EMIS Datareviews Series No. 13

Series Advisor: Dr. B. L. Weiss

PROPERTIES OF
Silicon Carbide

ELECTRONIC MATERIALS INFORMATION SERVICE

Other books in the EMIS Datareviews Series from INSPEC:

- No. 1   Properties of Amorphous Silicon (2nd Edition, 1989)
- No. 2   Properties of Gallium Arsenide (2nd Edition, 1990)
- No. 3   Properties of Mercury Cadmium Telluride (1987; out of print)
- No. 4   Properties of Silicon (1988)
- No. 5   Properties of Lithium Niobate (1989; out of print)
- No. 6   Properties of Indium Phosphide (1991)
- No. 7   Properties of Aluminium Gallium Arsenide (1993)
- No. 8   Properties of Lattice-Matched and Strained Indium Gallium Arsenide (1993)
- No. 9   Properties and Growth of Diamond (1994)
- No. 10   Properties of Narrow Gap Cadmium-based Compounds (1994)
- No. 11   Properties of Group III Nitrides (1994)
- No. 12   Properties of Strained and Relaxed Silicon Germanium (1995)

PROPERTIES OF
Silicon Carbide

Edited by

GARY L HARRIS

Materials Science Research Center of Excellence
Howard University, Washington DC, USA

Published by: INSPEC, the Institution of Electrical Engineers,
London, United Kingdom

© 1995: INSPEC, the Institution of Electrical Engineers

Apart from any fair dealing for the purposes of research or private study, or criticism or review, as permitted under the Copyright, Designs and Patents Act, 1988, this publication may be reproduced, stored or transmitted, in any forms or by any means, only with the prior permission in writing of the publishers, or in the case of reprographic reproduction in accordance with the terms of licences issued by the Copyright Licensing Agency. Inquiries concerning reproduction outside those terms should be sent to the publishers at the undermentioned address:

Institution of Electrical Engineers
Michael Faraday House,
Six Hills Way, Stevenage,
Herts. SG1 2AY, United Kingdom

While the editor and the publishers believe that the information and guidance given in this work is correct, all parties must rely upon their own skill and judgment when making use of it. Neither the editor nor the publishers assume any liability to anyone for any loss or damage caused by any error or omission in the work, whether such error or omission is the result of negligence or any other cause. Any and all such liability is disclaimed.

The moral right of the authors to be identified as authors of this work has been asserted by them in accordance with the Copyright, Designs and Patents Act 1988.

British Library Cataloguing in Publication Data

A CIP catalogue record for this book
is available from the British Library

ISBN 0 85296 870 1

Printed in England by Short Run Press Ltd., Exeter

Contents

Introduction by G.L. Harris … vii

Contributing Authors … xiii

Acknowledgements … xv

Abbreviations … xvi

1 BASIC PHYSICAL PROPERTIES

1.1 Density of SiC *G.L. Harris* … 3
1.2 Lattice parameters of SiC *G.L. Harris* … 4
1.3 Thermal conductivity of SiC *G.L. Harris* … 5
1.4 Acoustic velocity in SiC *G.L. Harris* … 7
1.5 Young's modulus of SiC *G.L. Harris* … 8
1.6 Miscellaneous properties of SiC *G.L. Harris* … 9

2 OPTICAL AND PARAMAGNETIC PROPERTIES

2.1 Optical absorption and refractive index of SiC *G.L. Harris* … 15
2.2 Phonos in SiC polytypes *J.A. Freitas Jr.* … 21
2.3 Photoluminescence spectra of SiC polytypes *J.A. Freitas Jr.* … 29
2.4 Impurities and structural defects in SiC determined by ESR *W.E. Carlos* … 42
2.5 ODMR investigations of defects and recombination processes in SiC
 T.A. Kennedy … 51

3 CARRIER PROPERTIES AND BAND STRUCTURE

3.1 Carrier mobilities and concentrations in SiC
 G.L. Harris, H.S. Henry and A. Jackson … 63
3.2 Effective masses in SiC *S. Yoshida* … 69
3.3 Band structure of SiC: overview *S. Yoshida* … 74
3.4 Pressure effects on band structure of SiC *I. Nashiyama* … 81

4 ENERGY LEVELS

4.1 Energy levels of impurities in SiC *I. Nashiyama* … 87
4.2 Deep levels in SiC *M.G. Spencer* … 93

5 SURFACE STRUCTURE, METALLIZATION AND OXIDATION

5.1	Surface structure and metallization of SiC *R. Kaplan and V.M. Bermudez*	101
5.2	Oxidation of SiC *J.J. Kopanski*	121

6 ETCHING

6.1	Introduction to etching of SiC *G.L. Harris*	133
6.2	Chemical etching of SiC *G.L. Harris*	134
6.3	Dry etching of SiC *K. Wongchotigul*	136
6.4	Electrochemical etching of SiC *J.S. Shor*	141

7 DIFFUSION OF IMPURITIES AND ION IMPLANTATION

7.1	Diffusion and solubility of impurities in SiC *G.L. Harris*	153
7.2	Ion implantation and anneal characteristics of SiC *K. Wongchotigul*	157

8 GROWTH

8.1	Bulk growth of SiC *S. Nishino*	163
8.2	Sublimation growth of SiC *A.O. Konstantinov*	170
8.3	Chemical vapour deposition of SiC *S. Nishino*	204
8.4	LPE of SiC and SiC-AlN *V.A. Dmitriev*	214

9 SiC DEVICES AND OHMIC CONTACTS

9.1	Ohmic contacts to SiC *G.L. Harris, G. Kelner and M. Shur*	231
9.2	Overview of SiC devices *G. Kelner and M. Shur*	235
9.3	SiC p-n junction and Schottky barrier diodes *G. Kelner and M. Shur*	238
9.4	SiC field effect transistors *G. Kelner*	247
9.5	SiC bipolar junction transistors and thyristors *M. Shur*	265
9.6	SiC optoelectronic devices *G. Kelner*	270
9.7	Potential performance/applications of SiC devices and integrated circuits *G. Kelner and M. Shur*	273

SUBJECT INDEX 275

Introduction

Semiconductor technology can be traced back to at least 150-200 years ago, when scientists and engineers living on the western shores of Lake Victoria produced carbon steel made from iron crystals rather than by 'the sintering of solid particles' [1]. Since that time, silicon and silicon-based materials have played a major role in the development of modern semiconductor device technology. The earliest pioneering work in SiC as a semiconductor material can be attributed to the Marconi company in 1907 [2]. They observed SiC electroluminescence for the first time. For almost 50 years, the work by the Marconi company appeared to go unnoticed. In 1955, Lely [3] developed a sublimation process using a growth cavity that produced rather pure platelets of SiC. The individual crystals were randomly sized and primarily hexagonal in shape. The process provided no control over the nucleation and orientation of these platelets. Since that time the interest in SiC as a semiconductor material can best be described as a roller-coaster ride with many peaks and valleys.

The first international conference on SiC was held in Boston, Massachusetts in 1959. Over 500 scientists attended the conference and 46 papers were presented. In 1968 and 1973, two other International Conferences were held, the last in Miami Beach. For almost ten years, the SiC effort in the USA, and around the world, went through another valley, except in the then USSR where the research effort remained fairly constant. This valley, or lack of interest, has been 'blamed on the lack of progress in crystal growth' [4]. Nishino [5] and Powell [6] made a large contribution to the renewed interest in SiC with the development of the heteroepitaxial growth of SiC on Si. This effort was then duplicated by Sasaki et al [7], Liaw and Davis [8] and Harris et al [9]. These recent results have led to three SiC Workshops hosted by North Carolina State University (1986, 1987, 1988), four International SiC Conferences co-sponsored by Howard University and Santa Clara University, yearly SiC workshops in Japan, an international high temperature semiconductor conference, the formation of several SiC companies, SiC-based products, new SiC solid solutions, and an overall renewed interest in this material.

Research on SiC in Japan has also been on the same roller-coaster. The work in the 1960s in Japan was centred around basic solid-state physics and crystal growth. The research in the 1970s almost completely died. In the 1980s, SiC made a comeback due largely to the project sponsored by the Ministry of International Trade and Industry (MITI) on 'Hardened ICs for Extreme Conditions', headed by the Electrotechnical Laboratory. In 1984, the Society of SiC was organized which has now become the Society of SiC and Related Wide Bandgap Semiconductors. The society has an annual meeting and the Institute of Space and Astronautical Science (ISAS) has organized an annual High Temperature Electronics meeting.

Recent topics in SiC research in Japan are micropipes in wafers, control of polytype in epitaxy, surface analysis of epitaxial layers, application for power devices, device simulation, and others [10].

As a crude measure of SiC activity, we have counted the number of articles in the Engineering Abstracts, Japanese Journal of Applied Physics, Applied Physics Letters, Materials Letters, Journal of Materials Science, Soviet Physics of Semiconductors, Soviet Physics Technical Letters, etc. for the years 1987 to 1994 (the 1994 figure is a projection based on one half of the year). The information is summarized in FIGURE 1. This data is a new unevaluated count of the number of publications dealing with silicon carbide. It provides a qualitative idea of the rate at which research in this area has grown. In 1987, only a few researchers and laboratories generated most of the published works but, in recent years, the activity at these laboratories has increased and many new research teams have become involved in SiC research throughout the US, Europe, Asia and Japan. This data does include the research activities in the former USSR where SiC activity has been very constant and at a very high level (several major research teams throughout the country). The relative research effort of semiconductors such as Si, Ge, III-V compounds and II-VI compounds is several orders of magnitude higher. However, their level of funding is also several orders of magnitude higher. The SiC effort has provided a larger 'bang for the buck'. This, I hope, will become evident after reviewing this book.

** 1994 data is a projection **

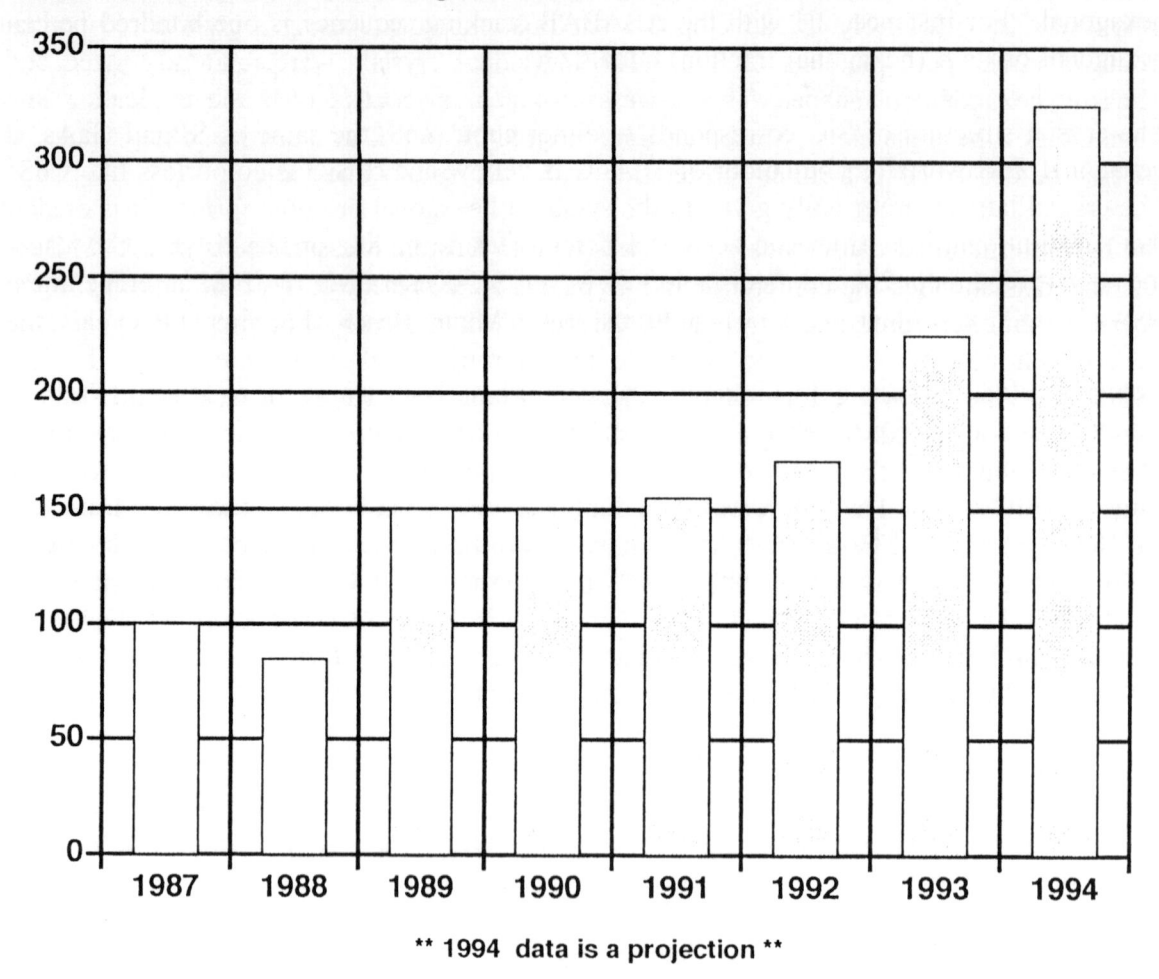

FIGURE 1 Silicon carbide publications (1987-1994)

The only chemically stable form of silicon and carbon is silicon carbide. The crystalline structure of SiC can be considered to consist of the close-packed stacking of double layers of Si and C atoms. One set of atoms (Si or C) is shifted along the main axis of symmetry by a quarter of the distance between the nearest similar layers. The bonding of silicon and carbon atoms is 88% covalent and 12% ionic with a distance between the Si and C atoms of 1.89 Å. Each Si or C atom is surrounded by four C or Si atoms in strong tetrahedral sp^3-bonds. The stacking of the double layers follows one of three possible relative positions. They are arbitrarily labelled A, B and C. One unique aspect of SiC is the stacking sequences of these three double layers which is the source of SiC's large number of crystallographic forms called polytypes. Polytypism is a one-dimensional polymorphism that is a result of the stacking sequence.

The only cubic form of SiC is called beta silicon carbide (β-SiC) which has a stacking sequence of ABCABCABC... Another convenient way of viewing these polytypes is the Ramsdell notation which is a number followed by a letter. The number represents the number of double layers in the stacking sequence and the letter represents crystal structure. For example, we have 3C for cubic or β-SiC. All hexagonal (H) and rhombohedral (R) types are commonly referred to as α-SiC with 6H the most common α-type (a more detailed discussion is given in Chapter 4). If the double layers have the same adjoining position, they are hexagonal. For instance, 2H with the ABABAB stacking sequence is one hundred percent hexagonal or its h (hexagonal fraction) equals one.

The crystal structure of SiC corresponds to either cubic ZnS, the same as Si and GaAs, or hexagonal ZnS wurtzite (the colour of 3C-SiC is yellow and α-SiC is colourless like CdS). The crystal lattices are usually given in the standard hexagonal or cubic forms. It is evident that substitutional impurities can occupy different sites depending on the polytype. Most of the polytypes, except 2H, are metastable. However, 3C does transform to 6H at temperatures above 2000 °C and other polytypes can transform at temperatures as low as 400 °C [11,12].

Numerous electronic and optoelectronic applications have been proposed based on SiC's basic electronic and optical properties. There are five primary applications of SiC-based material: 1) micro-structures, 2) optoelectronic devices, 3) high temperature electronics, 4) radiation hard electronics and 5) high power/high frequency devices. Micro-structure applications include X-ray masks and micro-machined structures such as speaker diaphragms and special micro-application tools. Optoelectronic applications would include substrates for the nitride family of devices, light emitting diodes, and UV detectors. Because of SiC's large bandgap almost all devices fabricated on SiC can be considered for high temperature applications. Nuclear reactor electronics, military systems and deep space electronics survivability are greatly improved with SiC. SiC's thermal conductivity and high field mobility will allow increased power density and high frequency operation. Many of these applications can be summarized in the 'Fruits of SiC Tree' in FIGURE 2.

Operable SiC devices have been fabricated successfully with 4H-SiC, 6H-SiC and 3C-SiC. For example, blue LEDs (light emitting diodes) were reported in 1977 that operated at elevated temperatures for virtually an unlimited lifetime with an overall efficiency of 2×10^{-5}. Recently [13], SiC blue LEDs have been fabricated with efficiencies greater than 2×10^{-4} and luminous intensity of 10 mcd at 20 mA with reported diode areas of 250 μm. Dmitriev et al [14] have produced a three colour (blue 470 nm, green 510 nm, red 650 nm) SiC single crystal display using epitaxial 6H and ion implantation into container-free liquid epitaxial material. The diameter of each diode is 300 - 500 μm with approximately 300 μm between emitters. There are commercial efforts in blue SiC LEDs in the US, Japan and Germany.

Research on 6H SiC MESFETs has also been conducted [15]. Generally speaking, these devices had gates on the order of 10 μm and showed good saturation. The 6H MESFETs reported drain breakdown voltages over 40 V with transconductances of approximately 2 - 4 ms mm^{-1}.

SiC inversion-mode MOSFETs have been reported by Davis et al [15]. The transconductance in this case was 0.27 ms mm^{-1} with a gate voltage of 24 V. This device operated well

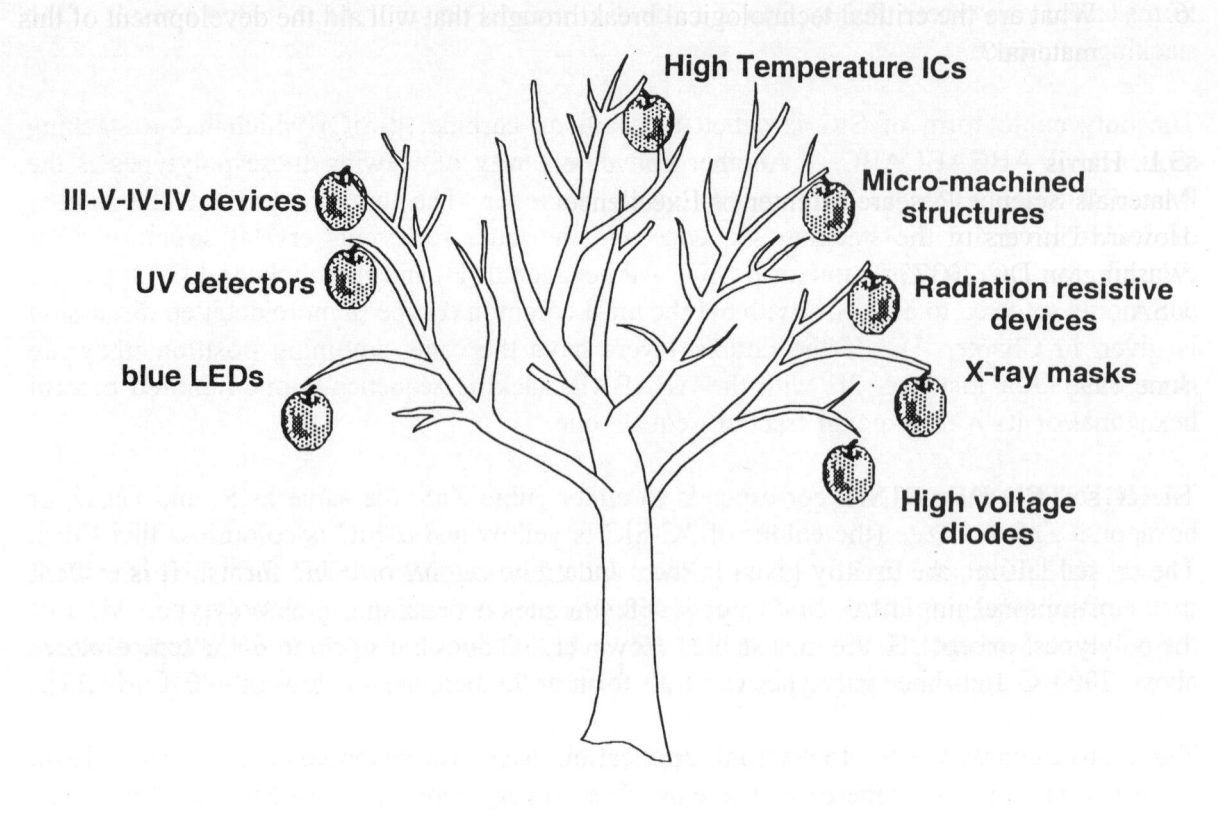

FIGURE 2 Fruits of the SiC tree

Fruits shown: High Temperature ICs, Micro-machined structures, Radiation resistive devices, X-ray masks, High voltage diodes, High voltage diodes, blue LEDs, UV detectors, III-V-IV-IV devices.

over 923 K with a transconductance, at this temperature, of 0.43 ms mm^{-1} at a gate voltage of 6 V.

Recently, a joint research project between Motorola Inc. and Cree Research yielded SiC FETs that operate at frequencies as high as 12.6 GHz [16]. The FETs were fabricated on 4H-SiC which has a higher mobility than 6H-SiC.

Many other SiC devices have been fabricated including UV detectors, β-SiC/Si solar cells, SiC memories, SiC inverters, etc. More details are provided in the device section of this book.

There are a number of important issues facing the SiC engineer and scientist. Let me end this introduction by proposing a few:

1. Can we grow large area defect-free boules of silicon carbide 3C, 4H and/or 6H?

2. Will heteroepitaxial polytype become a controlled process (3C on 6H, 4H and 3C on 6H, etc.)?

3. How can we achieve high temperature long term operation of SiC devices?

4. Can we develop solid solutions of $(SiC)_x AlN_{1-x}$ and other nitrides on SiC substrates?

5. What impact will SiC have on the LED flat plate market?

6. What are the critical technological breakthroughs that will aid the development of this material?

G.L. Harris
Materials Science Research Center of Excellence
Howard University
Washington DC 20375
USA

June 1995

SUGGESTED READING

[1] H.K. Henisch, R. Roy (Eds) [*Proc. Int. Conf. on Silicon Carbide*, University Park, Pennsylvania, USA, 20-23 Oct. 1968 (Pergamon Press Inc., 1969)]

[2] R.C. Marshall, J.W. Faust Jr., C.E. Ryan (Eds.) [*Proc. Int. Conf. on Silicon Carbide*, Miami Beach, Florida, USA, 17-20 Sept. 1973 (University of South Carolina Press, 1974)]

[3] J.I. Pankove (Ed.) [*Electroluminescence* (Springer-Verlag, 1977); ISBN 3-540-08127-5 (Berlin, Heidelberg, New York); ISBN 0-378-08127-5 (New York, Heidelberg, Berlin)]

[4] D. Emin, T.L. Aselage, C. Wood (Eds) [*Novel Refractory Semiconductors, MRS Soc. Symposia Proc.*, Anaheim, California, USA, 21-23 April 1987 (Materials Research Society, Pittsburgh, Pennsylvania, USA, 1987)]

[5] G.L. Harris, C.Y.-W. Yang (Eds) [*Proc. Int. Conf. on Amorphous and Crystalline Silicon Carbide*, Washington DC, USA, 10-11 Dec. 1987 (Springer-Verlag Berlin Heidelberg, 1989)]

[6] [*Ext. Abstr. Electrochem. Soc. (USA)* vol.89 (1989)]

[7] M.M. Rahman, C.Y.-W. Yang, G.L. Harris (Eds) [*Proc. Int. Conf. on Amorphous and Crystalline Silicon Carbide*, Santa Clara, California, USA, 15-16 Dec. 1988 (Springer-Verlag Berlin Heidelberg, 1989)]

[8] G.L. Harris, M.G. Spencer, C.Y.-W. Yang (Eds) [*Proc. Int. Conf. on Amorphous and Crystalline Silicon Carbide*, Howard University, Washington DC, USA, 11-13 April 1990 (Springer-Verlag Berlin Heidelberg, 1992)]

[9] A.A. Gippius, R. Helbig, J.P.F. Sellschop (Eds) [*Proc. Symposium C on Properties and Applications of SiC, Natural and Synthetic Diamond and Related Materials*, Strasbourg, France, 27-30 Nov. 1990 (North-Holland, Elsevier Science Publishers BV, the Netherlands, 1992)]

[10] C.Y.-W. Yang, M.M. Rahman, G.L. Harris (Eds) [*Proc. Int. Conf. on Amorphous and Crystalline Silicon Carbide*, Santa Clara, California, USA, 9-11 Oct. 1991 (Springer-Verlag Berlin Heidelberg, 1992)]

[11] M. Tajima (Ed.) [*Proc. Japanese High Temperature Electronics Meeting 1993, 1994 and 1995* ISAS, 3-1-1 Yoshnodai, Sagamihara 229, Japan]

[12] D.B. King, F.V. Thome [*Trans. Int. High Temperature Electronics Conf.*, Charlotte, North Carolina, 5-10 June 1994]

[13] Yu.M. Tairov, V.F. Tsvetkov [*Handbook of Electrical Engineering Materials*, in Russian, vol.3 (1988)]

[14]   [*Landolt-Bornstein, Group III: Crystal and Solid-State Physics* vol.17, Semiconductors (Springer-Verlag, 1982)]
[15]   W.J. Choyke [*NATO ASI Ser. E Appl. Sci. (Netherlands)* vol.185 (1990)]
[16]   M.G. Spencer, R.P. Devaty, J.A. Edmond, M. Asif Khan, R. Kaplan, M. Rahman (Eds) [*Proc. Conf. on Silicon Carbide and Related Materials*, Washington DC, USA, 1-3 Nov. 1993 (Institute of Physics Publishing Ltd., Bristol and Philadelphia, 1994)]
[17]   C.R. Abernathy, C.W. Bates Jr., D.A. Bohling, W.S. Hobson (Eds) [*Proc. Conf. on Chemical Perspectives of Microelectronic Materials*, Boston, Massachusetts, USA, 30 Nov.-3Dec. 1992 (Materials Res. Soc., Pittsburgh, Pennsylvania, USA, 1993)]

REFERENCES

[1]   [*Science (USA)* Sept. 22, 1979]
[2]   H.J. Round [*Electr. World (USA)* vol.19 (1907) p.309]
[3]   J.A. Lely [*Ber. Dtsch. Keram. Ges. (Germany)* vol.32 (1955) p.229]
[4]   R.B. Campbell [*IEEE Trans. Ind. Electron. (USA)* vol.IE-29 (1982) p.124]
[5]   S. Nishino, Y. Hazuki, H. Matsunami, T. Tanaka [*J. Electrochem. Soc. (USA)* vol.127 (1980) p.2674]
[6]   S. Nishino, J.A. Powell, H.A. Will [*Appl. Phys. Lett. (USA)* vol.42 (1983) p.460]
[7]   K. Sasaki, E. Sakuma, S. Misawa, S. Yoshida, S. Gonda [*Appl. Phys. Lett. (USA)* vol.45 (1984) p.72]
[8]   P. Liaw, R.F. Davis [*J. Electrochem. Soc. (USA)* vol.132 (1985) p.642]
[9]   G.L. Harris, K.H. Jackson, G.J. Felton, K.R. Osbourne, K. Fekade, M.G.Spencer [*Mater. Lett. (Netherlands)* vol.4 (1986) p.77]
[10]   M. Tajima [*Trans. Int. High Temp. Conf.* vol.I No.I-29 (1994)]
[11]   P. Krishna, R.C. Marshall, C.E. Ryan [*J. Cryst. Growth (Netherlands)* vol.8 (1971) p.129]
[12]   J.A. Powell, H.A. Will [*J. Appl. Phys. (USA)* vol.43 (1972) p.1400]
[13]   Personal communication, Cree Research Inc.
[14]   V.A. Dmitriev, Ya.Y. Morozenko, I.V. Popov, A.V. Suvorov, A.L. Syrkin, V.E. Chelnokov [*Sov. Tech. Phys. Lett. (USA)* vol.12 (1986) p.221]
[15]   R.F. Davis, G. Kelner, M. Shur, J.W. Palmour, J.A. Edmond [*Proc. IEEE (USA)* vol.79 (1991) p.677]
[16]   [*Electron. Des. (USA)* Nov.22 1993 p.26]

Contributing Authors

V.M. Bermudez	Naval Research Laboratory, Electronics Science and Technology Division, Washington, DC 20375-5347, USA	5.1
W.E. Carlos	Naval Research Laboratory, Electronics Science and Technology Division, Washington, DC 20375-5347, USA	2.4
V.A. Dmitriev	Cree Research Inc., 2810 Meridian Parkway, Suite 176, Durham, NC 27713, USA	8.4
J.A. Freitas Jr.	Sachs/Freeman Associates, Inc., Landover, MD 20785, USA	2.2, 2.3
G.L. Harris	Howard University, Materials Science Research Center of Excellence, Department of Electrical Engineering, Washington, DC 20059, USA	1.1-1.6, 2.1, 3.1, 6.1, 6.2, 7.1, 9.1
H.S. Henry	Howard University, Materials Science Research Center of Excellence, Department of Electrical Engineering, Washington, DC 20059, USA	3.1
A. Jackson	Howard University, Materials Science Research Center of Excellence, Department of Electrical Engineering, Washington, DC 20059, USA	3.1
R. Kaplan	Naval Research Laboratory, Electronics Science and Technology Division, Washington, DC 20375-5347, USA	5.1
G. Kelner	Naval Research Laboratory, Electronics Science and Technology Division, Washington, DC 20375-5347, USA	9.1-9.4, 9.6, 9.7
T.A. Kennedy	Naval Research Laboratory, Electronics Science and Technology Division, Washington, DC 20375-5347, USA	2.5
A.O. Konstantinov	A.F. Ioffe Physical Technical Institute, St. Petersburg, 194021, Russia	8.2
J.J. Kopanski	National Institute of Standards and Technology, Semiconductor Electronics Division, 225-A305, Gaithersburg, MD 20899, USA	5.2

I. Nashiyama	Japan Atomic Energy Research Institute, Radiation Engineering Division, Takasaki Radiation Chemistry Research Establishment, Watanuk-Machi-Takasaki, Gunma-Ken, Japan	3.4, 4.1
S. Nishino	Kyoto Institute of Technology, Department of Electronics and Information Science, Matsugasaki, Sakyo-ku, Kyoto 606, Japan	8.1, 8.3
J.S. Shor	Kulite Semiconductor Products, One Willow Tree Road, Leonia, NJ 07605, USA	6.4
M. Shur	University of Virginia, Department of Electrical Engineering, Charlottesville, VA 22903-2442, USA	9.1-9.3, 9.5, 9.7
M.G. Spencer	Howard University, Materials Science Research Center of Excellence, Department of Electrical Engineering, School of Engineering, Washington, DC 20059, USA	4.2
K. Wongchotigul	Howard University, Materials Science Research Center of Excellence, Department of Electrical Engineering, Washington, DC 20059, USA	6.3, 7.2
S. Yoshida	Electrotechnical Laboratory, 1-1-4 Umezona, Tsukuba-shi, Ibaraki 305, Japan	3.2, 3.3

Acknowledgements

It is a pleasure to acknowledge the work both of the contributing authors named on the previous pages and of the following experts in the field.

D.L. Barrett	Westinghouse Science & Technology Center, Pittsburgh, PA, USA
D.M. Brown	General Electric Corporate Research and Development Center, Schenectady, NY, USA
C.H. Carter Jr.	Cree Research, Inc., Durham, NC, USA
R.F. Davis	North Carolina State University, Raleigh, NC, USA
E. Duhart	Howard University, Materials Science Research Center of Excellence, Washington DC, USA
P.B. Klein	Naval Research Laboratory, Washington DC, USA
H. Matsunami	Kyoto University, Yoshidahonmachi, Sakyo, Kyoto, Japan
L.G. Matus	NASA - Lewis Research Center, Cleveland, OH, USA
J.W. Palmour	Cree Research, Inc., Durham, NC, USA
J.A. Powell	NASA - Lewis Research Center, Cleveland, OH, USA
V. Shields	Jet Propulsion Laboratory, Pasadena, CA, USA
X. Tang	Howard University, Materials Science Research Center of Excellence, Washington DC, USA

Abbreviations

The following abbreviations are used throughout the book.

α-SiC	hexagonal silicon carbide
A	acceptor
AC	alternating current
AES	Auger elctron spectroscopy
APB	anti-phase boundary
ASA	atomic sphere approximation
β-SiC	cubic silicon carbide
BE	bound exciton
BJT	bipolar junction transistor
BZ	Brillouin zone
CFLPE	container free liquid phase epitaxy
CL	cathodoluminescence
CMOS	complementary metal-oxide-semiconductor
C-V	capacitance-voltage
CVD	chemical vapour deposition
D	donor
DAP	donor-acceptor pairs
DAS	dimer-adatom stacking fault
DC	direct current
DLTS	deep level transient spectroscopy
DPB	double position boundary
EBIC	electron beam induced current
ECR	electron cyclotron resonance
EELS	electon energy loss spectroscopy
EHL	electron hole liquid
EL	electroluminescence
EMF	electro-motive force
EP	empirical pseudopotential
EPM	empirical pseudopotential method
EPR	electron paramagnetic resonance
ESD	electron stimulated desorption
ESR	electron spin resonance
ETB	empirical tight binding
FB	free-to-bound
FET	field effect transistor
G-V	conductance-voltage
HF	hyperfine

IC	integrated circuit
IDB	inversion domain boundary
IMPATT	impact ionization avalanche transient time
IR	infrared
I-V	current-voltage
JFET	junction field effect transistor
JFM	Johnson's figure of merit
KFM	Keyes' figure of merit
LA	longitudinal acoustic
LCAO	linear combination of atomic orbitals
LDA	local density approximation
LED	light emitting diode
LEED	low energy electron diffraction
LMTO	linear muffin-tin orbital
LO	longitudinal optical
LPE	liquid phase epitaxy
LTLPE	low temperature liquid phase epitaxy
LUC-MINDO	large unit cell-modified intermediate neglect of differential overlap
LVB	lower valence band
LZ	large zone
MBE	molecular beam epitaxy
MEIS	medium energy ion scattering
MESFET	metal semiconductor field effect transistor
MINDO	modified intermediate neglect of differential overlap
MIS	metal-insulator-semiconductor
MOCVD	metal-organic chemical vapour deposition
MOMBE	metal-organic molecular beam epitaxy
MOS	metal-oxide-semiconductor
MOSFET	metal-oxide-semiconductor field effect transistor
MOVPE	metal-organic vapour phase epitaxy
NBE	near band edge
NBE	nitrogen bound exciton
NEPM	non-local empirical pseudopotential method
NVRAM	non-volatile random access memory
ODMR	optically detected magnetic resonance
OLCAO	orthogonalised linear combination of atomic orbitals
OMVPE	organometallic vapour phase epitaxy
OPW	orthogonalised plane wave
PEC	photoelectrochemical
PL	photoluminescence
RAM	random access memory
RBS	Rutherford back-scattering

RF	radio frequency
RHEED	reflection high energy electron diffraction
RIE	reactive ion etching
RMS	root mean square
RS	Raman scattering
RT	room temperature
RTA	rapid thermal annealing
SCE	saturated calomel reference electrode
SCF	self-consistent field
SCR	silicon controlled rectifier
SCTB	self-consistent tight-binding
SE	sublimation epitaxy
SEM	scanning electron microscopy
SIMS	secondary ion mass spectrometry
SIS	semiconductor-insulator-semiconductor
STM	scanning tunnelling microscopy
SXS	soft X-ray spectroscopy
TA	transverse acoustic
TB	semi-empirical tight binding method
TEM	transmission electron microscopy
TL	thermoluminescence
TMA	thermomechanical analysis
TO	transverse optical
TPD	temperature programmed desorption
UHV	ultra high vacuum
UPS	ultraviolet photoemission spectroscopy
UV	ultraviolet
VPE	vapour phase epitaxy
XPS	X-ray photoemission spectroscopy
XRD	X-ray diffraction
ZPL	zero phonon line

CHAPTER 1

BASIC PHYSICAL PROPERTIES

1.1 Density of SiC
1.2 Lattice parameters of SiC
1.3 Thermal conductivity of SiC
1.4 Acoustic velocity in SiC
1.5 Young's modulus of SiC
1.6 Miscellaneous properties of SiC

1.1 Density of SiC

G.L. Harris

February 1995

A ROOM TEMPERATURE

Experimental measurements [1-3] place the density of SiC in the range from 3.166 to 3.24878 g cm^{-3} depending on the polytype. Most of these measurements were obtained by using X-ray data and calculating the density from

$$d = 4M/NV \tag{1}$$

where M is the gram formula weight (40.09715), V is the volume of the unit cell (i.e. the cube of the lattice parameter for 3C), N is Avogadro's constant (6.0221367×10^{23} per mole) and 4 is the number of formula units in the cell.

TABLE 1 Density of SiC at room temperature.

Density (g cm^{-3})	Polytype	Temperature (K)	Ref & comments
3.214	2H	293	[2]
3.166	3C	300	[4]
3.21427	3C	300	Using Eqn (1) and X-ray data in [1]
3.210	3C	300	[3]
3.211	6H	300	[3]
3.24878	6H	300	Using Eqn (1) and X-ray data in [1]

B DATA AT OTHER TEMPERATURES

Data at other temperatures can easily be obtained by using Eqn (1) and the lattice parameter measurements as a function of temperature from [1].

REFERENCES

[1] A. Taylor, R.M. Jones [*Proc. Conf. on Silicon Carbide*, Boston, USA, 1959 (Pergamon Press, New York, 1960) p.147]
[2] R.F. Adamsky, K.M. Merz [*Z. Kristallogr. (Germany)* vol.3 (1959) p.350]
[3] A.H. Mesquita de Gomes [*Acta Crystallogr. (Denmark)* vol.23 (1967) p.610]
[4] E.L. Kern, D.W. Hamil, H.W. Deem, H.D. Sheets [*Mater. Res. Bull. (USA)* vol.4 (1964) p.S25]

1.2 Lattice parameters of SiC

G.L. Harris

February 1995

The X-ray diffraction technique is the most commonly used method to determine the lattice parameters for SiC. Taylor and Jones [1] used this technique to perform some of the most complete lattice parameter determinations. The samples appeared to be very pure but the N_2 contamination was not measured which can lead to an uncertainty in the fourth decimal place. A detailed measurement of the lattice constants for 3C and 6H polytypes as a function of temperature can be found in [1].

TABLE 1 Lattice parameters of SiC.

Lattice parameter (a,c in Å)	Polytype	Temperature (K)	Ref
a = 4.3596	3C	297	[1]
a = 4.3582	3C	0	[1]
a = 3.0763	2H	300	[2]
c = 5.0480		300	[2]
a = 3.0730	4H	300	[3]
c = 10.053			
a = 3.0806	6H	297	[1]
c = 15.1173			
a = 3.080	6H	0	[1]
c = 15.1173			
a = 12.691	15R	300	[4]
α = 13°54′			
a = 17.683	21R	300	[4]
α = 9°58′			
a = 27.704	33R	300	[4]
α = 6°21′			

REFERENCES

[1] A. Taylor, R.M. Jones [in *Silicon Carbide - A High Temperature Semiconductor* Eds J.R. O'Connor, J. Smiltens (Pergamon Press, 1960) p.147]
[2] R.F. Adamski, K.M. Merz [*Kristallografiya (USSR)* vol.111 (1959) p.350]
[3] R.W.G Wyckoff [*Crystal Structures (USA)* vol.1 (1963) p.113]
[4] N.W. Thibault [*Am. Mineral. (USA)* vol.29 (1944) p.327]

1.3 Thermal conductivity of SiC

G.L. Harris

February 1995

A INTRODUCTION

Silicon carbide (SiC) is an attractive semiconductor material for high temperature electronic and electro-optic applications. From the device application point of view the thermal conductivity of SiC exceeds that of copper, BeO, Al_2O_3 and AlN. The actual value of the thermal conductivity can be varied with polytype and/or doping [1,2].

At very low temperatures, the thermal conductivity of 6H SiC could have a T^{-2} or T^{-3} temperature dependence [1]. For very pure or highly compensated material the temperature dependence is T^{-3}. Slack [3] has shown the T^{-2} temperature dependence is due to the heat flow associated with phonons. Doping of SiC shifts the thermal conductivity peak toward higher temperatures [1-3]. For the 4H and 6H samples reported the measured conductivity is perpendicular to the C direction.

B ROOM TEMPERATURE

The room temperature thermal conductivities of various polytypes are summarised in TABLE 1.

TABLE 1 Room temperature thermal conductivity of SiC.

Thermal conductivity, χ (W cm^{-1} K^{-1})	Polytype	Comments	Ref
3.2	3C	poly-3C	[1]
3.7	4H	-	[1]
3.6	6H	$N_N = 8 \times 10^{15}$ cm^{-3} at 300 K	[4]
3.6	6H	$N_N = 5 \times 10^{16}$ cm^{-3} at 300 K	[4]
3.6	6H	$N_N = 1 \times 10^{19}$ cm^{-3} at 300 K	[4]
2.31	6H	$N_{Al} = 5 \times 10^{19}$ cm^{-3} at 300 K	[4]
4.9	6H	-	[3]

N_N = nitrogen doping concentration
N_{Al} = aluminium doping concentration

C   TEMPERATURE RANGE

In the range of 8 to 40 K the variation of the thermal conductivity is listed in TABLE 2. The best fit is a function of the form:

$$K = AT^{-N}$$

and K is in units of $W\,cm^{-1}\,K^{-1}$ [1].

TABLE 2 Temperature dependence of thermal conductivity of SiC.

Temperature (K)	χ (W cm^{-1} K^{-1}) 6H	χ (W cm^{-1} K^{-1}) 4H	χ (W cm^{-1} K^{-1}) 3C
9	0.7	0.12	0.23
10	0.85	0.16	0.31
12	1.41	0.27	0.52
15	2.0	0.51	0.71
20	3.6	1.08	1.59
25	5.0	1.70	2.03
30	6.3	2.6	3.4
35	8.0	3.4	4.6
40	10.05	4.3	5.02

The changes with doping are discussed in [1] and [2].

REFERENCES

[1]   D. Morelli, J. Hermans, C. Beetz, W.S. Woo, G.L. Harris, C. Taylor [*Inst. Phys. Conf. Ser. (UK)* no.137 (1993) p.313-6]
[2]   I.I. Parafenova, Y.M. Tairov, V.F. Tsvetkov [*Sov. Phys.-Semicond. (USA)* vol.24 no.2 (1990) p.158-61]
[3]   G.A. Slack [*J. Appl. Phys. (USA)* vol.35 (1964) p.3460]
[4]   E.A. Burgemeister, W. Von Muench, E. Pettenpaul [*J. Appl. Phys. (USA)* vol.50 (1979) p.5790]

1.4 Acoustic velocity in SiC

G.L. Harris

February 1995

The acoustic velocity of SiC has been measured using both ultrasonic resonance [1] and ultrasonic pulse echo techniques. The latter has resulted in the appearance of CVD-SiC diaphragms in commercially available speakers [2].

TABLE 1 Acoustic velocity in SiC.

v (m s^{-1})	Polytype	Temperature (K)	Ref
13,300	6H	300	[1]
12,600	3C (poly)	297	[2]
13,730	4H	20	[3]
13,100	6H	5	[3]
13,260	6H	300	[3]
13,270	21R	300	[4]

REFERENCES

[1] G. Arlt, G.R. Schodder [*J. Acoust. Soc. Am. (USA)* vol.37 (1965) p.384]
[2] Y. Chinone, S. Ezaki, F. Fuijta, R. Matsumoto [*Springer Proc. Phys. (Germany)* vol. 43 (1989) p.198]
[3] S. Karmann, R. Helbig, R.A. Stein [*J. Appl. Phys. (USA)* vol.66 no.8 (1989) p.3922-4]
[4] D.W. Feldman, J.H. Parker, W.J. Choyke, L. Patrick [*Phys. Rev. (USA)* vol.170 (1968) p.698- and vol.173 (1968) p.787]

1.5 Young's modulus of SiC

G.L. Harris

February 1995

The Young's modulus of SiC has been measured using load-deflection measurements of suspended 3C-SiC diaphragms and free-standing 3C-SiC cantilever beams [1-4]. All of the samples were grown on Si. The Si substrates were masked and stripped from the back. The temperature coefficient for the Young's modulus was determined to be 3.2×10^{-5} GPa°C^{-1} [4].

TABLE 1 Young's modulus of SiC.

Thickness t (mm)	Young's modulus E (GPa)	Comments	Ref
3.13	392	297 K undoped	[2]
2.35	447	297 K undoped	[2]
1.29	442	297 K undoped	[2]
10	448	297 K undoped	[3]
10	694	297 K, p-type, 3C, Al-doped	[3]

REFERENCES

[1] L. Tang, M. Mehregany, L.G. Matus [*Appl. Phys. Lett. (USA)* vol.60 no.24 (1992) p.2992-4]
[2] L.G. Matus, L. Tang, M. Mehregany, D.J. Larkin, P.G. Neudeck [*Inst. Phys. Conf. Ser. (UK)* no.137 (1993) ch.3 p.185-8]
[3] K. Fekade, Q.M. Su, M.G. Spencer, M. Wuttig [*Inst. Phys. Conf. Ser. (UK)* no.137 (1993) ch.3 p.189-92]
[4] C.M. Su, K. Fekade, M.G. Spencer, M. Wuttig [Elastic and Anelastic Properties of CVD Epitaxy 3C-SiC, submitted to Appl. Phys. Lett. (USA)]

1.6 Miscellaneous properties of SiC

G.L. Harris

February 1995

A INTRODUCTION

In this Datareview, we will list other basic physical properties of SiC which either have not been extensively studied or for which there is good agreement.

B DATA

The data for the various SiC polytypes are summarised in TABLE 1.

TABLE 1 Basic physical properties of SiC.

Physical property	Value	Experimental conditions	Ref
Diffusion length of minority carriers L_p	0.03 - 0.144 µm	$(N_D - N_A) = 10^{16} - 10^{17}$ cm^{-3}, actual value depends on quenching cycle, 6H	[1]
	0.2 - 0.4 µm	$(N_D - N_A) = 10^{16} - 10^{17}$ cm^{-3}, sandwich sublimation growth method, 6H	[2,3]
	0.5 - 1.0 µm	$(N_D - N_A) = 1.5 - 20 \times 10^{17}$ cm^{-3}, from ion implantation devices, 6H	[4]
$L_n + L_p$	0.4 - 1.5 µm	$(N_D - N_A) = 6 \times 10^{16} - 10^{18}$ cm^{-3}, see [13] for doping and temperature dependence, 6H	[13]
Dielectric constant $\varepsilon(0)$	9.75	300 K, 3C	[5,6]
$\varepsilon(\infty)$	6.52	300 K, 3C	[5,6]
$\varepsilon(0)$	9.66	300 K, 6H($\perp$ c-axis)	[5,6]
	10.3	300 K, 6H($\parallel$ c-axis)	
$\varepsilon(\infty)$	6.52	300 K, 6H($\perp$ c-axis)	[5,6]
	6.70	300 K, 6H($\parallel$ c-axis)	
Work function	4.52 eV	300 K, 6H, {0001} faces	[7]
	$\phi_B = C_2 \phi_m + C_3$	ϕ_B = work function on 6H SiC, ϕ_m = work function on metal, $C_2 = 0.6$, $C_3 = -1.82$, 6H, 300 K	[8,9]
	4.533 eV	300 K, 3C, based on W calculations	[9]
	4.116 eV	300 K, 3C, based on Mo calculations	[9]
Breakdown field E_B	2 - 3 x 10^6 V cm^{-1}	300 K, 6H, p = 10^{17} - 10^{18} cm^{-3}	[10]
	10640 x $N_D^{0.142}$ V cm^{-1}	Based on experimental data, 300 K, 6H	[11]
	8185 x $N_D^{0.142}$ V cm^{-1}	Scaled from 6H data, 300 K, 3C	[11]
Saturation velocity V_d	2.0 x 10^7 cm s^{-1}	Epitaxial layer, 6H	[10]

1.6 Miscellaneous properties of SiC

TABLE 1 continued

Physical property	Value	Experimental conditions	Ref
Electro-optic coefficient r_{41}	2.7×10^{-12} m V^{-1}	Measured at 633 nm, 300 K, β-SiC	[12]
Elastic coefficient, stiffness			
Cubic C_{11}	3.52×10^{12} dyn cm^{-2}		[14]
C_{12}	1.2×10^{12} dyn cm^{-2}		
C_{44}	2.329×10^{12} dyn cm^{-2}	Calculated	
C_{11}	2.89×10^{12} dyn cm^{-2}		[23]
C_{12}	2.34×10^{12} dyn cm^{-2}		
C_{44}	0.55×10^{12} dyn cm^{-2}		
Hexagonal-6H			
C_{11}	5.0×10^{12} dyn cm^{-2}		[15]
C_{12}	0.92×10^{12} dyn cm^{-2}		
C_{44}	1.68×10^{12} dyn cm^{-2}		[16]
C_{33}	5.64×10^{12} dyn cm^{-2}		
C_{33}	5.5×10^{12} dyn cm^{-2}		
C_{33}	5.6×10^{12} dyn cm^{-2}		
$C_{66} = (C_{11}-C_{12})/2$	2.04×10^{12} dyn cm^{-2}		
Hexagonal-4H			
C_{44}	6.0×10^{12} dyn cm^{-2}		[16]
21R			
C_{44}	5.6×10^{12} dyn cm^{-2}		[17]
Elastic coefficient, applicance-6H			
S_{11}	2.03×10^{13} cm^2 dyn^{-1}	Calculated	[15]
S_{12}	0.421×10^{13} cm^2 dyn^{-1}		
S_{44}	5.95×10^{13} cm^2 dyn^{-1}		
S_{11}	2.03×10^{13} cm^2 dyn^{-1}		[16]
S_{11}	2.04×10^{13} cm^2 dyn^{-1}		
Hexagonal-4H			
S_{11}	2.14×10^{13} cm^2 dyn^{-1}		[16]
Piezoelectric coefficient-6H			
e_{33}	0.2 C m^{-2}		[18]
e_{15}	0.08 C m^{-2}		
Electromechanical coupling coefficient-6H			
K_{31}	$< 4 \times 10^{-3}$		[19]
K_{33}	$< 4 \times 10^{-3}$		
Piezoresistive effect π_{11}	-142×10^{12} cm^2 dyn^{-1}		[20]
Gauge factor G_f	-31.8 maximum (3C) -29.4 (6H)		[21] [22]

1.6 Miscellaneous properties of SiC

TABLE 1 continued

Physical property	Value	Experimental conditions	Ref
Debye temperature β α	1430 K 1200 K		[24] [25]
Thermal EMF Q	$-70\,\mu V\,°C^{-1}$ $-110\,\mu V\,°C^{-1}$	293 K 1273 K	[26]
Specific heat β α	$0.17\,cal\,g^{-1}\,°C^{-1}$ $0.22\,cal\,g^{-1}\,°C^{-1}$ $0.28\,cal\,g^{-1}\,°C^{-1}$ $0.30\,cal\,g^{-1}\,°C^{-1}$ $0.27\,cal\,g^{-1}\,°C^{-1}$ $0.35\,cal\,g^{-1}\,°C^{-1}$	20 °C 200 °C 1000 °C 1400 - 2000 °C 700 °C 1550 °C	[27] [28]
Bulk modulus-6H	14.01	10^6 psi, 20 °C	[28]
Spectral emissivity $\varepsilon(\lambda)$	0.94 (at $\lambda = 0.9\,\mu m$)	1800 °C	[29]
Magnetic susceptibility χ (6H)	10.6×10^{-6} g-mol	1300 °C	[30]

C   CONCLUSION

Various physical properties of several of the polytypes of SiC have been listed together with the relevant experimental conditions such as temperature, lattice direction and carrier density.

REFERENCES

[1]   V.S. Ballandovich, G.N. Violina, [*Sov. Phys.-Semicond. (USA)* vol.15 no.8 (1981) p.959-60]

[2]   E.N. Mokhov, M.G. Ramm, R.G. Verenchikova, G.A. Lomakina [*Abstract All-Union Conf. in Wide-Gap Semiconductors*, Leningrad, 1979 (in Russian) p.51]

[3]   M.G. Ramm, E.N. Mokhov, R.G. Verenchikova [*Izv. Akad. Nauk SSSR Neorg. Mater. (USSR)* vol.15 (1979) p.2233]

[4]   E.V. Kalinina, G.F. Kholujanov [*Inst. Phys. Conf. Ser. (UK)* no.137 (1994) ch.6 p.675-6]

[5]   L. Patrick, W.J. Choyke [*Phys. Rev. B (USA)* vol.2 (1970) p.2255]

[6]   W.J. Choyke [*NATO ASI Ser. E, Appl. Sci. (Netherlands)* vol.185 (1990) and references therein]

[7]   J.A. Dillion, R.E. Schlier, H.E. Farnsworth [*J. Appl. Phys. (USA)* vol.30 (1959) p.675]

[8]   P. Chattopadhyay, A.N. Daw [*Solid-State Electron. (UK)* vol.28 no.8 (1985) p.831-6]

[9] C. Jacob, S. Nishino, M. Mehregany, J.A. Powell, P. Pirouz [*Inst. Phys. Conf. Ser. (UK)* no.137 (1994) ch.3 p.247-50]

[10] W.V. Munch, I. Pfaffeneder [*J. Appl. Phys. (USA)* vol.48 (1977) p.4831]

[11] B.J. Baliga [*Springer Proc. Phys. (Germany)* vol.71 (1992) p.305-13]

[12] X. Tang, K.G. Irvine, D. Zhang, M.G. Spencer [*Springer Proc. Phys. (Germany)* vol.71 (1992) p.206-9]

[13] M.M. Anikin, A.A. Lebedev, S.N. Pyatko, V.A. Soloview, A.M. Strelchuk [*Springer Proc. Phys. (Germany)* vol.56 (1992) p.269-73]

[14] K.B. Tolpygo [*Sov. Phys.-Solid State (USA)* vol.2 no.10 (1961) p.2367-76]

[15] G. Arlt, G.R. Schodder [*J. Acoust. Soc. Am. (USA)* vol.37 no.2 (1965) p.384-6]

[16] S. Karmann, R. Helbig, R.A. Stein [*J. Appl. Phys. (USA)* vol.66 no.8 (1989) p.3922-4]

[17] D.W. Feldman, J.H. Parker, W.J. Choyke, L. Patrick [*Phys. Rev. (USA)* vol.170 (1968) p.698- and vol.173 (1968) p.787]

[18] H.J. van Dall et al [*Sov. Phys.-Solid State (USA)* vol.12 (1963) p.109-27]

[19] Landolt-Bornstein [*Numerical Data and Functional Relationships in Science and Technology, New Series*, vol.2, Ed. K.H. Hellwege (Springer-Verlag, Berlin, Germany, 1969) p.61]

[20] I.V. Raptskaya, G.E. Rudashevskii, M.G. Kasaganova, M.I. Islitsin, M.B. Reifman, E.F. Fedotova [*Sov. Phys.-Solid State (USA)* vol.9 (1968) p.2833]

[21] J.S. Shor, D. Goldstein, A.D. Kurtz [*Springer Proc. Phys. (Germany)* vol.56 (1992)]

[22] L. Bemis, J.S. Shor, A.D. Kurtz [*Inst. Phys. Conf. Ser. (UK)* no.137 (1994) ch.7 p.723-6]

[23] [*Gmelins Handbuch der Anorganischen Chemie, 8th edition, Silicium* Part B (Weinheim, Verlag Chemie, GmbH, 1959)]

[24] P.B. Pickar et al [Engineering Sciences Lab., Dover, NJ, Research on Optical Properties of Single Crystal of Beta Phase Silicon Carbide. Summary Tech. Rept. no. OR8394, May 5, 1965-June 5, 1966. Contract no. DA-28-017-AMC-2002, A, Oct. (1966) 85p. AD-641 198]

[25] G.A. Slack [*J. Appl. Phys. (USA)* vol.35 (1964) p.3460]

[26] R.G. Breckenridge [Union Carbide Corp., Parma Res. Lab., Thermoelectric Materials, bi-monthly PR no. 8, Mar. 28 - May 28, 1960. Contract no. Nobs-77066, June 15, 1960. AD-245 092]

[27] E.L. Kern, D.W. Hamill, H.W. Deem, H.D. Sheets [*Mater. Res. Bull. (USA)* vol.4 (1969) p.S25]

[28] P.T.B. Shaffer [*Handbook of High Temperature Materials - No. 1. Materials Index* (Plenum Press, 1964) p.107]

[29] J.L. Durand, C.K. Houston [Martin Co., Martin-Marietta Corp., Orlando, FL. Infrared Signature Characteristics. ATL-TR-66-8. Contract no. AF 08 635 5087. Jan. 1966. 174 p. AD 478 597]

[30] D. Das [*Indian J. Phys. (India)* vol.40 (1966) p.684]

CHAPTER 2

OPTICAL AND PARAMAGNETIC PROPERTIES

2.1　Optical absorption and refractive index of SiC
2.2　Phonons in SiC polytypes
2.3　Photoluminescence spectra of SiC polytypes
2.4　Impurities and structural defects in SiC determined by ESR
2.5　ODMR investigations of defects and recombination processes in SiC

2.1 Optical absorption and refractive index of SiC

G.L. Harris

February 1995

A INTRODUCTION

This Datareview outlines the data on optical absorption and refractive index in the various polytypes of SiC. The optical transitions give rise to the characteristic colour of each polytype. Values for both the ordinary and extraordinary refractive indices versus wavelength are given.

B OPTICAL ABSORPTION

The optical absorption in SiC can, in general, be characterized by intraband and interband absorption components. The interband transitions in n-type polytypes other than 3C are responsible for the well-known colours of nitrogen-doped samples. The intraband absorption is of the free-carrier type and results in sub-bandgap transitions that are found in most forms of SiC. Biedermann [1] has measured the optical absorption bands at room temperature along the $E \perp c$ and $E \parallel c$ directions for 4H, 6H, 8H and 15R n-type SiC. These are the most anisotropical used types of SiC. These polytypes are uniaxial and are strongly dichroic [2]. The surface is normally perpendicular to the c-axis. This gives rise to the green colour in 6H, the yellow colour in 15R, and the green-yellow colour in 4H polytypes. These bands responsible for the colour are attributed to optical transitions from the lowest conduction band to other sites of increased density of states in the higher, empty bands [3], thus producing the various colours of nitrogen-doped 6H, 15R and 4H. Cubic SiC changes form a pale canary yellow to a greenish yellow when the material is doped. The yellow arises from a weak absorption in the blue region (see [3-6]). The shift towards the green for doped β-SiC is due to the free-carrier intraband absorption, which absorbs red preferentially [3].

The absorption edges of the seven polytypes of SiC have been studied by Choyke [4]. Solangi and Chaudhry [6] and Nishino et al [7] have measured the absorption coefficient of β-SiC. The shapes of the absorption curves are characteristic of indirect transitions. In FIGURE 1, the square root of the absorption coefficient is plotted vs. photon energy. These actual curves were supplied by [5]. In this case, $E \perp c$, the stronger absorption curves were used at liquid nitrogen temperature.

2.1 Optical absorption and refractive index of SiC

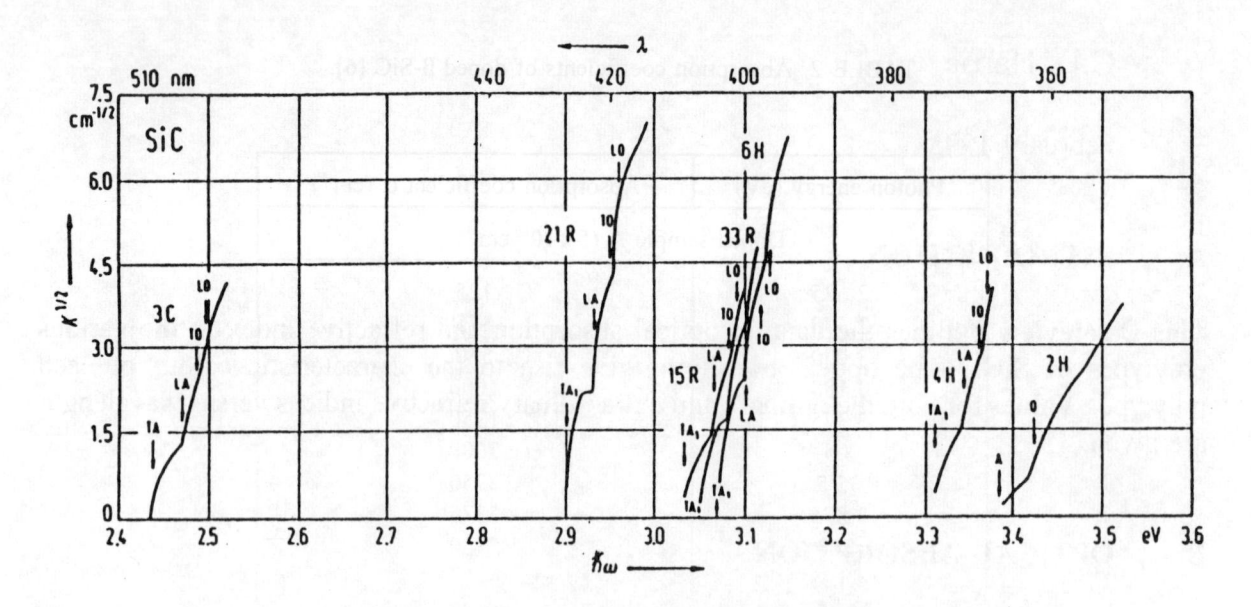

FIGURE 1 Absorption edge of several polytypes at 4.2 K; light polarised E⊥c [5].

TABLE 1 Absorption coefficients of β-SiC [2].

Sample A (relatively pure)	
Photon energy (eV)	Absorption coefficient α (cm^{-1})
2.25	47
2.375	63
2.50	88
2.625	150
2.75	360
2.875	407
3.0	698
3.125	1007
3.25	1400
3.375	1800
3.5	2190

The arrows on FIGURE 1 mark the principal phonons and a detailed discussion is provided in the later sections of this book. These indirect interband transitions reflect on the simultaneous absorption or emission of phonons. The mechanism of exciton transition for an indirect semiconductor may involve a bound exciton which may recombine without phonon emission during the transition. This is explained in detail in [4,5].

The absorption coefficient as a function of wavelength in β-SiC grown on Si is also provided in [5,6].

2.1 Optical absorption and refractive index of SiC

TABLE 2 Absorption coefficients of doped β-SiC [6].

Photon energy (eV)	Absorption coefficient α (cm⁻¹)
Doped sample A (5×10^{16} cm^{-3})	
2.625	125
2.375	375
3.000	550
3.188	1000
3.375	1250
3.5	2000
3.75	2750
3.875	3375
4.00	3750
4.125	4325
4.25	5000
4.375	5450
Doped sample B (6.9×10^{16} cm^{-3})	
2.438	125
2.50	250
2.625	300
2.75	500
2.875	750
3.00	1125
3.188	1500
3.375	2000
3.550	2720
3.675	3300
3.75	3700
3.825	4300
4.0	4950

C   REFRACTIVE INDEX OF SiC

Refractive indices for several polytypes of silicon carbide have been measured [8-12]. The 6H polytype of SiC has been measured in the most detail [8]. For the hexagonal form the c axis is assumed to be perpendicular to the surface. Thus, a normal incidence wave can be used to measure transmission and/or reflection. This normal incidence wave is often called the ordinary ray. A prism with its sides perpendicular to the c-axis is used to determine both n_o and n_e in the normal way (n_o = ordinary ray, n_e = extraordinary ray). A short summary of the data is presented here; for additional information please refer to [8,9].

2.1 Optical absorption and refractive index of SiC

TABLE 3 Refractive index of SiC polytypes.

Refractive index n	Wavelength λ (nm)	Material type	Ref
2.7104	467	3C	[8]
2.6916	498		
2.6823	515		
2.6600	568		
2.6525	589		
2.6446	616		
2.6264	691		
n empirical fit: $n(\lambda) = 2.55378 + (3.417 \times 10^4)/\lambda^2$ ($\lambda = 467 - 691$ nm)			
n_o 2.7121	435.8	2H prism edge	[10]
n_e 2.7966	parallel to c-axis		
n_o 2.7005	450.3		
n_e 2.7883			
n_o 2.6686	500.7		
n_e 2.7470			
n_o 2.6480	546.1		
n_e 2.7237			
n_o 2.6461	551.1		
n_e 2.7215			
n_o 2.6295	601.5		
n_e 2.7029			
n_o 2.6173	650.9		
n_e 2.6892			
n_o empirical fit: $n_o(\lambda) = 2.5513 + (2.585 \times 10^4)/\lambda^2 + (8.928 \times 10^8)/\lambda^4$ $n_e(\lambda) = 2.6161 + (2.823 \times 10^4)/\lambda^2 + (11.49 \times 10^8)/\lambda^4$			
n_o 2.7186	467	4H prism edge parallel to c-axis	[9]
n_e 2.7771			
n_o 2.6980	498		
n_e 2.7548			
n_o 2.6881	515		
n_e 2.7450			
n_o 2.6655	568		
n_e 2.7192			
n_o 2.6588	589		
n_e 2.7119			
n_o 2.6508	616		
n_e 2.7033			
n_o 2.6335			
n_e 2.6834			
$n_o(\lambda) = 2.5610 + (3.4 \times 10^4)/\lambda^2$ $n_e(\lambda) = 2.6041 + (3.75 \times 10^4)/\lambda^2$			

TABLE 3 continued

Refractive index n	Wavelength λ (nm)	Material type	Ref
n_o 2.7074 n_e 2.7553	467	6H prism edge parallel to c-axis	[9]
n_o 2.6870 n_e 2.7331	498		
n_o 2.6789 n_e 2.7236	515		
n_o 2.6557 n_e 2.6979	568		
n_o 2.6488 n_e 2.6911	589		
n_o 2.6411 n_e 2.6820	616		
n_o 2.6243 n_e 2.6639	691		
$n_o(\lambda) = 2.5531 + (3.34 \times 10^4)/\lambda^2$ $n_e(\lambda) = 2.5852 + (3.68 \times 10^4)/\lambda^2$			
n_o 2.7081 n_e 2.7609	467	15 R prism edge parallel to c-axis minimum deflection method	[9]
n_o 2.6894 n_e 2.7402	498		
n_o 2.6800 n_e 2.7297	515		
n_o 2.6572 n_e 2.7043	568		
n_o 2.6503 n_e 2.6968	589		
n_o 2.6429 n_e 2.6879	616		
n_o 2.6263 n_e 2.6676	691		
$n_o(\lambda) = 2.558 + (3.31 \times 10^4)/\lambda^2$ $n_e(\lambda) = 2.5889 + (3.74 \times 10^4)/\lambda^2$			
n_o 2.639	633	from electro-optic coefficient measurement β-SiC, 300 K	[14]

D CONCLUSION

The interband transitions in SiC polytypes give rise to the characteristic colours of green (6H), yellow (15R) and green-yellow (4H). Cubic SiC exhibits a pale yellow colour when undoped and greenish yellow in the doped state. The variations of refractive index with wavelength for the 3C, 2H, 4H, 6H and β-SiC polytypes are listed.

REFERENCES

[1] E. Biedermann [*Solid State Commun. (USA)* vol.3 (1965) p.343]
[2] L. Patrick, W.J. Choyke [*Phys. Rev. (USA)* vol.186 no.3 (1969) p.775-7]
[3] W.J. Choyke [*NATO ASI Ser. E, Appl. Sci. (Netherlands)* vol.185 (1990) p.563-87]
[4] W.J. Choyke [*Mater. Res. Soc. Symp. Proc. (USA)* vol.97 (1987) p.207-19]
[5] W.J. Choyke [*Mater. Res. Bull. (USA)* vol.4 (1969) p.5141-52]
[6] A. Solangi, M.I. Chaudhry [*J. Mater. Res. (USA)* vol.7 no.3 (1992) p.539-54]
[7] S. Nishino, H. Matsunami, T. Tanaka [*Jpn. J. Appl. Phys. (Japan)* vol.14 (1975) p.1833]
[8] P.T.B. Schaffer, R.G. Naum [*J. Opt. Soc. Am. (USA)* vol.59 (1969) p.1498]
[9] P.T.B. Schaffer [*Appl. Opt. (USA)* vol.10 (1971) p.1034]
[10] J.A. Powell [*J. Opt. Soc. Am. (USA)* vol.62 (1972) p.341]
[11] V. Rehn, J.L. Stanford, V.O. Jones [*Proc. 13th Int. Conf. Phys. Semicond.*, Rome, 1976, Ed. F.G. Fumi (Typografia Marves, Rome, 1976) p.985]
[12] W.J. Choyke, E.D. Palik [*Handbook of Optical Constants of Solids* (Academic Press, 1985) p.587-95]
[13] W.J. Choyke [*NATO ASI Ser. E, Appl. Sci. (Netherlands)* vol.185 (1990) p.863, and references therein]
[14] X. Tang, K.G. Irvine, D. Zhang, M.G. Spencer [*Springer Proc. Phys. (Germany)* vol.71 (1992) p.206-9]

2.2 Phonons in SiC polytypes

J.A. Freitas Jr.

July 1994

A INTRODUCTION

The technique of Raman scattering (RS) to study vibrational spectra in the numerous polytypes of SiC will be described. An explanation of the various notations used to describe the stacking sequences in these polytypes will then be given. Section C discusses the various optical phonons studied by RS and the concept of a common phonon spectrum for all polytypes will be introduced. Raman studies are also used to assess crystalline structure and quality of epitaxial layers of SiC on Si and SiC substrates. Section D outlines several other excitations of interest, e.g. polaritons, plasmons, and electronic RS, as well as impurity and defect recognition in irradiated and ion implanted material.

B RAMAN SPECTRA AND PHONON DISPERSION IN SiC POLYTYPES

Raman scattering (RS) is one of the most commonly used techniques to study vibrational phenomena in solids [1]. Inelastic light scattering in crystals is susceptible to selection rules originating from wavevector (**q**) conservation [1,2]. The magnitudes of the incident (**q**$_i$) and the scattered (**q**$_s$) radiation wavevectors are much smaller than that of a general vector (**q**) in the Brillouin Zone (BZ) (q_i, q_s << q). Hence, in order to conserve **q** the created (Stokes) or annihilated (anti-Stokes) phonon must have a wavevector of magnitude $q \approx 0$, i.e. near the centre of the BZ (Γ-point). Thus, first order RS can investigate only phonons with $q \approx 0$. Phonons with larger **q** may be observed in second order (two phonon) Raman spectra. The $q \approx 0$ restriction can also be removed by introducing impurities or defects into the material or by fabricating crystals with a much larger lattice constant along the growth direction than that of the corresponding single crystals (superlattice). A technique more suited to the study of phonon dispersion is neutron scattering, since the thermal neutron wavevector is about the same order of magnitude as a general BZ vector [3]. However, neutron scattering spectroscopy has some serious limitations, such as poor resolution and frequency determination. In addition, neutron scattering experiments require large samples. In contrast, RS can employ micro-Raman spectrometers with a probing laser spot size of ~1 µm [4].

SiC has fascinated material scientists not only for its extreme mechanical, thermal and electronic properties but also because of its unusual structural properties. The basic unit of SiC consists of a covalently bonded tetrahedron of Si (or C) atoms with a C (or Si) at the centre. The identical polar layers of Si_4C (or C_4Si) are continuously stacked and the permutation of stacking sequences allows an endless number of different one-dimensional orderings (polytypes) without variation in stoichiometry.

The stacking sequence of these polytypes can be described by the 'ABC' notation, where A, B and C represent the three sites available in one sublattice. For example (...ABCABC...) and (...ABABAB...) stand for cubic (3C or β-SiC) and simple hexagonal (wurtzite) respectively. The intermixing of these two simple forms can also occur, generating a large number of ordered structures (α-SiC). The most commonly used notation for identifying polytypes is due to Ramsdell [5]. In this notation, the number of layers in the stacking direction (c-axis), before the sequence is repeated, is combined with the letter representing the Bravais lattice type, i.e. cubic (C), hexagonal (H) or rhombohedral (R). The arbitrariness of the layer designation in the Ramsdell notation can be avoided by using the notation suggested by Jagodzinski [6]. In this later notation, any one layer in the stacking sequences can be described as having a local cubic (k) or hexagonal (h) environment (at least with respect to the positions of the centroids of the immediate neighbours). As a result of the tetrahedral unit acentricity in sequences of the type hhhh the layers are offset and rotated by 180 degrees. Consequently one observes two stacking operations for SiC: (a) layer-translation only, yielding (kkkkkk) (or ...ABCABC...), and (b) layer-translation and rotation, yielding (hhhh) (or ...ABABAB...).

Zhadanov [7] visualised the SiC stacking sequences by considering the sequence of non-basal tetrahedral planes. The observed zig-zag of layers ([111] direction for cubic, and [0001] direction for hexagonal and rhombohedral) was represented by a notation which denotes the number of consecutive layers without rotation, the total stacking sequence, and a subscript indicating the sequence repetition. A detailed discussion of polytypism, polytypic transformation and notations has been reported by Jepps and Page [8].

TABLE 1 lists some of the simple SiC polytypes along with the four notations briefly discussed here. The percentage of hexagonality, also listed in TABLE 1, can be determined from the Jagodzinski notation. For example, the 6H polytype is represented by $(hkk)_2$, and therefore it has six biplanes composed of two with hexagonal coordination and four with cubic coordination. Thus, the 6H polytype has 33 % of hexagonality character. The number of atoms (n_a) per unit cell for hexagonal and rhombohedral polytypes also can be determined from the number of biplanes (n_b) in the polytype by: $n_a = n_b \times n$, where n = 2 or 2/3 for the hexagonal and rhombohedral polytypes, respectively.

TABLE 1 Seven of the most simple SiC polytypes with four notations (R = Ramsdell notation, J = Jagodzinski notation, Z = Zhadanov notation). They are listed by increasing percent hexagonality.

R	ABC notation	J	Z	% of hexagon-ality	Space group	No. of atoms per unit cell
3C	ABC	(k)	(∞)	0	$T_d^2(F\bar{4}3m)$	2
8H	ABACBABC	$(khkk)_2$	(44)	25	$C_{6v}^4(P6_3mc)$	16
21R	ABCACBACABCBACBCABACB	$(hkkhkkk)_3$	$(34)_3$	29	$C_{3v}^5(R3m)$	14
6H	ABCACB	$(hkk)_2$	(33)	33	$C_{6v}^4(P6_3mc)$	12
15R	ABACBCACBABCBAC	$(hkhkk)_3$	$(32)_3$	40	$C_{3v}^5(R3m)$	10
4H	ABAC	$(hk)_2$	(22)	50	$C_{6v}^4(P6_3mc)$	8
2H	ABAB	$(h)_2$	(11)	100	$C_{6v}^4(P6_3mc)$	4

C PHONONS IN SiC

Cubic SiC crystallises in the zinc blende structure (space group T_d^2; $F\bar{4}3m$), which has two atoms per unit cell, and thus three optical modes are allowed at the centre of the BZ. Since SiC is a polar crystal, the optical modes are split into one non-degenerate longitudinal optical phonon (LO) and two degenerate transverse optical phonons (TO). The first and second order RS studies of cubic SiC were discussed in detail by Olego and co-workers [9-11]. The first and second order spectra measured as a function of hydrostatic pressure reveal the mode-Grüneisen parameters of the LO and TO modes at Γ and of several optical and acoustical phonons at the edge of the BZ [9,10]. The data analysis also provides information about the relaxation mechanisms of the Γ optical phonons. Information about the transverse effective charge was obtained from a study of the pressure [10] and temperature [11] dependences of the long-wavelength optical phonons.

2H-SiC, the rarest polytype, has the wurtzite structure and belongs to the space group C_{6v}^4 ($P6_3mc$). This uniaxial crystal has four atoms per unit cell, and consequently has nine long-wavelength optical modes. Group theory predicts the following Raman active lattice phonons, near the centre of the BZ: an A_1 branch with phonon polarisation in the uniaxial direction, a doubly degenerate E_1 branch with phonon polarisation in the plane perpendicular to the uniaxial direction, and two doubly degenerate E_2 branches. The A_1 and E_1 phonons are also infrared active, while E_2 is only Raman active. In uniaxial crystals, in addition to the long-range electrostatic forces responsible for the longitudinal-transverse splitting, we observe the short-range interatomic forces which exhibit the anisotropy of the force constants. The predominance of one of the two independent forces dictates the split between the different long-wavelength optical modes. For a comprehensive discussion of Raman selection rules, tensors and anisotropy see, for example, [2,12].

The most common and one of the most studied SiC polytypes is 6H. Like 2H and other hexagonal polytypes, 6H-SiC belongs to the space group C_{6v}^4, but has 12 atoms per unit cell (see TABLE 1), leading to 36 phonon branches, 33 optical and 3 acoustic. Therefore the number of phonons observed by RS is greater for 6H than for 2H-SiC, resulting in an additional complication in distinguishing between several normal modes with the same symmetry. The assignment of the Raman spectra of 6H-SiC may be simplified by using the concept of a standard large zone (LZ) introduced by Patrick [13]. Since the SiC polytypes are characterised by a one-dimensional stacking sequence of planes in the axial direction (c-axis), the standard LZ for 6H-SiC extends to $6\pi/c$, where c is the unit cell axial dimension. Note that the reciprocal lattice vector is $2\pi/c$; thus we obtain pseudomomentum vectors $q=0$, $2\pi/c$, $4\pi/c$ and $6\pi/c$, all equivalents to $q=0$ in the BZ. This procedure, referred to as 'zone-folding' (or reduced zone scheme), is discussed extensively in the RS studies of superlattices (see for example [14]). Feldman et al [15] have used Raman spectroscopy to study lattice vibration in 6H-SiC. They define a reduced momentum $x = q/q_{max}$ which establishes the points in the phonon dispersion curves (for both optic and acoustic branches) accessible by RS. They observed that all long-wavelength modes are found on the LZ axis, and classified the modes in two categories, namely: (a) the strong modes at $x=0$, which show angular dependence, and (b) the weak modes, with no angular dependence, and with identification that depends upon their assignment to values of $x=q/q_{max}$ ($x\neq 0$) in the LZ. The weak modes were conveniently classified according to their atomic motions as axial (parallel to c: extraordinary) or planar (perpendicular to c: ordinary). Feldman et al [16] extended their

2.2 Phonons in SiC polytypes

RS studies to other polytypes and verified the concept of a common phonon spectrum (within 2%) for all polytypes [13]. They also were able to construct the phonon dispersion curves for SiC from their first-order RS data from various polytypes (except 3C-SiC, since this polytype is isotropic), namely: 4H, 6H, 15R and 21R. The phonon dispersion curves show discontinuities in the LZ representation (observed as doublet peaks in the Raman spectra), which result from the dependence of the force constant on the types of layers (hexagonal or cubic) in the polytype structure [17]. Discontinuities in the phonon dispersion curves are the main effect of layered structures (for a detailed discussion of 'mini-gaps', see [14] p.98). Recently Nakashima et al [18] used Raman measurements from 8H and 27R polytypes to fill out the phonon dispersion curve reported by Feldman et al [16].

All the previously discussed Raman studies have been carried out with small single-crystal samples grown via the very high temperature Lely sublimation process [19]. Since large samples are required for practical applications, monocrystalline Si has been generally chosen as the substrate for the growth of the β-SiC [20-23]. However, the mismatches in the coefficients of thermal expansion ($\approx 8\%$ at 473 K) and lattice parameters ($\approx 20\%$) have compromised the film quality. Room temperature RS can be used conveniently to verify the crystalline structure and quality of the film, as well as the reduction of residual strain (of free-standing films) under annealing treatments [24]. Biaxial stress in SiC/Si and its dependence upon film thickness and buffer layer was examined in detail by Feng et al [25,26].

In an attempt to reduce the concentration of all of the various defects commonly observed in the 3C-SiC/Si heteroepitaxial system, some research groups investigated the growth of 3C-SiC on 6H-SiC single crystal substrates [27-30]. As in the 3C-SiC/Si system, RS was successfully used for crystallographic identification of the deposited film [31]. This was made possible because of evident differences between 3C- and 6H-SiC Raman spectra, originating from the 6H uniaxial symmetry and zone folding. In the spectral range of the strong modes ($x = q/q_{max} = 0$; 760 to 980 cm^{-1}), by using a convenient RS configuration, the 6H-SiC Raman spectrum will show an A_1-LO (~ 964 cm^{-1}) and two E_2 (788 and 766 cm^{-1}) phonons, while the 3C-SiC Raman spectrum will present an LO (972 cm^{-1}) and TO (796 cm^{-1}) phonon. The differences in the phonon frequencies (5 cm^{-1} or more) and number of lines (see TABLE 2 for more details) provide an easy way of distinguishing between the two polytypes [24,31]. More significant differences are observed at the low frequency region (140 to 510 cm^{-1}, acoustic phonon range), where a number of phonons (weak modes) can be observed in 6H-SiC, but not in 3C. These weak modes, listed in TABLE 3 for 4H, 6H and 15R polytypes, can be observed at the Γ point because of the folding of the LZ into the real BZ. As discussed previously, in the case of 6H-SiC, the LZ is six times larger than the reduced BZ. Consequently each phonon branch can be observed at four different energy values at the centre of the BZ.

2.2 Phonons in SiC polytypes

TABLE 2 Frequencies of 3C- and 6H-SiC polytypes modes. The anisotropy is given by TO(1) - TO(2).

Polytype	E_1 (cm^{-1})	E_2 (cm^{-1})	TO(2) (cm^{-1})	LO(2) (cm^{-1})	TO(1) (cm^{-1})	LO(1) (cm^{-1})	TO(1) - TO(2) (cm^{-1})	Ref
3C	-	-	-	-	796	972	0	[16,17,30]
	-	-	-	-	796.2	972.7		[10]
6H	766	788	788	964	797	970	9	[16,17]
	768	789		967	796			[30]

TABLE 3 Symmetry and frequency of phonon modes for 4H-, 6H- and 15R-SiC polytypes. $x = q/q_{max}$ is the phonon reduced momentum. N.O. stands for not observed.

Polytype	$x = q/q_{max}$	Optical (cm^{-1}) Axial	Optical (cm^{-1}) Planar	Acoustic (cm^{-1}) Axial	Acoustic (cm^{-1}) Planar	Ref
4H	0.5		{E$_2$}776		{E$_2$}204	[16]
			{E$_2$}N.O.		{E$_2$}196	[16]
	1.0	{A$_1$}838	{E$_1$}N.O.	{A$_1$}610	{E$_1$}266	[16]
6H	0.33		{E$_2$}788(789)		{E$_2$}149(150)	[15]([17])
			{E$_2$}N.O.		{E$_2$}145(140)	[15]([17])
	0.67	{A$_1$}889	{E$_1$}777	{A$_1$}508	{E$_1$}241	[15]
		{A$_1$}N.O.	{E$_1$}769	{A$_1$}504	{E$_1$}236	[15]
	1.0		{E$_2$}766(768)		{E$_2$}262(266)	[15]([17])
15R	0.4	{A$_1$}938	{E}785(786)	{A$_1$}337	{E}172(173)	[16]([17])
		{A$_1$}932	{E}N.O.	{A$_1$}331	{E}167(167)	[16]([17])
	0.8	{A$_1$}860	{E}769(770)	{A$_1$}577	{E}256(256)	[16]([17])
		{A$_1$}N.O.	{E}N.O.	{A$_1$}569	{E}254(255)	[16]([17])

D MISCELLANEOUS ELEMENTARY EXCITATIONS IN SiC

Inelastic light scattering is not restricted to vibrational phenomena; indeed RS has been used in the study of other excitations in the SiC polytype system, e.g. polaritons, plasmons, and electronic RS. In polar crystals TO phonons and photons with approximately the same energy and wave vector may couple strongly. This mixed propagating excitation, called the polariton, can no longer be described as a phonon or a photon, and its dispersion is determined by the material dielectric constant. In the case of a uniaxial crystal [32], the orientation of the electric field (E) of the incident and the scattered light with respect to the crystal axis must be taken into account. Since the crystal refractive index and the dielectric constant for the ordinary and extraordinary rays are different, the polariton is anisotropic. Azhnyuk et al [33] carried out a detailed study of near-forward Raman scattering (inelastic scattering experiment performed at small angles with respect to the incident beam direction) in 6H-SiC. They were able to observe the extraordinary (A_1-branch) and the ordinary (E_1-branch) polaritons. They also investigated the effect of free-electrons on the polariton dispersion curve. It was observed that for nitrogen-doped (N-doped) samples the polariton shift decreased with increasing free charge carrier concentration (plasmon). The interaction between plasmon and polariton is commonly called plasmariton.

In a heavily N-doped 6H sample (6×10^{19} cm^{-3}) Klein et al [34] observed an asymmetric broadening and a shift of the A_1(LO) phonon which were attributed to the overdamped coupling between LO phonon and plasmon modes [35]. The interaction between these two excitations occurs via their macroscopic electric fields when the frequency of oscillation of a free-carrier plasma is close to that of the LO phonon. The dependence of the LO phonon-overdamped plasmon coupled modes on carrier concentration was reported by Yugami et al [36] in 3C-SiC films, where the carrier concentrations varied from 6.9×10^{16} to 2×10^{18} cm^{-3}. They verified that the carrier concentrations obtained from RS were in fairly good agreement with the Hall measurement values, and that the Faust-Henry coefficient [35] for the 3C-SiC (C = +0.35) was close to the value reported for 6H-SiC (C = +0.39) [34].

Low temperature Raman measurements performed on N-doped 6H-SiC show electronic transitions of E_2 symmetry at 13.0, 60.3 and 62.6 meV, which were assigned to 1s(A_1) and 1s(E) valley-orbit transitions at the three inequivalent N-donor sites [37] (one hexagonal and two cubics, see TABLE 1). The fourth peak at 78.8 meV was interpreted as a phonon mode in the phonon gap region [37,38]. These interpretations were confirmed by an RS study of the 15R-SiC polytype which has two hexagonal and three cubic sites [37]. Colwell and Klein [37] argued that the observation of both the E_2 symmetry of the Raman measurements and a single transition per site is associated with the conduction-band minima lying along the edges of the BZ. This is consistent with suggestions by Junginger and van Haeringen [39] and Herman et al [40].

The formation and annihilation of defects induced in semiconductors under irradiation or ion implantation is still a subject of both academic and technological interest. Room temperature Raman scattering can conveniently be used to identify a substitutional impurity (localised vibrational mode) [41], or structural defect (breaking of the translational symmetry of the crystal) [42]. The lattice recovery under heat treatment can easily be verified by monitoring the reduction of the phonon linewidth and frequency shifts [42]. Although Raman scattering has been extensively used to study lattice damage and recovery in single and compound semiconductors, only a few studies have been reported in the SiC system. Raman scattering studies of H-, D- and He-ion implanted single crystals of 6H-SiC indicate that high-frequency local modes associated with the presence of H appear only after annealing at about 900 °C [43]. However, no feature associated with D was observed. The lattice seems to recover after annealing at about 1500 °C, since only the vibrational spectrum of unimplanted material was observed [43]. Rahn et al [44] observed that at higher ion implantation doses (10^{20} to 10^{22} H ions cm^{-3}) new Raman lines appear in their spectra. Their polarisation experiments suggest that these lines are associated with two different types of bonded carbon interstitial defects.

E CONCLUSION

The differences between the various SiC polytypes in terms of stacking sequence have been discussed using the four standard notations. Percentage hexagonality and number of atoms per unit cell have been listed. Raman scattering is the most widely used technique to study optical modes in the various polytypes. The 6H-SiC polytype has 36 phonon branches and resort is made to the concept of a standard large zone to simplify interpretation of the spectra. Studies covering a wide range of polytypes have established a common phonon spectrum for all polytypes. Growth of large samples, for practical applications, has been accomplished by epitaxy on both Si and SiC substrates. Raman scattering is then used to assess both the structure and quality of the epitaxial layers and to differentiate between polytypes. Phonons and photons can interact in polar crystals to produce polaritons whose dispersion is fixed by the dielectric constants. Interactions of phonons with plasmon (free charge carrier concentration) modes also occur. Room temperature Raman scattering can be used to differentiate between impurities and defects in irradiated or ion implanted/annealed samples.

REFERENCES

[1] R.S. Krisnhan, G.W. Chantry, R.A. Cowley, C.E. Hathaway, P. Lallemand [*The Raman Effect* vol.1, Ed. A. Anderson (M. Dekker Inc., New York, USA, 1971)]
[2] R. Loudon [*Adv. Phys. (UK)* vol.13 (1964) p.423-82]
[3] H. Bilz, W. Kress [*Phonon Dispersion Relations in Insulators* (Springer, Berlin, Heidelberg, 1979)]
[4] See, for example, J.A. Freitas Jr., J.S. Sanghera, U. Strom, P.C. Pureza, I.D. Aggarwal [*J. Non-Cryst. Solids (Netherlands)* vol.140 (1992) p.166-71]
[5] R.S. Ramsdell [*Am. Mineral. (USA)* vol.32 (1947) p.64-82]
[6] H. Jagodzinski [*Acta Crystallogr. (Denmark)* vol.2 (1949) p.201-15]
[7] G.R. Zhadanov [*C. R. Acad. Sci. (France)* vol.48 (1945) p.39-55]
[8] N.W. Jepps, T.F. Page [*Prog. Cryst. Growth Charact. Mater. (UK)* vol.7 (1983) p.259-307]
[9] D. Olego, M. Cardona [*Phys. Rev. B (USA)* vol.25 (1982) p.1151-60]
[10] D. Olego, M. Cardona, P. Vogl [*Phys. Rev. B (USA)* vol.25 (1982) p.3878-88]
[11] D. Olego, M. Cardona [*Phys. Rev. B (USA)* vol.25 (1982) p.3889-96]
[12] C.A. Arguello, D.L. Rousseau, S.P.S. Porto [*Phys. Rev. (USA)* vol.181 (1969) p.1351-63]
[13] L. Patrick [*Phys. Rev. (USA)* vol.167 (1968) p.809-13]
[14] B. Jusserand, M. Cardona [*Light Scattering in Solids V*, Topics in Applied Physics, Eds M. Cardona, G. Güntherodt (Springer, Berlin, Heidelberg, 1989) p.49-152]
[15] D.W. Feldman, J.H. Parker Jr., W.J. Choyke, L. Patrick [*Phys. Rev. (USA)* vol.170 (1968) p.698-704]
[16] D.W. Feldman, J.H. Parker Jr., W.J. Choyke, L. Patrick [*Phys. Rev. (USA)* vol.173 (1968) p.787-93]
[17] C.H. Hodges [*Phys. Rev. (USA)* vol.187 (1969) p.994-9]
[18] S. Nakashima, H. Katahama, Y. Nakakura, A. Mitsuishi [*Phys. Rev. B (USA)* vol.33 (1986) p.5721-9]
[19] J.A. Lely [*Ber. Dtsch. Keram. Ges. (Germany)* vol.32 (1955) p.229-51]
[20] S. Nishino, J.A. Powell, H.A. Will [*Appl. Phys. Lett. (USA)* vol.42 (1983) p.460-2]

[21] A. Addamiano, P.H. Klein [*J. Cryst. Growth (Netherlands)* vol.70 (1984) p.291-4]
[22] K. Sasaki, E. Sakuma, S. Misana, S. Yoshida, S. Gonda [*Appl. Phys. Lett. (USA)* vol.45 (1984) p.72-3]
[23] P.H. Liaw, R.F. Davis [*J. Electrochem. Soc. (USA)* vol.132 (1985) p.642-50]
[24] J.A. Freitas Jr., S.G. Bishop, A. Addamiano, P.H. Klein, H.J. Kim, R.F. Davis [*Mater. Res. Soc. Symp. Proc. (USA)* vol.46 (1985) p.581-6]
[25] Z.C. Feng, A.J. Mascarenhas, W.J. Choyke, J.A. Powell [*J. Appl. Phys. (USA)* vol.64 (1988) p.3176-86]
[26] Z.C. Feng, W.J. Choyke, J.A. Powell [*J. Appl. Phys. (USA)* vol.64 (1988) p.6827-35]
[27] R.W. Bartlett, R.A. Mueller [*Mater. Res. Bull. (USA)* vol.4 (1969) p.341]
[28] P. Rai-Choudhury, N.P. Formigoni [*J. Electrochem. Soc. (USA)* vol.116 (1969) p.1140]
[29] H.S. Kong, B.L. Jiang, J.T. Glass, G.A. Rozgonyi, K.L. More [*J. Appl. Phys. (USA)* vol.63 (1988) p.2645]
[30] H.S. Kong, J.T. Glass, R.F. Davis [*J. Mater. Res. (USA)* vol.4 (1989) p.204-14]
[31] H. Okumura et al [*J. Appl. Phys. (USA)* vol.61 (1987) p.1134-6]
[32] S.P.S. Porto, B. Tell, T.C. Damen [*Phys. Rev. Lett. (USA)* vol.16 (1966) p.450-2]
[33] Yu.M. Azhnyuk, V.V. Artamonov, M.Ya. Valakh, A.P. Litvinchuk [*Phys. Status Solidi B (Germany)* vol.135 (1986) p.75-84]
[34] M.V. Klein, B.N. Ganguly, P.J. Colwell [*Phys. Rev. B (USA)* vol.6 (1972) p.2380-8]
[35] W.L. Faust, C.H. Henry [*Phys. Rev. Lett. (USA)* vol.17 (1966) p.1265]
[36] H. Yugami et al [*J. Appl. Phys. (USA)* vol.61 (1987) p.354-8]
[37] P.J. Colwell, M.V. Klein [*Phys. Rev. B (USA)* vol.6 (1972) p.498-515]
[38] P.J. Colwell, W.D. Compton, M.V. Klein, L.B. Schein [*Proc. 10th Int. Conf. on the Phys. of Semiconductors*, Cambridge, MA, Eds S.P. Keller, J.C. Hensel, F. Stern (USAEC Div. of Technical Information Extension, Oak Ridge, Tenn., 1970) p.484-5]
[39] H.G. Junginger, W. van Haeringen [*Phys. Status Solidi (Germany)* vol.37 (1970) p.709]
[40] F Herman, J.P. van Dyke, R.L. Kortum [*Mater. Res. Bull. (USA)* vol.4 (1969) p.S167]
[41] R.J. Nemanich [*Mater. Res. Soc. Symp. Proc. (USA)* vol.69 (1986) p.23-37]
[42] G. Burns, F.H. Dacol, C.R. Wie, E. Burstein, M. Cardona [*Solid State Commun. (USA)* vol.62 (1987) p.449-54]
[43] L.A. Rahn, P.J. Colwell, W.J. Choyke [*Proc. 3rd Int. Conf. on Light Scattering in Solids*, Campinas, SP, Brazil, Eds M. Balkanski, R.C.C. Leite, S.P.S. Porto (Flammarion Sc., Paris, 1975) p.607-11]
[44] L.A. Rahn, P.J. Colwell, W.J. Choyke [*Bull. Am. Phys. Soc. (USA)* vol.21 (1976) p.408]

2.3 Photoluminescence spectra of SiC polytypes

J.A. Freitas Jr.

March 1995

A INTRODUCTION

Photoluminescence (PL) spectroscopy is recognized as a powerful, sensitive, non-destructive technique for the detection and identification of impurities and other defects in semiconductors. Utilizing PL spectroscopy involves the measurement and interpretation of the spectral distribution of recombination radiation emitted by the sample. Electrons and holes which are optically excited across the forbidden energy gap usually become localized or bound at an impurity or defect before recombining, and the identity of the localized centre to which they are bound can often be determined from the PL spectrum. Usually, relatively sharp-line, near band edge spectra arise from the recombination of electron-hole pairs which form bound excitons (BE) [1] at impurity sites, or the free-to-bound (FB) transitions [2] which involve the recombination of free electrons (holes) with holes (electrons) bound at neutral acceptors (donors), while lower energy and broader PL bands arise from the recombination of carriers localized at deep traps. Information about crystalline structure and quality can be inferred from the spectral energy position and/or from thermal quenching studies. There are very few constraints on the materials that can be characterized by PL, and a broad scope of phenomena such as excitation and recombination mechanisms, impurities and structural defects can be investigated [3].

B DONORS

In 1964, Choyke et al [4] reported a seminal work on the optical characterization of cubic SiC grown by the so-called Lely technique. Their low temperature absorption measurements revealed the indirect character of the bandgap, with an exciton energy gap at 2.390 eV. They also reported the low temperature PL spectrum, characterized by five well-resolved lines, including a 2.379 eV zero phonon line (ZPL) (recombination without phonon emission) located 10 meV below the exciton energy gap (E_{gx} [5]), and its phonon replicas TA, LA, TO and LO at 46, 79, 94 and 103 meV, respectively, at the low energy side of the ZPL line. On the basis of comparison with PL spectra of nitrogen-doped 6H and 15R crystals grown in the same furnace run [6,7], the authors attribute the five-line PL spectrum in the cubic crystals to the recombination of excitons bound at neutral nitrogen donors (nitrogen replacing the carbon atom: N_c) [4]. Investigation of the Zeeman and uniaxial splitting of these five-line spectra [8] determined that the luminescence signature does arise from the recombination of excitons bound to isolated neutral donors (i.e. not complexes). Dean et al [9] studied the two electron satellite transitions associated with this five-line donor-bound exciton spectrum and inferred a highly accurate value of 53.6 ± 0.5 meV for the donor ionization energy. These results are consistent with but do not provide proof of the attribution to N_c donors.

PL spectra assigned to nitrogen donors in the SiC system were first reported for the 6H polytype [6]. The observed PL spectrum, consisting of about 50 lines, was attributed to exciton recombination after capture by neutral nitrogen impurity, resulting in a four-particle complex (the donor ion, an electron and an exciton [10]). The presence of nitrogen in 6H SiC was unambiguously established by electron spin resonance (ESR) studies reported previously by Woodbury and Ludwig [11]. In cubic SiC, the N_c ESR spectrum was reported ten years later [12,13]. The N_c 6H PL spectrum differs from that in 3C-SiC in its complexity (larger number of resolved lines), which originates from the larger 6H unit cell and from the carbon site inequivalency [14,15]. In [6], the reported values for E_{gx} and the three ZPLs are 3.024 eV, 3.008 eV, 2.993 eV and 2.991 eV, respectively. Excitons can also be captured by an ionized nitrogen donor, resulting in a three-particle complex (the donor ion and an exciton [10]). The PL spectrum of the three-particle nitrogen-exciton complex in the 6H polytype [16] shows that the exciton binding energies are an order of magnitude greater than in the case of four-particle complexes. By combining the PL data from four- and three-particle complexes, Hamilton et al [16] estimated ionization energies of 0.17, 0.20 and 0.23 eV, for the three inequivalent nitrogen donors.

The four-particle and three-particle complexes have also been observed in the 15R [7] and 33R [17] polytypes: however, in both materials one of the ZPLs is missing. The missing line may be caused by a small exciton binding energy, which may produce an unresolved doublet, or by concealing a low intensity line in between many phonon replicas. Surprisingly, the three-particle spectrum is absent from the PL spectra of 3C [4], 2H [18], 4H [19] and 21R [20] polytypes. Although there is not a clear explanation, it is possible that the binding energy of the hole to a neutral (N_c^o) or ionized (N_c^+) nitrogen donor is very sensitive to the donor environment, such that in some polytypes the binding of a hole in the three-particle complex is extremely small and the hole is not bound [21]. The PL spectra corresponding to phonon emission for the seven polytypes discussed here have as a common characteristic six strong lines with about the same displacement from the ZPLs (nearly the same energies in all polytypes) [22,23], with the exception of the 2H SiC [24]. These six 'principal' phonons are those with the same **k** value as the conduction band minima in the large zone [17]. Note that the PL spectra reflect the position of the conduction band minima, and hence phonon energies measured in the PL spectra do not coincide with Raman scattering measurements, which show a common set of dispersion curves for the axial direction of all polytypes [25]. The energies of the principal phonons observed in the PL spectra of six polytypes are listed in TABLE 1. The optical absorption and mechanism of excitation and recombination in the seven SiC polytypes discussed here were reviewed in [26].

TABLE 1 Principal phonon energies (meV) in six polytypes derived from the luminescence spectrum (from [26]). The notation (TA) for transverse acoustic, (LA) for longitudinal acoustic, (TO) for transverse optical, and (LO) for longitudinal optical phonons is accurate for the cubic polytype, but not for the hexagonal and rhombohedral polytypes. For details see Datareview 2.2.

Phonon branch	3C	4H	6H	21R	15R	33R
TA_1	46.3	46.7	46.3	46.5	46.3	46.3
TA_2	-	51.4 53.4	53.5	53.0	51.9	52.3
LA	79.5	76.9 78.8	77.0	77.5	78.2	77.5
TO_1	94.4	95.0	94.7	94.5	94.6	94.7
TO_2	-	-	95.6	-	95.7	95.7
LO	102.8	104.0 104.3	104.2	104	103.7	103.7

2.3 Photoluminescence spectra of SiC polytypes

Free and bound excitons in cubic SiC were investigated by Nedzvetskii et al [27]. From their study of the dependence of the absorption and luminescence spectra on temperature they determined for the exciton binding energy (E_x), the bound exciton binding energy to N_c, and the fundamental energy gap the values of 13.5 meV, 9.1 meV and $E_g = 2.4018$ eV (at 4.2 K), respectively. The values reported for E_x, E_{gx}, and phonon energy are in good agreement with results reported by Choyke et al [4]. In [27], it was also observed that the phonon-assisted luminescence of bound and free excitons obeyed the same momentum selection rule. This observation was recently confirmed by Moore et al [28]. The free exciton (intrinsic) PL spectrum in the 6H polytype was partially reported in [6]; however, the exciton ionization energy for the 6H as well as for the 4H polytype was inferred from electroabsorption measurements [29,30]. The intrinsic PL spectra of 6H and 4H polytypes were reported later by Ikeda and Matsunami [31]. This work also included results on the 3C and 15R polytypes. The exciton binding energies reported by Ikeda and Matsunami, which were inferred from the temperature dependence of the free exciton luminescence intensity, were about half the previously reported values for 3C, 6H, and 4H. This observation made the authors assume the same trend for the 15R polytype. They explained their experimental discrepancy in E_x by the existence of a competing mechanism associated with the presence of a defect band observed on their PL spectra. Different values for E_x have also been observed by wavelength modulated absorption spectroscopy [32]. IR absorption measurements have been used to access the ionization energy for donors in different polytypes [33,34]. Recently, high resolution IR transmission measurements of high quality bulk 3C-SiC samples yielded a nitrogen binding energy of 54.2 meV and valley-orbit splitting of 8.34 meV [35]. E_g, E_{gx}, E_x, and N_c ionization energies for six polytypes are listed in TABLE 2.

TABLE 2 Low temperature (≤ 5 K) energy bandgap (E_g), exciton energy gap (E_{gx}), exciton binding energy (E_x) and ionization energy of the N_c (E_i) for six polytypes. The energy bandgap is given by $E_g = E_{gx} + E_x$.

Polytype	E_g (eV)	E_{gx} (eV)	E_x (meV)	E_i (eV)
3C	2.403	2.390 [26]	13.5 [27]	0.054 [35]
4H	3.285	3.265 [26]	20.0 [30]	0.124 [31]
				0.066 [31]
6H	3.101	3.023 [26]	78.0 [30]	0.17 [7]
				0.20 [7]
				0.23 [7]
15R	2.946	2.906 [26]	40.0 [31]	0.14 [7]
				0.16 [7]
				0.16 [7]
				0.20 [7]
21R		2.853 [26]		
33R		3.003 [26]		$0.15 \leq E_i \leq 0.23$ [17]

PL studies of cubic SiC under high excitation show the formation of complexes consisting of two [36] or more excitons [37] bound to a neutral N_c donor. At lower temperature and higher excitation level the multiple bound exciton state progresses to an electron hole plasma, finally achieving an electron-hole liquid (EHL) state [38]. The EHL binding energy was found to be 19.5 ± 4 meV, and the critical temperature about 41 K [38]. Evidence for condensation of

2.3 Photoluminescence spectra of SiC polytypes

an electron hole plasma in the 4H and 15R SiC polytypes has also been reported [39]. Excitons and EHL in the cubic SiC have been extensively studied by Russian workers [40-45].

Although good quality SiC polytypes have been grown previously, the small crystal dimensions and quality control have been the main difficulties to be overcome before this material can realise fully its potential. The development of a chemical vapour deposition (CVD) technique made possible the epitaxial deposition of large and relatively thick, crack-free cubic SiC films on Si substrates [46-49]. The sublimation process for growing higher purity α-SiC single crystal, developed by Lely [50], was modified by adding a seed [51-53], wherein only one large crystal of a single polytype is grown. Currently, high quality epitaxial layers of 3C or 6H polytypes are deposited by CVD on these α-SiC substrates [54-57].

The earlier PL studies performed in Lely grown materials provide a baseline characterization for CVD SiC films. Several workers [58-62] have reported the observation at liquid helium temperature of characteristic five-line near-band edge PL spectra in nominally undoped films of CVD-grown cubic SiC on Si. They are basically equivalent to the five-line spectra first observed by Choyke et al [4] in Lely-grown crystals of cubic SiC. The observed increase of line-width and red-shift in peak position were attributed to a high impurity and structural defect concentration, as well as strain due to the Si-SiC lattice mismatch [58]. The bi-axial strain effect was studied in supported, partially supported, and free-standing films (un-annealed and annealed) [58]. A detailed study of bi-axial strain and its effect upon increasing film thickness was subsequently published by Choyke et al [62]. They observed that for films thicker than 3 μm the film stress decreased slightly with increasing film thickness. However, the material continues to improve with increasing thickness, since the band-edge emission intensity increases in comparison with the defect band PL intensities. Although 3C- and 6H-SiC films have been deposited previously on 6H-SiC substrates [54-57], only recently have films with PL spectra similar to those of high quality Lely-grown material been reported [63,64].

One of the advantages of the CVD growth technique is to conveniently allow, during growth, the introduction of n-type and p-type dopants into monocrystalline SiC films. Although there has been some investigation into the incorporation of different donor-like impurities in SiC [65], up to now there have been no reports of any optical evidence of donor impurities other than nitrogen [66,59]. Although careful examination of the nitrogen bound exciton (NBE) spectra from CVD films of cubic SiC reveals that most n-type samples exhibit weak shoulders on the low-energy side of the ZPL and each of the phonon replicas, the thermal quenching energy of these features is found to be about twice that observed for the NBE. These observations suggest a donor with a binding energy of about 100 meV [61]. The 15-20 meV shallow donor, commonly observed in Hall measurements of cubic SiC films deposited on Si, seems to be associated with a structural defect or with donor complexes, rather than a single isolated substitutional impurity [67,68].

2.3 Photoluminescence spectra of SiC polytypes

C ACCEPTORS AND DONOR-ACCEPTOR PAIRS

Low temperature absorption measurements performed in p-type 3C-, 4H-, 6H- and 15R-SiC doped with boron and 6H-SiC doped with Al show extrinsic absorption features which have been attributed to excitons bound to neutral acceptors [69-72]. Electroabsorption measurements performed in boron-doped 6H polytype show that the observed effective microfield is considerably stronger than the expected values generated by the substitutional boron acceptor, suggesting that the excitons may be bound to dipole complexes, which may include the N-donor [73].

Although acceptor-doped SiC polytypes have been intensively studied by PL, cathodoluminescence (CL) and electroluminescence, only some recently reported PL and preliminary Zeeman studies show evidence of an exciton bound to a neutral-acceptor in 3C- and 6H-SiC films [74]. The observation of a four-particle acceptor complex reported by Clemen et al [74] was realized in samples with low N content and low Al-doping levels. In general, the background concentration of N_c is about 10^{17} cm^{-3} and the Al doping about the same level or one order of magnitude higher. Samples containing these donor (D) and acceptor (A) levels have their PL spectra dominated by recombination processes involving donor-acceptor pairs (DAP).

In the DAP recombination mechanism electrons bound at neutral donors recombine radiatively with holes bound at neutral acceptors. If an isolated neutral donor and acceptor are separated by a reasonably small distance r, the electron and hole can recombine with emission of a photon of energy E(r) given by [75]:

$$E(r) = E_g - (E_D + E_A) + (e^2/\varepsilon r) - (e^2/\varepsilon)(a^5/r^6) \tag{1}$$

where E_g is the forbidden gap energy, E_D and E_A are the isolated donor and acceptor binding energies, respectively, ε is the static dielectric constant and a is the effective van der Waals coefficient for the interaction between a neutral donor and a neutral acceptor. Note that Eqn (1) is derived assuming the validity of the effective mass approximation. The peak position of the luminescence band depends on the density of pairs as a function of r, on the transition probability of a pair of a given r and on the excitation intensity. If the separation between donor and acceptor is very large, E(r) in Eqn (1) tends to the limiting value

$$E_{min} = E_g - (E_D + E_A) \tag{2}$$

which can be estimated from the low energy tail of the distant DAP band [76].

Choyke and Patrick reported a well resolved sharp line spectra between 2.10 and 2.35 eV in Al-doped Lely-grown crystals of cubic SiC [77]. They attributed these sharp lines to the recombination of close DAPs, where the N replaces C and Al replaces Si (Type II spectrum). These lines were resolved up to shell m = 80, and an extrapolation to infinite separation yielded an accurate value of the minimum photon energy, $E_{min} = 2.0934$ eV. If we use the value of $E_g = 2.403$ eV, at 4.2 K [77], we obtain from Eqn (2):

$$E_D(N) + E_A(Al) = 310 \text{ meV} \tag{3}$$

2.3 Photoluminescence spectra of SiC polytypes

For large shell numbers (m > 200) the DAP spectrum has a broad peak due to unresolved distant pairs. At temperatures above 77 K a new peak with higher emission energy than the former appears. This latter peak is attributed to free-electrons recombining with holes bound to deep Al acceptors (free-to-bound transition) [78]. With $E_g = 2.402$ eV at 77 K, and assuming that the electron kinetic energy increases faster than E_g decreases, one obtains about 257 meV for $E_A(Al)$ [79]. From Eqn (3) we obtain $E_D(N) = 53$ meV, which is in excellent agreement with results obtained from independent experiments [9]. Recently, Freitas et al [61] reported discrete close-pair PL spectra in Al-doped CVD cubic SiC films equivalent in every detail to the DAP spectra reported in [77]. Their study of the temperature dependence of the N-Al close pair PL yielded a thermal activation energy of 54 meV for the nitrogen donors, which coincides with the 54 meV binding energy determined from the spectral energies of the sharp-line close pair spectra.

Although Al seems to be the most studied p-type impurity in the cubic polytype, detailed PL characterizations have also been reported on material doped with B and Ga acceptors [80-84]. Al, B and Ga acceptor impurities are also radiative centres in other SiC polytypes. PL studies of α-SiC doped with these impurities reveal the existence of site-dependent impurity levels originating from the lattice site inequivalence [84,85]. Recently, there has been some discussion about the validity of previously reported values for the ionization energy of B in 6H-SiC. Suttrop et al [86] have shown strong experimental evidence that the ionization energy of the isolated B acceptor is 0.32 eV. They also reported that the commonly observed yellow emission is associated with a boron-related deep centre (D-centre) with a 0.58 eV ionization energy. These results have been confirmed by Reinke et al [87], who have suggested that B occupies a Si site in 6H-SiC. Such results raise questions about the previously reported values for the B ionization energy in other SiC polytypes. The ionization energies of some acceptors in the most common SiC polytypes are listed in TABLE 3.

TABLE 3 Ionization energy (eV) of some acceptor impurities in the most common SiC polytypes.

Impurity/ Polytype	Al	B	Be	Ga	Sc
3C	0.257 [61]	0.735 [80]	-	0.344 [83]	-
4H	0.191 [85]	0.647 [85]	-	0.267 [31]	-
6H	0.249 [85]	0.723 [85]	0.4 [88]	0.333 [85]	0.24 [89]
	0.239 [85]	0.698 [85]	0.6 [88]	0.317 [85]	
		0.32 [86]			
15R	0.236 [85]	0.700 [85]	-	0.320 [85]	-
	0.230 [85]	-	-	0.311 [85]	-
	0.223 [85]	-	-	0.305 [85]	-
	0.221 [85]	0.666 [85]	-	0.300 [85]	-
	0.206 [85]	-	-	0.282 [85]	-

D   TRANSITION METALS

Impurity trace analyses have shown that transition metals such as titanium and vanadium are pervasive impurities in Lely and/or modified-Lely grown SiC polytypes [90]. Since transition metals are unstable and have multiple charge states they can be electrically active deep levels.

2.3 Photoluminescence spectra of SiC polytypes

The identification and control of the background contamination level of transition metal impurities is required for the satisfactory performance of devices based on the SiC material system.

Ti was the first transition metal impurity to be identified in the SiC system. The initial controversy over the assigment of the 'ABC spectrum' in 6H-SiC [91,92] was resolved by Hagen and van Kemenade [93]. In their study of samples doped with the ^{15}N isotope they found that the ABC-spectrum was unchanged, thereby showing that it is not N-related. Later Kemenade and Hagen studied 6H crystals doped with the five different Ti isotopes, and concluded that the ABC-spectrum was due to Ti [94]. The PL spectrum of Ti in SiC has been observed in 4H, 6H, 15R and 33R polytypes. These results are explained by a model in which the silicon-substitutional, isoelectronic neutral Ti(3d^o) impurities bind excitons [95]. In 6H- and 4H-SiC polytypes, the presence of a deep electrically active donor resulting from the (N_c-Ti_{Si}) nearest-neighbour pair [96,97] has been observed.

In the last few years Schneider and co-workers have performed a number of experiments on various SiC polytypes which exhibit a characteristic infrared emission in the 1.3 to 1.5 µm spectral range [98]. They have assigned this emission band to vanadium impurities substituting the various silicon sites in the lattice. In their extensive work they found three charge states of vanadium which act as an electrically amphoteric deep level in SiC. They also suggest that vanadium may have an important role in the minority-carrier lifetime in SiC-based optoelectronic devices [98,99]. Recently, trace amounts of vanadium impurities have been detected in 3C-SiC grown by the modified-Lely technique [100].

E RADIATION- AND IMPLANTATION-INDUCED DEFECTS

The success of the non-equilibrium CVD growth technique for the homo- and hetero-epitaxial growth of SiC has been the major thrust of the last decade. Many of the potential applications of SiC in semiconductor electronics depend upon ion-implantation techniques for device fabrication. Several authors [101-118] have reported studies of the lattice damage induced by ion-implantation or by fast-particle irradiation, and of lattice damage recovery, after the seminal work of Makarov [119].

H and He backscattering studies of lattice recovery after post-ion implantation annealing procedures show that the lattice damage is largely repaired by a 1200 °C anneal [120]. PL studies of a number of bulk [102] and thin film [59] SiC samples implanted with a variety of ions and annealed with increasing temperature have shown that two defects, referred to as D_1 and D_2, continue to dominate the luminescence spectra even for annealing temperatures above 1600 °C. In addition, the PL spectra did not show any feature that could be attributed, unambiguously, to the particular implanted ion-species. The stability of these centres at high annealing temperature suggests that they may be defect complexes. Although there has been some association of D_1 with excess carbon vacancies in 6H-SiC films [111], a definite microscopic model for these centres does not yet exist. The D_1 and D_2 centres have been observed as native defects in as-grown 3C-SiC films deposited on Si substrates, and their relative intensity to the edge emission increases with increasing annealing temperature [59].

It is important to mention that up to the present time hydrogen and deuterium are the only impurities which have been unambiguously associated with specific PL bands of implanted and annealed SiC. The observation of C-H and C-D bond-stretching modes leads to a model of the luminescence centre, namely an H or D atom bonded to a C atom at a Si vacancy [121]. H centres, similar to H spectra of implanted and annealed Lely samples, have been detected in 6H-, 4H- and 15R-SiC as-grown CVD films [122].

Impurity-defect complexes are observed in fast-particle irradiated samples, where the minimal lattice damage does not require the annealing procedure. One can observe a number of lines that undergo progressive changes with annealing temperature below 1000 °C, without interference of the D_1 and D_2 spectra. There has been a great deal of electron paramagnetic resonance (EPR) work on these defects in 3C- and 6H-SiC [123]. The large number, complexity and sample dependence of both the PL and the EPR spectra make it very difficult to associate the EPR spectra and the PL spectra with the same centre.

We should mention that antisite defects may be the most common type for as-grown materials, since both lattice atoms belong to the same column of the periodic table. Recent calculations reported by Bernholc et al [124] predict formation energies as low as ~3 eV for both of the antisite defects, Si_C and C_{Si}, in cubic SiC. However, theoretical studies [124,125] suggest that both types of isolated antisite defects in 3C-SiC are electrically inactive.

F   CONCLUSION

Photoluminescence spectra of several of the SiC polytypes have been reviewed. Nitrogen has been identified as a donor impurity but no other donor impurities have been identified from optical measurements. A donor binding energy of ~100 meV is suggested with the shallow donor seen in electrical measurements assigned to a structural defect or donor complex. Doping with boron, gallium and aluminium produces acceptors, and ionization energies are given, but these are normally bound to donors giving rise to donor-acceptor pairs in photoluminescence spectra. Transition metal elements, e.g. Ti and V, are common in SiC and Ti has been identified with the ABC-spectrum. Vanadium on silicon sites produces IR emission at 1.3 to 1.5 µm and may influence minority-carrier lifetime in SiC-based devices. Ion implantation damage can be reduced by a 1200 °C anneal but two defects, possibly defect complexes, still dominate the photoluminescence spectrum even after 1600 °C annealing. Antisite defects may be the most common type in as-grown material but theory suggests that they will be electrically inactive.

ACKNOWLEDGEMENT

The author wishes to thank P.B. Klein for many helpful discussions and for critically reading this manuscript.

REFERENCES

[1] P.J. Dean, P.C. Herbert [*Excitons*, Solid State Science, vol.14, Ed. K. Cho (Springer, Berlin, 1979) p.55]

[2] D.J. Ashen, P.J. Dean, D.T. Hurle, J.B. Mullin, A.M. White [*J. Phys. Chem. Solids (UK)* vol.36 (1975) p.1041]

[3] S.G. Bishop, J.A. Freitas Jr. [*J. Cryst. Growth (Netherlands)* vol.106 (1990) p.38-46]

[4] W.J. Choyke, D.R. Hamilton, L. Patrick [*Phys. Rev. A (USA)* vol.133 (1964) p.1163-6]

[5] $E_{gx} = E_g - E_x$, where E_g is the energy gap, and E_x is the exciton binding energy.

[6] W.J. Choyke, L. Patrick [*Phys. Rev. (USA)* vol.127 (1962) p.1868-77]

[7] L. Patrick, D.R. Hamilton, J.W. Choyke [*Phys. Rev. (USA)* vol.132 (1963) p.2023-31]

[8] R.L. Hartman, P.J. Dean [*Phys. Rev. B (USA)* vol.2 (1970) p.951-9]

[9] P.J. Dean, W.J. Choyke, L. Patrick [*J. Lumin. (Netherlands)* vol.15 (1977) p.299-314]

[10] M.A. Lampert [*Phys. Rev. Lett. (USA)* vol.1 (1958) p.1868]

[11] H.H. Woodbury, G.W. Ludwig [*Phys. Rev. (USA)* vol.124 (1961) p.1083]

[12] Yu.M. Altaiskii, I.M. Zaritskii, V.Ya. Zevin, A.A. Konchits [*Sov. Phys.-Solid State (USA)* vol.12 (1971) p.2453]

[13] The ESR spectrum exhibited by unpaired spins of electrons bound to donors can provide an identification of the donor species through the interaction of the paramagnetic electrons with the nuclear magnetic moments of the donor atoms. This central hyperfine interaction splits the ESR spectrum into 2I+1 lines, where 'I' is the nuclear spin, providing in many cases an identification of the donor. For more details see for example Datareview 2.4.

[14] There are 12 atoms per unit cell (consequently 36 phonon branches) and three inequivalent carbon sites in the 6H polytype (two cubic and one hexagonal). Since the nitrogen replaces carbon equally on each of the carbon sites, one may observe three series of lines, each comprised of a zero-phonon line (ZPL) and its phonon replicas, totalling three ZPLs and 108 phonon replicas, disregarding phonon degeneracy.

[15] L. Patrick [*Phys. Rev. (USA)* vol.127 (1962) p.1878]

[16] D.R. Hamilton, W.J. Choyke, L. Patrick [*Phys. Rev. (USA)* vol.131 (1963) p.127-33]

[17] W.J. Choyke, D.R. Hamilton, L. Patrick [*Phys. Rev. A (USA)* vol.139 (1965) p.1262-74]

[18] L. Patrick, D.R. Hamilton, W.J. Choyke [*Phys. Rev. (USA)* vol.143 (1966) p.526-36]

[19] L. Patrick, W.J. Choyke, D.R. Hamilton [*Phys. Rev. A (USA)* vol.137 (1965) p.1515-20]

[20] D.R. Hamilton, L. Patrick, W.J. Choyke [*Phys. Rev. A (USA)* vol.138 (1965) p.1472-6]

[21] W.J. Choyke, L. Patrick, D.R. Hamilton [*7th Int. Conf. on the Phys. of Semic.* Eds C.Benoit, A. La Guillaume (Dunod, Paris, 1964) p.751-8]

[22] The cubic SiC polytype has the conduction-band minima at X [9], and hence the two TA and TO momentum conserving phonons should be degenerate, which is consistent with the observation of 4 rather than 6 phonons in the four-particle PL spectrum.

[23] R. Kaplan, R.J. Wagner, H. Kim, R.F. Davis [*Solid State Commun. (USA)* vol.55 (1985) p.67]

2.3 Photoluminescence spectra of SiC polytypes

[24] The 2H-SiC polytype has distinct spectra for the radiation electric field parallel (extraordinary: axial) and perpendicular (ordinary: planar) to the optic axis. These different phonon energies are conveniently listed as acoustic and optical branches [18].

[25] The large zone scheme and a common set of dispersion curves for the uniaxial crystal direction were discussed in Datareview 2.2, Section B (p.21 of this book).

[26] W.J. Choyke [*Mater. Res. Bull. (USA)* vol.4 (1969) p.141-52]

[27] D.S. Nedzvetskii, B.V. Novikov, N.K. Prokofeva, M.B. Reifman [*Sov. Phys.-Semicond. (USA)* vol.2 (1969) p.914-9]

[28] W.J. Moore, R. Kaplan, J.A. Freitas Jr., Y.M. Altaiskii, V.L. Zuev, L.M. Ivanova [*Mater. Res. Soc. Symp. Proc. (USA)* vol.339 (1994) to be published]

[29] V.I. Sankin [*Sov. Phys.-Solid State (USA)* vol.17 (1975) p.1191-2]

[30] G.B. Dubrovskii, V.I. Sankin [*Sov. Phys.-Solid State (USA)* vol.17 (1975) p.1847-8]

[31] M. Ikeda, H. Matsunami [*Phys. Status Solidi A (Germany)* vol.58 (1980) p.657-63]

[32] R.G. Humphrey, D. Bimberg, W.J. Choyke [*Solid State Commun. (USA)* vol.39 (1981) p.163-7]

[33] W. Suttrop, G. Pensl, W.J. Choyke, A. Dormen, S. Leibenzeder, R. Stein [*Springer Proc. Phys. (Germany)* vol.71 (1992) p.129-35]

[34] W. Gotz et al [*J. Appl. Phys. (USA)* vol.73 (1993) p.3332]

[35] W.J. Moore, P.J. Lin-Chung, J.A. Freitas Jr., Y.M. Altaiskii, V.L. Zuev, L.M. Ivanova [*Phys. Rev. (USA)* vol.48 (1993) p.12289-91]

[36] B.V. Novikov, M.M. Pimonenko [*Sov. Phys.-Solid State (USA)* vol.13 (1973) p.2323-5]

[37] P.J. Dean, D.C. Herbert, D. Bimberg, W.J. Choyke [*Phys. Rev. Lett. (USA)* vol.37 (1976) p.1635-8]

[38] D. Bimberg, M.S. Skolnick, W.J. Choyke [*Phys. Rev. Lett. (USA)* vol.40 (1978) p.56-60]

[39] M.S. Skolnick, D. Bimberg, W.J. Choyke [*Solid State Commun. (USA)* vol.28 (1978) p.865]

[40] I.S. Gorban, G.N. Mishinova [*Sov. Phys.-Solid State (USA)* vol.25 (1983) p.142-3]

[41] I.S. Gorban, G.A. Gubanov, V.G. Lysenko, A.A. Pletyushkin, V.B. Timofeev [*Sov. Phys.-Solid State (USA)* vol.26 (1984) p.1385-9]

[42] V.A. Vakulenko, I.S. Gorban, V.A. Gubanopov, A.A. Pletyushkin [*Sov. Phys.-Solid State (USA)* vol.26 (1984) p.2139]

[43] V.D. Kulakovskii, V.A. Gubanov [*Sov. Phys.-JEPT (USA)* vol.61 (1985) p.550]

[44] V.D. Kulakovskii, V.A. Gubanov [*Sov. Phys.-JEPT (USA)* vol.27 (1985) p.1359]

[45] I.S. Gorban, A.P. Krokhmal [*Sov. Phys.-Solid State (USA)* vol.28 (1986) p.1279]

[46] S. Nishino, Y. Hazuki, H. Matsunami, T. Tanaka [*J. Electrochem. Soc. (USA)* vol.127 (1980) p.2674-80]

[47] S. Nishino, J.A. Powell, H.A. Will [*Appl. Phys. Lett. (USA)* vol.42 (1983) p.460-2]

[48] H.P. Liaw, R.F. Davis [*J. Electrochem. Soc. (USA)* vol.131 (1984) p.3014]

[49] A. Addamiano, P.H. Klein [*J. Cryst. Growth (Netherlands)* vol.70 (1980) p.291]

[50] J.A. Lely [*Ber. Dtsch. Keram. Ges. (Germany)* vol.32 (1955) p.229-36]

[51] G. Ziegler, P. Lanig, D. Theis, C. Weyrich [*IEEE Trans. Electron Devices (USA)* vol.ED-30 (1983) p.277-81]

[52] C.H. Carter Jr., L. Thang, R.F. Davis [*4th National Review Meeting on the Growth and Characterization of SiC*, Raleigh, NC, USA, June 19, 1987 (unpublished)]

[53] K. Koga, Y. Ueda, T. Nakata, T. Yamaguchi, T. Niina [*J. Vac. Soc. Jpn. (Japan)* vol.30 (1987) p.886-90]

[54] H.S. Kong, J.T. Glass, R.F. Davis [*Appl. Phys. Lett. (USA)* vol.49 (1986) p.1074]
[55] H.S. Kong, J.T. Glass, R.F. Davis [*J. Mater. Res. (USA)* vol.4 (1989) p.204]
[56] H. Matsunami, K. Shibahara, N. Kuroda, S. Nishino [*Springer Proc. Phys. (Germany)* vol.34 (1989) p.34-9]
[57] H.S. Kong, J.T. Glass, R.F. Davis [*Appl. Phys. Lett. (USA)* vol.64 (1988) p.2672]
[58] J.A. Freitas Jr., S.G. Bishop, A. Addamiano, H.J. Kim, R.F. Davis [*Mater. Res. Soc. Symp. Proc. (USA)* vol.46 (1985) p.581-6]
[59] J.A. Freitas Jr., S.G. Bishop, J.A. Edmond, J. Ryu, R.F. Davis [*J. Appl. Phys. (USA)* vol.61 (1987) p.2011-6]
[60] H. Okumura et al [*Jpn. J. Appl. Phys. (Japan)* vol.127 (1987) p.L116-8]
[61] J.A. Freitas Jr., S.G. Bishop, P.E.R. Nordquist Jr., M.L. Gipe [*Appl. Phys. Lett. (USA)* vol.52 (1988) p.1695-7]
[62] W.J. Choyke, Z.C. Feng, J.A. Powell [*J. Appl. Phys. (USA)* vol.64 (1988) p.3163-75]
[63] J.A. Powell et al [*Appl. Phys. Lett. (USA)* vol.56 (1990) p.1353-5]
[64] J.A. Powell et al [*Appl. Phys. Lett. (USA)* vol.56 (1990) p.1442-4]
[65] See for example H. Kim, R.F. Davis [*J. Electrochem. Soc. (USA)* vol.133 (1986) p.2350-7]
[66] J.A. Freitas Jr., W.E. Carlos, S.G. Bishop [*Springer Proc. Phys. (Germany)* vol.56 (1989) p.135-42]
[67] S.G. Bishop et al [*Springer Proc. Phys. (Germany)* vol.34 (1987) p.90-8]
[68] W.J. Moore [*Springer Proc. Phys. (Germany)* vol.56 (1987) p.156-60]
[69] I.S. Gorban', A.P. Krokhmal [*Sov. Phys.-Solid State (USA)* vol.28 (1986) p.1279-81]
[70] I.S. Gorban', A.P. Krokhmal [*Sov. Phys.-Solid State (USA)* vol.20 (1986) p.39-41]
[71] I.S. Gorban', A.P. Krokhmal [*Sov. Phys.-Solid State (USA)* vol.19 (1977) p.733-5]
[72] I.S. Gorban', A.P. Krokhmal [*Sov. Phys.-Solid State (USA)* vol.15 (1973) p.1213-4]
[73] E.I. Radonova, V.I. Sankin, V.I. Sokolov [*Sov. Phys.-Semicond. (USA)* vol.15 (1981) p.140-3]
[74] L.L. Clemen, W.J. Choyke, R.P. Devaty, J.A. Powell, H.S. Khong [*Springer Proc. Phys. (Germany)* vol.71 (1992) p.105-15]
[75] J.J. Hopfield, D.G. Thomas, M. Gershenzon [*Phys. Rev. Lett. (USA)* vol.10 (1963) p.162]
[76] D.G. Thomas, M. Gershenzon, F.A. Trumbore [*Phys. Rev. A (USA)* vol.133 (1964) p.269]
[77] V.A. Kiselev, B.V. Novikov, M.M. Pimonenko, E.B. Sadrin [*Sov. Phys.-Solid State (USA)* vol.13 (1973) p.926]
[78] G. Zanmarchi [*J. Phys. Chem. Solids (UK)* vol.29 (1968) p.1727-36]
[79] W.J. Choyke, L. Patrick [*Proc. 3rd Int. Conf. on Silicon Carbide* Eds R.C. Marshall, J.W. Faust Jr., C.E. Ryan, Miami Beach, Florida, 17-20 September, 1973 (University of South Carolina Press, Columbia, 1974) p.261-83]
[80] S. Yamada, H. Kuwabara [*Proc. 3rd Int. Conf. on Silicon Carbide* Eds R.C. Marshall, J.W. Faust Jr., C.E. Ryan, Miami Beach, Florida, 17-20 September, 1973 (University of South Carolina Press, Columbia, 1974) p.305-12]
[81] H. Kuwabara, S. Yamada [*Phys. Status Solidi A (Germany)* vol.30 (1975) p.739-46]
[82] H. Kuwabara, S. Yamada, S. Tsunekawa [*J. Lumin. (Netherlands)* vol.12/13 (1976) p.531-6]
[83] H. Kuwabara, K. Yamanaka, S. Yamada [*Phys. Status Solidi A (Germany)* vol.37 (1976) p.K157-61]

[84] A. Suzuki, H. Matsunami, T. Tanaka [*J. Electrochem. Soc. (USA)* vol.124 (1977) p.241-6]
[85] M. Ikeda, H. Matsunami, T. Tanaka [*Phys. Rev. B (USA)* vol.22 (1980) p.2842-54]
[86] W. Suttrop, G. Pensl, P. Lanig [*Appl. Phys. A, Solids Surf. (Germany)* vol.51 (1990) p.231-7]
[87] J. Reinke, S. Greulich-Weber, J.-M. Spaeth, E.N. Kalabukhova, S.N. Lukin, E.N. Mokhov [*Inst. Phys. Conf. Ser. (UK)* no.137 (1994) p.211-14]
[88] Yu.P. Maslokovets, E.N. Mokhov, Yu.A. Vodakov, G.A. Lomakina [*Sov. Phys.-Solid State (USA)* vol.10 (1968) p.634]
[89] Yu.M. Tairov, I.I. Khlebnikov, V.F. Tsvetkov [*Phys. Status Solidi (Germany)* vol.25 (1974) p.349]
[90] P.A. Glasow [*Springer Proc. Phys. (Germany)* vol.34 (1989) p.13-34]
[91] P.J. Dean, R.L Hartman [*Phys. Rev. B (USA)* vol.5 (1972) p.4911-24]
[92] L. Patrick [*Phys. Rev. B (USA)* vol.7 (1973) p.1719-21]
[93] S.H. Hagen, A.W.C. van Kemenade [*J. Lumin. (Netherlands)* vol.6 (1974) p.131]
[94] A.W.C. van Kemenade, S.H. Hagen [*Solid State Commun. (USA)* vol.14 (1974) p.1331-3]
[95] L. Patrick, W.J. Choyke [*Phys. Rev. B (USA)* vol.10 (1974) p.5091-4]
[96] K. Maier, J. Schneider, W. Wilkening, S. Leibenzeder, R. Stein [*Mater. Sci. Eng. B (Switzerland)* vol.11 (1992) p.27]
[97] K. Maier, H.D. Muller, J. Schneider [*Mater. Sci. Forum (Switzerland)* vol.83-87 (1992) p.1183]
[98] J. Schneider, H.D. Muller, K. Maier, W. Wilkening, F. Fuchs [*Appl. Phys. Lett. (USA)* vol.56 (1990) p.1184-6]
[99] J. Schneider, K. Maier [*Physica B (Netherlands)* vol.185 (1993) p.199-206]
[100] M.G. Spencer [private communication (1994)]
[101] I.I. Geitsi, A.A. Nesterov, L.S. Sminorv [*Sov. Phys.-Semicond. (USA)* vol.4 (1970) p.744]
[102] W.J. Choyke, L. Patrick [*Phys. Rev. B (USA)* vol.4 (1971) p.6 ; vol.4 (1971) p.1843-7]
[103] L. Patrick, W.J. Choyke [*J. Phys. Chem. Solids (UK)* vol.34 (1973) p.565]
[104] I.S. Gorban, V.A. Kravets, G.N. Mishinova, K.V. Nazarenko [*Sov. Phys.-Semicond. (USA)* vol.10 (1976) p.1254]
[105] N.V. Kodrau, V.V. Makarov [*Sov. Phys.-Semicond. (USA)* vol.11 (1977) p.569]
[106] P.J. Dean, D. Bimberg, W.J. Choyke [*Inst. Phys. Conf. Ser. (UK)* vol.46 (1979) p.447]
[107] Y.A. Vodakov, E.N. Mokhov, M.G. Ramn, A.D. Roenkov [*Krist. Tech. (Germany)* vol.14 (1979) p.729]
[108] M. Ikeda, H. Matsunami, T. Tanaka [*Jpn. J. Appl. Phys. (Japan)* vol.19 (1980) p.1201]
[109] N.V. Kodrau, V.V. Makarov [*Sov. Phys.-Semicond. (USA)* vol.15 (1981) p.813]
[110] N.V. Kodrau, V.V. Makarov [*Sov. Phys.-Semicond. (USA)* vol.15 (1981) p.960]
[111] Y.A. Vodakov, G.A. Lomakina, E.N. Mokhov [*Sov. Phys.-Solid State (USA)* vol.24 (1982) p.780]
[112] V.M. Gusev, K.D. Demakov, V.G. Stolyarova [*Radiat. Eff. (UK)* vol.69 (1983) p.307]
[113] Y. M. Suleimanov, A.M. Grekov, V.M. Grekov [*Sov. Phys.-Solid State (USA)* vol.25 (1983) p.1060]

[114] V.I. Lvin, Y.M. Tairov, V.F. Tsvetkov [*Sov. Phys.-Semicond. (USA)* vol.18 (1984) p.747]
[115] E.E. Violina, K.D. Demakov, A.A. Kal'nin, F. Nolbert, E.N. Potapov, Y.M. Tairov [*Sov. Phys.-Solid State (USA)* vol.26 (1984) p.960]
[116] J.A. van Vechten [*J. Cryst. Growth (Netherlands)* vol.71 (1985) p.326]
[117] Y.M. Suleimanov, V.M. Grekhov, K.D. Demakov, I.V. Plyuto [*Sov. Phys.-Solid State (USA)* vol.27 (1985) p.1910]
[118] Y.A. Vodakov, G.A. Lomakina, E.N. Mokhov, V.G. Oding, M.G. Ramn, V.I. Sokolov [*Sov. Phys.-Semicond. (USA)* vol.20 (1986) p.900]
[119] V.V. Makarov [*Sov. Phys.-Solid State (USA)* vol.9 (1967) p.2 ; vol.9 (1967) p.457]
[120] R.R. Hart, H.L. Dunlap, O.J. Marsh [*Radiation Effects in Semiconductors* Eds Gordon and Breach (N.Y., 1971) p.405-10]
[121] W.J. Choyke [*Inst. Phys. Conf. Ser. (UK)* vol.31 (1977) p.58]
[122] L.L. Clemen, R.P. Devaty, W.J. Choyke, A.A. Burk Jr., D.J. Larkin, J.A. Powell [*Proc. Amorphous and Crystalline Silicon Carbide V*, Ed. M.G. Spencer (Washington, DC, USA, November 1993)]
[123] For details see Datareview 2.4.
[124] J. Bernholc, S.A. Kajira, C. Wang, A. Antonelli [*Mater. Sci. Eng. B (Switzerland)* vol.11 (1992) p.265]
[125] P.J. Lin-Chung, Y. Li [*Mater. Sci. Forum (Switzerland)* vol.10-12 (1986) p.1247]

2.4 Impurities and structural defects in SiC determined by ESR

W.E. Carlos

January 1994

A INTRODUCTION

Electron spin resonance (ESR) or, alternatively, electron paramagnetic resonance (EPR) is based on the resonant absorption of electromagnetic energy by unpaired electrons in a magnetic field. In a typical apparatus the frequency is held constant and the magnetic field swept with a small AC modulation field superimposed to permit phase-sensitive detection of the absorption. In a commercial spectrometer magnetic fields are < 1 T and the electromagnetic irradiation is in the microwave range with 9-10 GHz being the most common frequency range, although frequencies over 100 GHz have been used in some work reviewed here. The unpaired electrons of interest in most studies of semiconductors are localized at either impurities or structural defects in the crystal lattice. Primarily through the hyperfine interaction of the unpaired electronic spin with nuclear spins, information about the chemical nature of the defect can be obtained. For example, one might determine the identity of an impurity and the lattice site it occupies. Rotational studies of single crystals resolve further details about the symmetry of the defect. Studies of the evolution of the ESR signal with measurement temperature and annealing processes, as well as other in-situ and ex-situ stimuli, such as light, provide information on the relationship of a defect to the host lattice.

ESR has been used to characterize a number of impurity-related and structural defects in several polytypes of SiC. Most centres observed in SiC can be described by a simple spin-Hamiltonian:

$$\mathcal{H} = \beta \mathbf{BgS} + \sum \mathbf{IAS} + \mathbf{SDS} \tag{1}$$

where β is the Bohr magneton for the electron and the first term describes the electronic Zeeman interaction between the magnetic field, **B**, and the electronic spin, **S**, through the tensor, **g**. The second term represents the hyperfine interactions between the electronic spin and nuclear spins, **I**, by **A**. The third term gives the coupling of electronic spins by the crystal field tensor, **D**, and is zero if $S = \frac{1}{2}$, as is the case for most centres discussed in this section. A more general spin Hamiltonian includes nuclear Zeeman and quadrupolar terms.

Nitrogen, which is a residual shallow donor in all polytypes, is the most extensively studied impurity in both Lely grown crystals and CVD grown films. Other impurities which have been investigated include acceptors and transition metals. Radiation-induced structural defects have been studied in 3C and 6H polytypes, initially in Lely grown crystals and more recently in CVD grown films.

B THE NITROGEN DONOR

Clearly, the most important impurity in all polytypes of SiC is nitrogen, which appears to primarily substitute for carbon and act as a shallow donor. There is some dispute over

2.4 Impurities and structural defects in SiC determined by ESR

whether nitrogen is the only residual shallow donor or whether there is a second donor level due to a structural defect. While this is the motivation for much of the more recent work on 3C films, we will focus on what ESR can tell us about N in SiC. ^{14}N is the only nearly 100% abundant isotope with I=1 which yields a 'fingerprint' three line spectrum or multiple three line spectra. The spectra of N (or any other substitutional impurity) in many of the SiC polytypes are complicated by the existence of multiple inequivalent carbon sites. It is perhaps best to start with the 3C polytype which has a simple zinc blende crystal structure in which all C (or Si) sites are equivalent even though work on N in the 6H polytype predates the 3C work by more than a decade.

Altaiskii et al [1] were the first to report on the ESR spectrum of nitrogen donors in Lely grown 3C single crystals. In that study the three-line spectrum is partially resolved. Later work by Carlos et al [2] showed that a well resolved three-line spectrum could be obtained for the CVD films as shown in FIGURE 1. The g-tensor and hyperfine interaction are isotropic with g = 2.0050 and A = 0.11 mT. Both of these parameters are in good agreement with expectations from simple effective-mass-approximation calculations for this polytype [5].

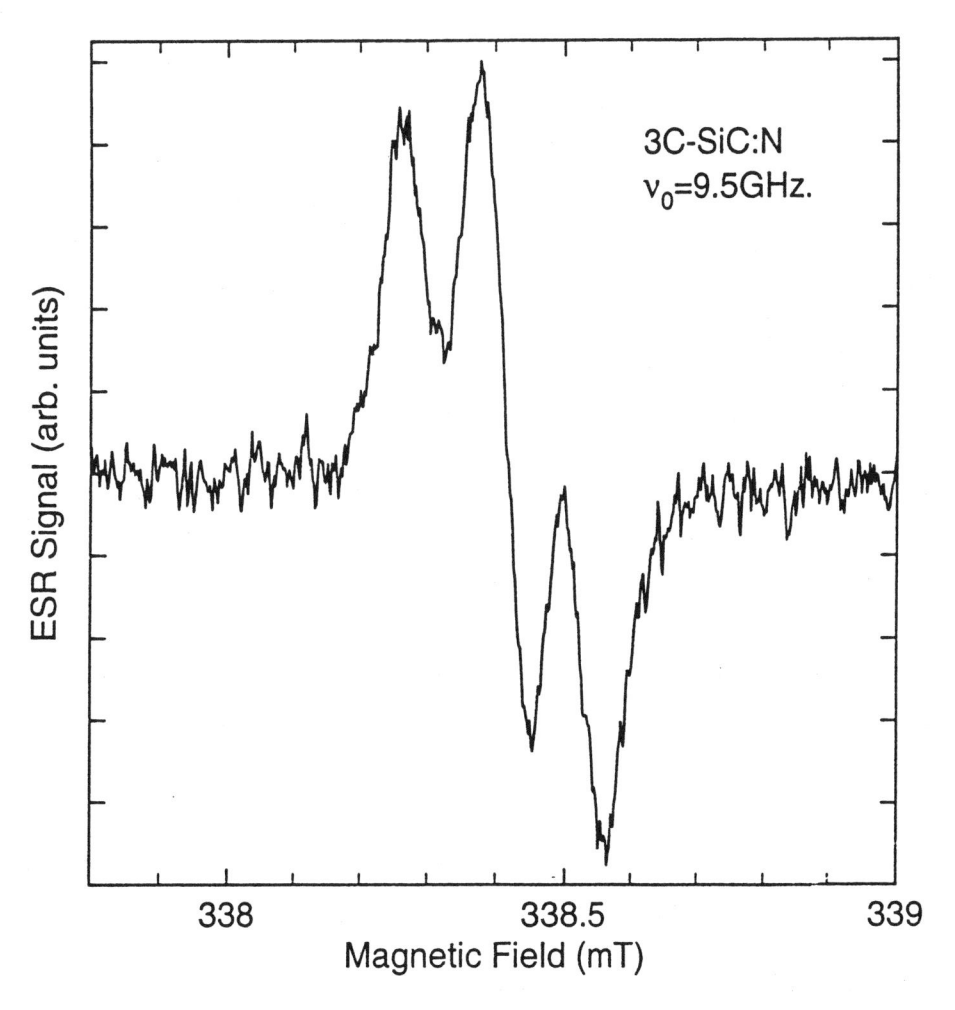

FIGURE 1 The three-line ESR spectrum of the nitrogen donor in 3C-SiC taken at 20 K. The lineshape is temperature independent below about 30 K.

2.4 Impurities and structural defects in SiC determined by ESR

The temperature dependence of the nitrogen spectrum in the 3C films has been investigated by Okumura and co-workers [3,4] and Carlos and co-workers [2,5]. Both groups find that the lineshape is essentially constant for T<30K, but that the lines broaden and converge into a single line at higher temperatures as is illustrated in FIGURE 2. Both changes in observed strength of the hyperfine interaction and in the linewidths are consistent with previous observations for donors such as phosphorus in Si [6]. The observed hyperfine interaction decreases with temperature as electrons move from the symmetric combination of hydrogenic 1S orbitals to partially populate the antisymmetric combinations which have no central hyperfine interaction. The observed line splitting is then a weighted average of that due to the symmetric and antisymmetric states. Carlos et al [5] used this to estimate the central-cell correction for the N-donor to be about 8 meV which is consistent with the results of photoluminescence measurements [7].

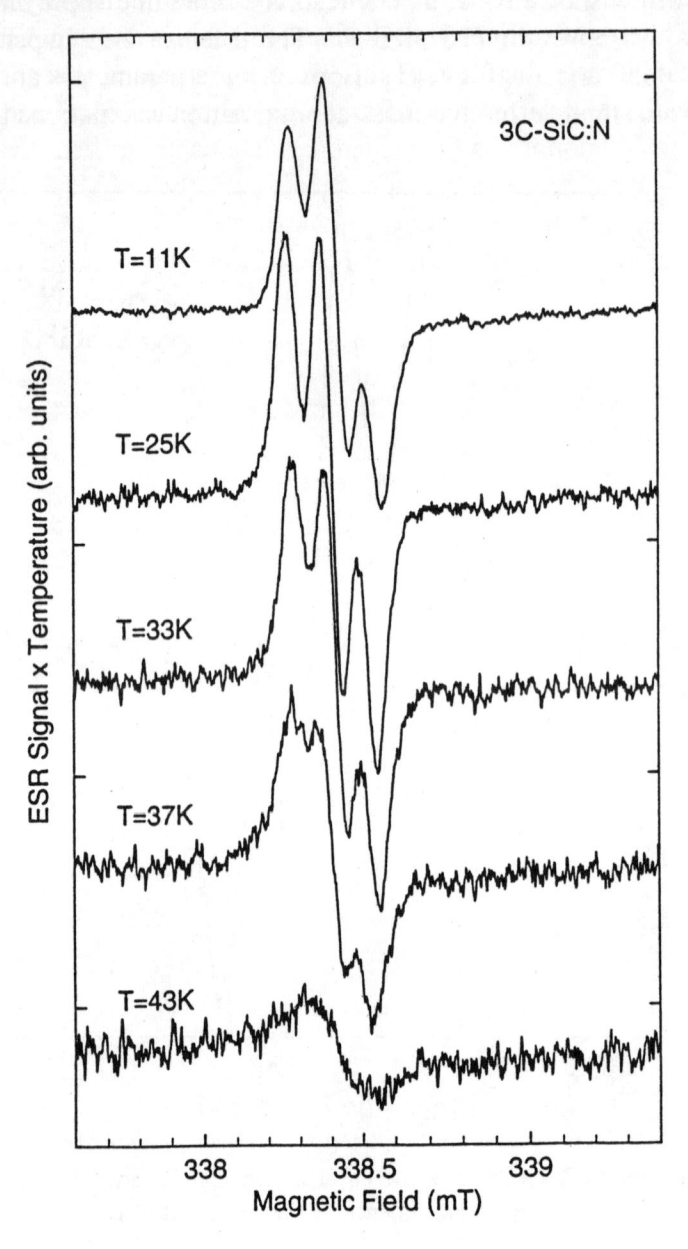

FIGURE 2 The evolution of the ESR lineshape of nitrogen in 3C-SiC with temperature. The intensity has been scaled by the measurement temperature to compensate for the changes in magnetization.

2.4 Impurities and structural defects in SiC determined by ESR

An underlying question in many of the ESR measurements of the lineshape of the N-donor and its temperature dependence is the possible existence of an additional broad line at a similar g-value to the three-line spectrum. Most authors agree that the three lines in the isolated N-donor spectrum should have symmetric lineshapes and, since there is a slight asymmetry to the full spectrum, this leads to the conclusion that there is another relatively broad ESR line shifted slightly from the three-line spectrum. This could be due to a second donor, a conduction electron spin resonance or a structural defect. A weak signal, possibly due to a Si vacancy, is also observed at higher temperatures (T > 50 K) in some samples.

The first ESR measurements on SiC were performed on 6H polytype, Lely-grown single crystals [8-10] and the signal due to residual N-donors was observed. The 6H polytype has three inequivalent C sites, two with cubic symmetry and one with hexagonal symmetry, and so in principle there should be three triplets due to N-donors in this polytype. Woodbury and Ludwig [9] observed two sets of lines with different intensities and suggested the more intense was due to the unresolved combination of the two cubic sites and that the weaker triplet was due to the hexagonal site. However infrared, photoluminescence and Raman data were interpreted to indicate that donors on the hexagonal site had a significantly smaller (~100 meV versus ~160 meV) binding energy than those on the cubic sites [11]. Therefore, in material with significant compensation one would only expect to see the ESR spectra of carriers trapped on the two cubic sites. Subsequent ESR measurements have shown a line with an anisotropic linewidth in addition to the other two triplets in samples with relatively high (but non-metallic) carrier concentration levels [12-14]. The two sets of lines with similar hyperfine interactions are then interpreted as being due to N-donors on the two cubic sites and the additional line as due to donors on the hexagonal site with a weak anisotropic hyperfine interaction [12,13].

The spectra of the N-donor in the 4H [15,16] and 15R [17] polytypes have been analyzed based on results from the 6H polytype. The 4H polytype has one cubic and one hexagonal C site while the 15R has two hexagonal sites and three cubic sites. In both cases triplets are associated with cubic sites and anisotropically broadened lines (with presumably much weaker hyperfine interactions) are associated with the hexagonal sites. In TABLE 1 we summarize the results on the four polytypes mentioned here as well as earlier less detailed results from Veinger [18] and from Deigen et al [19] for the 8H, 21R and 27R polytypes. In those cases we presume that the N-hyperfine interactions are those of the lowest energy N sites.

C ACCEPTORS

Next to nitrogen the most studied impurity in SiC has been boron [9,20,21]. A single boron-related centre has been observed by Woodbury and Ludwig [9] and by Zubatov and co-workers [20]. This was initially modelled as B_C [9] based on a weak hyperfine interaction which was consistent with four neighbouring Si atoms and the 4.7% natural abundance of ^{29}Si. Using isotopically enriched material Zubatov and co-workers showed that the largest hyperfine interaction for this centre, which they call K3, was with a single carbon atom. After considering a $B_{Si} + V_C$ model [20,21], they settled on the B_{Si} [21] with the boron atom Jahn-Teller shifted away from one of the C atoms. Approximately half of the unpaired spin is on that C atom with ~1% on each of the three neighbouring Si atoms giving the weak ^{29}Si hyperfine interaction first observed by Woodbury and Ludwig and ~4% is on the boron atom.

2.4 Impurities and structural defects in SiC determined by ESR

TABLE 1 Parameters of the N-donor.

Polytype	Site sym.	g-tensor	N-hyperfine	Ref
3C	Cubic	2.0050	0.11 mT	[5]
4H	Cubic	2.0043, 2.0013	1.82 mT	[15,16]
	Hexagonal	2.0055, 2.0010	n.r.	
6H	Cubic	2.0037, 2.0030	1.20 mT	[12,13]
	Cubic	2.0040, 2.0026	1.20 mT	
	Hexagonal	2.0048, 2.0028	n.r.	
15R	Cubic	2.0035, 2.0030	1.07 mT	[17]
	Cubic	2.00415, 2.0028	1.19 mT	
	Cubic	2.00385, 2.0033	1.19 mT	
	Hexagonal	2.0035, 2.0028	n.r.	
	Hexagonal	2.0031, 2.00225	n.r.	
8H		g = 2.017	0.75 mT	[18]
		g = 2.017	0.96 mT	
21R		?	1.07 mT	[19]

Note: The principal axis of the g-tensors of the hexagonal and rhombohedral polytypes is the c-axis, with the parallel component listed first and the perpendicular component second. In some cases, indicated by n.r., hyperfine interaction was not resolved.

Zubatov et al [21] showed that the K3 centre has very similar ESR parameters and temperature dependence in the 3C, 2H, 4H and 6H polytypes with some slight differences due to inequivalent sites in the 4H and 6H polytypes. The details of the low temperature spectra are given in TABLE 2. As the sample warms up ($T \approx 70$ K) the static Jahn-Teller distortion gives way to a dynamic Jahn-Teller effect and the motion of the boron atom causes first the ^{13}C and ^{29}Si hyperfine to be lost and finally the weak ^{11}B hyperfine is no longer resolved. The low temperature at which this occurs is the basis of the argument for dropping V_C from the model for the K3 centre [21]. The case of boron in SiC is then somewhat of a puzzle. The donor-acceptor pair luminescence clearly is due to boron on the same sublattice as the donor, presumably N [22], i.e., the C sublattice, with no evidence of B_{Si}. However, the ESR shows only B_{Si} with no evidence of B_C. The somewhat unsatisfying answer may be that B substitutes for both Si and C and that ESR is more sensitive to B_{Si} and photoluminescence has only been able to clearly resolve features due to B_C.

Aluminium centres have not been observed by ESR; however, the Al_{Si} acceptor has been observed by optically detected magnetic resonance (ODMR) [23]. In that work, which will be discussed in detail in Datareview 2.5, a triplet of lines was assigned to Al on the three inequivalent Si sites in 6H SiC with the g-value for the resonances following $\cos\theta$ dependencies, where θ is the angle between the c-axis and the magnetic field. They find g-values for the field parallel to the c-axis of 2.412, 2.400 and 2.325 and discuss possible assignments of these with the inequivalent Si sites. Very little has been done with gallium doped SiC. Vodakov et al [24] mention observing an ESR centre in Ga doped 6H SiC very similar to the boron related centre, K3, although it completely disappeared for $T > 50$ K. Baranov et al [25] have used ODMR to study Ga acceptors in the 6H polytype and, again, that work is discussed in detail in Datareview 2.5. They obtain similar g-tensors to the Al acceptors and are able to resolve hyperfine interactions. Specifically, they resolve two sites with (1) $g_\parallel = 2.27$, $A_\parallel(^{69}Ga) = 3.2$ mT, $A_\parallel(^{71}Ga) = 4.1$ mT and (2) $g_\parallel = 2.21$, $A_\parallel(^{69}Ga) = 3.9$ mT,

2.4 Impurities and structural defects in SiC determined by ESR

$A_{\parallel}(^{71}Ga) = 4.9$ mT with $g_{\perp} = 0.6$ for both. The hyperfine interactions indicate 5 - 6 % localization on the Ga atom, consistent with a shallow acceptor.

TABLE 2 The B_{Si} centre in several SiC polytypes.

Polytype	g_z, g_x, g_y	A_z, A_x, A_y (mT)	Nuclei	T_{max} (K)
3C	2.0020, 2.0057, 2.0063	6.26, 1.90, 1.90	1-^{13}C	50
		1.05, 0.89, 0.89	3-^{29}Si	
		0.07, 0.21, 0.21	1-^{11}B	
2H	2.0020, 2.0070, 2.0070	0.16, 0.22, 0.22	1-^{11}B	70
4H	2.0064, 2.0028, 2.0059	0.70, 6.08, 1.70	1-^{13}C	35
		0.86, 1.00, 0.86	3-^{29}Si	
		0.20, 0.09, 0.09	1-^{11}B	
	2.0018, 2.0070, 2.0070	6.06, 1.70, 1.70	1-^{13}C	70
		1.04, 0.88, 0.88	3-^{29}Si	
		0.20, 0.12, 0.12	1-^{11}B	
6H	2.0059, 2.0021, 2.0062	1.70, 6.05, 1.70	1-^{13}C	65
		1.02, 1.04, 1.02	3-^{29}Si	
		0.18, 0.10, 0.10	1-^{11}B	
	2.0062, 2.0028, 2.0059	1.65, 6.07, 1.65	1-^{13}C	35
		1.05, 1.00, 1.05	3-^{29}Si	
		0.21, 0.12, 0.12	1-^{11}B	
	2.0020, 2.0068, 2.0068	6.16, 1.72, 1.72	1-^{13}C	95
		1.04, 0.85, 0.85	3-^{29}Si	
		0.19, 0.14, 0.14	1-^{11}B	

Note: Data in TABLE 2 is based on [21]. Dynamic effects prevent resolution of all the hyperfine structure above T_{max}.

D TRANSITION METAL IMPURITIES

Transition metals such as titanium and vanadium are common trace impurities in SiC, especially in Lely-grown single crystals. Titanium substitutes for Si and the Ti-N complex has been observed using ESR by Vainer et al [26] in the 6H polytype. Maier and co-workers [27,28] have observed isolated Ti on both Si sites and the Ti-N complex in the 4H polytype. Maier and co-workers observed a highly anisotropic resonance with $g = g_{\parallel}\cos\theta$ ($g_{\parallel} = 1.706$) in the 4H (but not the 6H) polytype which they identified as the isolated Ti on the hexagonal Si site. Another less anisotropic resonance was identified as isolated Ti on the cubic site. A strong electron-phonon interaction gives the g tensor of the former an observable isotopic dependence. Hyperfine structure due to ^{47}Ti, ^{49}Ti and ^{29}Si was also resolved for this case. The magnetic moments of the two I > 0 nuclei are nearly identical although ^{47}Ti has I = 5/2 and ^{49}Ti has I = 7/2 and both groups [26,27] find $A \approx 1$ mT for H $\parallel$ c. Maier and co-workers [28] also observed a Ti-N complex and resolved ^{14}Ni hyperfine splitting was ~0.1 mT for both polytypes. Vainer et al observed three distinct resonances with similar symmetries which they attributed to the three inequivalent Si sites in the 6H material, while Maier and co-workers were not able to resolve two distinct resonances for the cubic and hexagonal sites in the 4H material. The ODMR studies of Lee et al [29] on the isolated Ti_{Si} centre in the 4H, 6H and 15R polytypes will be covered in Datareview 2.5.

Vanadium also substitutes for Si. Both the neutral (S=½) and the negatively charged (S=1) states have been observed in the 6H and 4H polytypes [27,30]. ^{51}V is nearly 100% abundant with I=7/2 and so octets of lines serve as a 'fingerprint' for V although the S=½ spectra are lost above T≈20K due to vibronic coupling. In the 6H polytype, the cubic sites have weakly anisotropic resonances near g=2, while the hexagonal site is strongly anisotropic with $g = g_\parallel \cos\theta$, where $g_\parallel = 1.749$, much like the one Ti centre. In n-type samples the S=1 state is observed, with g≈2 and a strong crystal field splitting of the spectrum of V on the hexagonal site in either 4H or 6H polytype, due to the axial electric field gradient.

E   STRUCTURAL DEFECTS

The first work on structural defects in 3C and 6H SiC was by Balona and Loubser [31]. They studied electron- and neutron-irradiation induced defects in Lely-grown single crystals. Itoh and co-workers [32-36] have recently investigated electron and proton irradiated 3C films. The latter authors were able to clearly identify ^{13}C and ^{29}Si hyperfine structure for two centres which they labelled T1 and T5 and to assign these a negatively charged Si vacancy and a positively charged C vacancy, respectively. The authors had labelled several features T2, T3 and T4 which now appear to be parts of the hyperfine structure of these two centres. Balona and Loubser had suggested that a centre with parameters similar to the T1 centre was due to a carbon divacancy; the more complete hyperfine data indicates that the Si vacancy model is correct. The details of the parameters for the two centres are given in TABLE 3. ESR parameters of several other centres which Balona and Loubser observed in various 6H samples after proton or electron irradiation are also included. Their results seem to be quite sample dependent and given the improvements in crystal quality might merit repetition.

TABLE 3   Radiation-induced defects in SiC.

Polytype	Centre	Model	g-tensor	Hyperfine (mT)	Ref
3C	T1	V_{Si}^-	2.0029	$A_\parallel = 2.86$ (^{13}C) $A_\perp = 1.18$ (^{13}C) $A \approx -0.29$ (^{29}Si)	[31,32]
3C	T5	V_C^+	$g_1 = 2.0020$ $g_2 = 2.0007$ $g_3 = 1.9951$	$A_\parallel = 2.02$ (^{29}Si) $A_\perp = 1.48$ (^{29}Si)	[33]
6H	A	V_C^0 (S=1)	g = 2.0020	A = 0.45	[30]
6H	B	V_C^-	$g_\parallel = 2.0032$ $g_\perp = 2.0051$	$A_\parallel = 0.3$ $A_\perp = 0.71$	[30]
6H	C	$(V_C + V_C)^+$	$g_\parallel = 2.0050$ $g_\perp = 2.0037$	A = 0.41	[30]
6H	D	?	g = 2.0026		[30]
6H	E1	V_C^+	g = 2.0034	A = 0.24	[30]
	E2	V_C^+	$g_\parallel = 2.0033$ $g_\perp = 2.0028$	A = 0.48	[30]

Note: g and A tensors designated ∥ and ⊥ have principal axes along the Si-C bond directions while T5 has the g-tensor aligned along the <100> directions.

F   CONCLUSION

Electron spin resonance reveals the unpaired electrons associated with impurities or structural defects and can be used to identify the lattice site positions of these features. Nitrogen is shown to substitute for carbon and acts as a shallow donor. The various ESR triplets due to nitrogen in several SiC polytypes give information on the lattice sites occupied. For the acceptor boron, ESR shows it to occupy Si sites only, in disagreement with DAP photoluminescence measurements which show only boron on carbon sites. It may be that boron substitutes on both sites and the two techniques have sensitivity for only one particular lattice site. The aluminium acceptor is not observed in ESR but gallium has been noted in one report. Transition metals, Ti and V, have been identified by ESR both isolated on Si sites and in Ti-N complexes. Several charged vacancy defects have been assigned from ESR spectra in irradiated samples.

REFERENCES

[1]   Yu.M. Altaiskii, I.M. Zaritskii, V.Ya. Zevin, A.A. Konchits [*Sov. Phys.-Solid State (USA)* vol.12 (1971) p.2453-4]

[2]   W.E. Carlos et al [*Mater. Res. Soc. Symp. Proc. (USA)* vol.97 (1987) p.253-8]

[3]   H. Okumura et al [*Jpn. J. Appl. Phys. (Japan)* vol.27 (1988) p.1712-7]

[4]   I. Nashiyama, H. Okumura, E. Sakuma, S. Misawa, K. Endo, S. Yoshida [*Amorphous and Crystalline Silicon Carbide III* Eds G.L. Harris, M.G. Spencer, C.Y. Yang (Springer-Verlag, Berlin, 1992) p.149-54]

[5]   W.E. Carlos, J.A. Freitas, J. Pazik [to be published]

[6]   D.J. Lépine [*Phys. Rev. B (USA)* vol.2 (1970) p.2429-39]

[7]   P.J. Dean, W.J. Choyke, L. Patrick [*J. Lumin. (Netherlands)* vol.15 (1977) p.299-314]

[8]   J.S. van Wieringen [*Semiconductors and Phosphors* Eds M. Schön, H. Welker (Interscience Publishers, Inc., New York, 1958) p.367-70]

[9]   H.H. Woodbury, G.W. Ludwig [*Phys. Rev. (USA)* vol.124 (1961) p.1083-9]

[10] G.E.G. Hardeman [*J. Phys. Chem. Solids (UK)* vol.24 (1963) p.1223-33]

[11] O.V. Vakulenko, Yu.A. Marazuev, B.M. Shotov [*Sov. Phys.-Solid State (USA)* vol.18 (1976) p.1795-7]

[12] E.N. Kalabukhova, N.N. Kabdin, S.N. Lukin [*Sov. Phys.-Solid State (USA)* vol.29 (1987) p.1461-2]

[13] E.N. Kalabukhova, N.N. Kabdin, S.N. Lukin, T.L. Petrenko [*Sov. Phys.-Solid State (USA)* vol.30 (1988) p.1457-8]

[14] O.V. Vakulenko, V.S. Lysyi [*Sov. Phys.-Solid State (USA)* vol.30 (1988) p.1446-7]

[15] E.N. Kalabukhova, S.N. Lukin, B.D. Shanina, L.V. Artamonov, E.N. Mokhov [*Sov. Phys.-Solid State (USA)* vol.32 (1990) p.482-6]

[16] Yu.A. Vodakov, E.N. Kalabukhova, S.N. Lukin, A.A. Lepneva, E.N. Mokhov, B.D. Shanina [*Sov. Phys.-Solid State (USA)* vol.33 (1991) p.1869-75]

[17] E.N. Kalabukhova, N.N. Kabdin, S.N. Lukin, E.N. Mokhov, B.D. Shanina [*Sov. Phys.-Solid State (USA)* vol.31 (1989) p.378-83]

[18] A.I. Veinger [*Sov. Phys.-Semicond. (USA)* vol.3 (1969) p.52-6]

[19] M.F. Deigen, I.M. Zaritskii, L.A. Shul'man [*Sov. Phys.-Solid State (USA)* vol.12 (1971) p.2343-6]

[20] A.G. Zubatov, V.G. Stepanov, Yu.A. Vodakov, E.N. Mokhov [*Sov. Phys.-Solid State (USA)* vol.24 (1982) p.504-5; *Sov. Tech. Phys. Lett. (USA)* vol.8 (1982) p.120-1]

[21] A.G. Zubatov, I.M. Zaritskii, S.N. Lukin, E.N. Mokhov, V.G. Stepanov [*Sov. Phys.-Solid State (USA)* vol.27 (1985) p.197-201]

[22] S. Yamada, H. Kuwabara [*Silicon Carbide - 1973* Eds R.C. Marshall, J.W. Faust Jr., C.E. Ryan (Univ. of South Carolina Press, Columbia, SC, 1974) p.305-12]

[23] Le Si Dang, K.M. Lee, G.D. Watkins, W.J. Choyke [*Phys. Rev. Lett. (USA)* vol.45 (1980) p.390-4]

[24] Yu.A. Vodokov, A.G. Zubatov, E.N. Mokhov, V.G. Stepanov [*Sov. Tech. Phys. Lett. (USA)* vol.8 (1982) p.122-3]

[25] P.G. Baranov, V.A. Vetrov, N.G. Romanov, V.I. Sokolov [*Sov. Phys.-Solid State (USA)* vol.27 (1985) p.2085-6]

[26] V.S. Vainer, V.A. Il'in, V.A. Karachinov, Yu.M. Tairov [*Sov. Phys.-Solid State (USA)* vol.28 (1986) p.201-4]

[27] K. Maier, J. Schneider, W. Wilkening, S. Leibenzeder, R. Stein [*Mater. Sci. Eng. B (Switzerland)* vol.11 (1992) p.27-30]

[28] K. Maier, H.D. Müller, J. Schneider [*Mater. Sci. Forum (Switzerland)* vol.83-87 (1992) p.1183-94]

[29] K.M. Lee, Le Si Dang, G.D. Watkins, W.J. Choyke [*Phys. Rev. B (USA)* vol.32 (1985) p.2273-84; *Solid State Commun. (USA)* vol.37 (1981) p.551-4]

[30] J. Schneider et al [*Appl. Phys. Lett. (USA)* vol.56 (1990) p.1184-6]

[31] L.A. de S. Balona, J.H.N. Loubser [*J. Phys. C, Solid State Phys. (UK)* vol.3 (1970) p.2344-51]

[32] H. Itoh, N. Hayakawa, I. Nashiyama, E. Sakuma [*J. Appl. Phys. (USA)* vol.66 (1989) p.4529-31]

[33] H. Itoh, M. Yoshikawa, I. Nashiyama, S. Misawa, H. Okumura, S. Yoshida [*IEEE Trans. Nucl. Sci. (USA)* vol.37 (1990) p.1732-8]

[34] H. Itoh, M. Yoshikawa, I. Nashiyama, S. Misawa, H. Okumura, S. Yoshida [*J. Electron. Mater. (USA)* vol.21 (1992) p.707-10]

[35] H. Itoh, M. Yoshikawa, I. Nashiyama, S. Misawa, H. Okumura, S. Yoshida [*Amorphous and Crystalline Silicon Carbide III* Eds G.L. Harris, M.G. Spencer, C.Y. Yang (Springer-Verlag, Berlin, 1992) p.143-8]

[36] I. Nashiyama et al [*Amorphous and Crystalline Silicon Carbide and Related Materials II* Eds M.M. Rahman, C.Y.-W. Yang, G.L. Harris (Springer-Verlag, Berlin, 1989) p.100-5]

2.5 ODMR investigations of defects and recombination processes in SiC

T.A. Kennedy

November 1993

A INTRODUCTION

Optically-Detected Magnetic Resonance (ODMR) has provided magnetic and hyperfine parameters for excited states of defects in SiC. Since ODMR combines EPR with photoluminescence, it gives a link between the information provided by each experiment. In SiC, most of the publications describe donor- and acceptor-resonances detected on distant donor-acceptor-pair (DAP) recombination. As in the other experiments, interesting effects occur due to the different symmetries which arise from the polytypism of SiC.

ODMR depends on the existence of spin selection rules for the recombination of an electron and hole. Since most of the published results in SiC concern DAP recombination, the data can be understood with the spin Hamiltonian for a weakly coupled donor-acceptor pair:

$$\mathcal{H} = \beta \mathbf{B} g^D S^D + \beta \mathbf{B} g^A S^A + J S^D S^A \tag{1}$$

where the first term describes the Zeeman spitting of the donor states, the second the Zeeman splitting of the acceptor states, and the third the exchange coupling between the donor and acceptor. For distant donor-acceptor pairs, this coupling is much smaller than the Zeeman interaction for either the electron or hole. This exchange interaction can broaden the ODMR lines beyond the values observed for isolated impurities by EPR. Under the assumption that both the donor and acceptor states are doublets, this Hamiltonian leads to four excited states (see FIGURE 1). Two of these states are allowed to recombine radiatively, but two cannot. Hence, the population of the forbidden states builds up. When microwaves are applied and the magnetic resonance condition is fulfilled, population is transferred to the allowed states which recombine to produce ODMR.

The ODMR results have been placed into three groups. First, the results on donors and acceptors are reviewed. Second, the ODMR of Ti in SiC is described. Finally, the work on native and radiation-induced defects is presented.

B DONORS

Cubic (3C) SiC contains sites of cubic symmetry with the common donor, nitrogen, expected to occupy the carbon-site. Hence an isotropic g-value is expected and a single hyperfine (HF) splitting is expected. An isotropic ODMR at $g = 2.00$ was reported in thermally quenched material, but no HF interaction was resolved [1]. The ODMR was observed as a decrease in the photoluminescence. More often, the ODMR of donors has been observed as an enhancing signal on DAP recombination in SiC doped with acceptors (see TABLE 1). In 3C material doped with Al, an isotropic resonance at $g = 2.0065$ was observed again without hyperfine splitting (see FIGURE 2) [2]. The linewidths observed in ODMR are larger than those in

2.5 *ODMR investigations of defects and recombination processes in SiC*

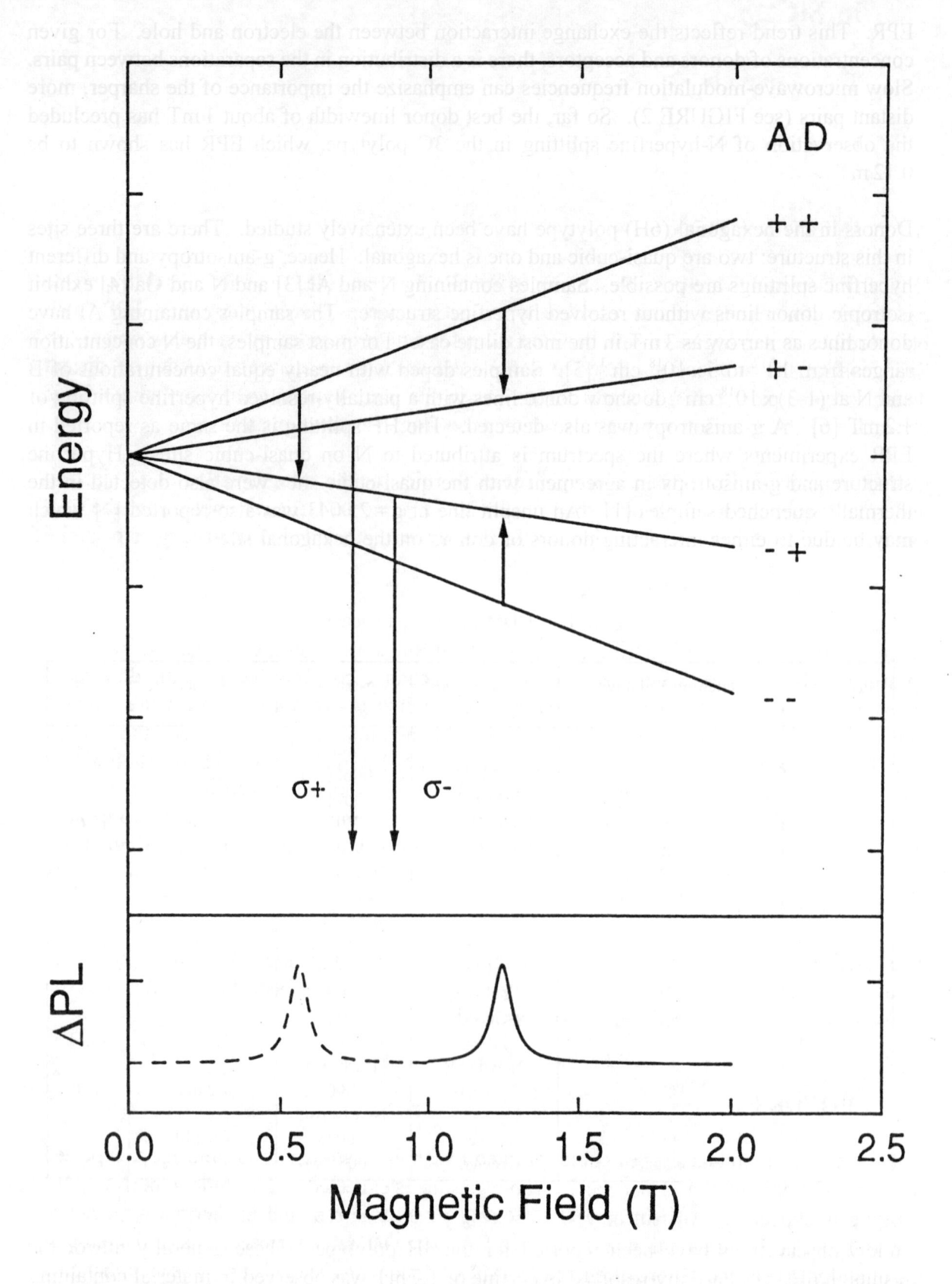

FIGURE 1 ODMR processes. The top shows the energy levels for donor (D) and acceptor (A) spins in a magnetic field. The two middle states can decay by emitting a circularly polarized photon. The small arrows indicate microwave-induced transitions from non-emitting states to emitting states. The bottom shows the change in the emission at magnetic resonance for the acceptor (dashed) and donor (solid).

2.5 ODMR investigations of defects and recombination processes in SiC

EPR. This trend reflects the exchange interaction between the electron and hole. For given concentrations of donors and acceptors, there is a distribution in the separations between pairs. Slow microwave-modulation frequencies can emphasize the importance of the sharper, more distant pairs (see FIGURE 2). So far, the best donor linewidth of about 1 mT has precluded the observation of N-hyperfine splitting in the 3C polytype, which EPR has shown to be 0.12 mT.

Donors in the hexagonal (6H) polytype have been extensively studied. There are three sites in this structure: two are quasi-cubic and one is hexagonal. Hence, g-anisotropy and different hyperfine splittings are possible. Samples containing N and Al [3] and N and Ga [4] exhibit isotropic donor lines without resolved hyperfine structure. The samples containing Al have donor lines as narrow as 3 mT in the most dilute case. For most samples, the N concentration ranges from 10^{17} to 5×10^{18} cm^{-3} [5]. Samples doped with nearly equal concentrations of B and N at $(1-3) \times 10^{18}$ cm^{-3} do show donor lines with a partially-resolved hyperfine splitting of 1.2 mT [6]. A g-anisotropy was also detected. The HF splitting is the same as reported in EPR experiments where the spectrum is attributed to N on quasi-cubic sites. Hyperfine structure and g-anisotropy in agreement with the quasi-cubic sites were also detected in the thermally quenched samples [1]. An unsplit line at g = 2.0043 was also reported [1] which may be due to either interacting donors or donors on the hexagonal site.

TABLE 1 ODMR assigned to donors.

Polytype	Quenching/Enhancing	Emission band	g-tensor (error)	^{14}N Hyperfine	Ref
3C	Q	?	2.00	Not res.	[1]
	E	2.13 eV	2.0065(15) Isotropic	Not res.	[2]
4H	E	N-Al DAP N-B DAP	2.00	1.7 mT	[5,6]
	Q	?	2.00 Nearly iso.	1.7 mT	[1]
	(Cross relaxation)	Green Ti	2.003(1) Nearly iso.		[11]
6H	E	2.65 eV N-Al DAP	2.004(2) Isotropic	Not res.	[3,7]
	E	'Yellow' N-Ga DAP	2.00 Isotropic	Not res.	[4,5]
	E	2.1 eV N-B DAP	2.004 Nearly iso.	1.2 mT	[5,6]
	Q	2.36 eV	2.00 Nearly iso.	1.2 mT	[1]
	Q	2.36 eV	2.0043	Unsplit	[1]
	(Cross relaxation)	Green Ti	2.0035(5)	1.23 mT	[11]

A few observations have been reported for the 4H polytype. These generally mirror the results in 6H [5]. Partially resolved hyperfine of 1.7 mT was observed in material containing N and B [5,6].

- 53 -

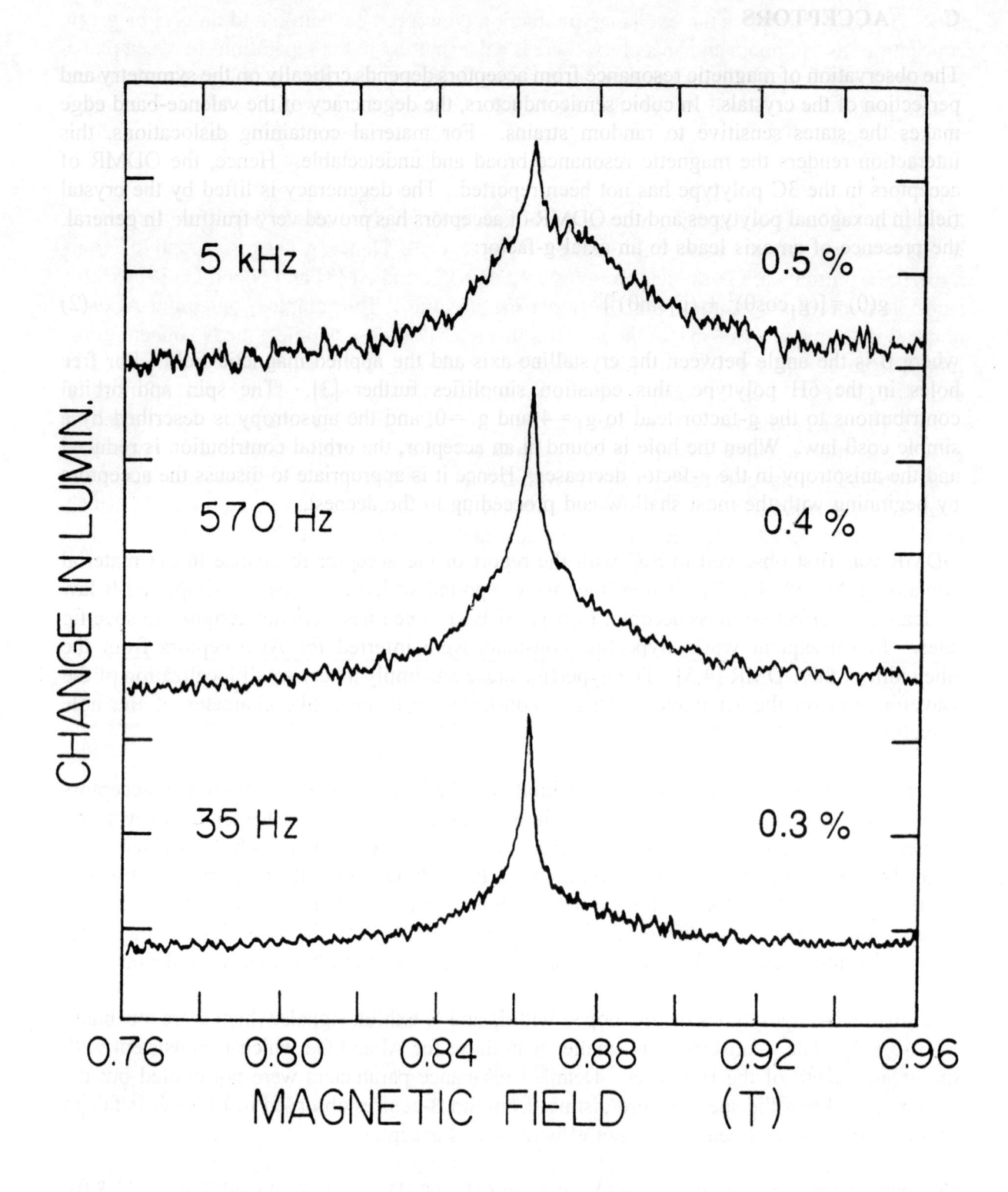

2.5 ODMR investigations of defects and recombination processes in SiC

FIGURE 2 ODMR of donors in cubic SiC for three modulation frequencies. The lower frequencies emphasize the contributions from more distant donor-acceptor pairs. Since the exchange interaction is weaker for distant pairs, the linewidth is smaller.

C   ACCEPTORS

The observation of magnetic resonance from acceptors depends critically on the symmetry and perfection of the crystals. In cubic semiconductors, the degeneracy of the valence-band edge makes the states sensitive to random strains. For material containing dislocations, this interaction renders the magnetic resonance broad and undetectable. Hence, the ODMR of acceptors in the 3C polytype has not been reported. The degeneracy is lifted by the crystal field in hexagonal polytypes and the ODMR of acceptors has proved very fruitful. In general, the presence of an axis leads to an axial g-factor:

$$g(\theta) = [(g_{\parallel}\cos\theta)^2 + (g_{\perp}\sin\theta)^2]^{1/2} \qquad (2)$$

where θ is the angle between the crystalline axis and the applied magnetic field. For free holes in the 6H polytype, this equation simplifies further [3]. The spin and orbital contributions to the g-factor lead to $g_{\parallel} = 4$ and $g_{\perp} = 0$, and the anisotropy is described by a simple $\cos\theta$ law. When the hole is bound at an acceptor, the orbital contribution is reduced and the anisotropy in the g-factor decreases. Hence it is appropriate to discuss the acceptors by beginning with the most shallow and proceeding to the deepest.

ODMR was first observed in SiC with the report of the acceptor resonance in 6H material containing Al and N [3,7]. Three lines were reported which followed the simple $\cos\theta$ law indicating an effective-mass acceptor (see TABLE 2). The lines were not assigned to specific sites. In subsequent work, hyperfine constants were inferred for Al acceptors from the linewidth of the ODMR [4,5]. The hyperfine constants imply a very small localization of the wavefunction on the Al nuclei. This is consistent with the p-like character of the hole wavefunction.

6H crystals doped with Ga and N also exhibit an ODMR which can be assigned to acceptors [4,5]. Two signals were reported with a large g-anisotropy (see TABLE 2). However, the perpendicular g-value is finite with a value of 0.6. This is consistent with the greater depth of the Ga acceptor compared to the Al acceptor. Partially resolved HF structure was observed for the Ga nuclei. The localization of the wavefunction is still small, but about twice that for Al-acceptors. From the difference in hyperfine coupling for the two signals, the more localized centre was assigned to the quasi-cubic sites, which are believed to be deeper.

Both 6H and 4H polytypes of SiC doped with B and N exhibit signals which were attributed to B [5,6,8]. The g-anisotropy is smaller than those for Al and Ga which is consistent with the greater depth of the B centres. Detailed resonance parameters were not quoted but the authors stated that the spectra were distinct from the B-centres observed in EPR. It is fair to conclude that these B-centres are not effective-mass acceptors.

Complex, anisotropic spectra were observed in 6H and 4H polytypes doped with Sc [5,8,9]. The g-factors parallel to the crystal axis are smaller than those that are perpendicular to it (see TABLE 2). This anisotropy is opposite to that of the shallow acceptors. Many spectra are observed whose parameters indicate that the Sc-acceptors, like the B-centres, are not effective-mass-like.

2.5 ODMR investigations of defects and recombination processes in SiC

TABLE 2 ODMR assigned to acceptors. All spectra are axial with respect to the C_6 axis. All the ODMR reported here are enhancing signals.

Polytype	Dopant	Emission	Label/site sym.	g-tensor (error) g∥	g⊥	Hyperfine	Ref
4H	B	N-B DAP		(Similar to 6H)			[6]
	Sc			(Similar to 6H)			[5,8]
	Al	Cross relax. Green, Ti		2.35(2)	0	Not res.	[11]
6H	Al	2.65 eV N-Al DAP	I	2.412(2)	0	Not res.	[3,7]
	Al	2.65 eV N-Al DAP	II	2.400(2)	0	Not res.	[3,7]
	Al	2.65 eV N-Al DAP	III	2.325(2)	0	Not res.	[3,7]
	Al		i			^{27}Al 0.6 mT (estimated)	[4,5]
	Al		ii			^{27}Al 0.7 mT (estimated)	[4,5]
	Ga	Yellow N-Ga DAP	Hexagonal	2.27	0.6(2)	^{69}Ga 3.2 mT	[4,5,8]
	Ga	Yellow N-Ga DAP	Cubic	2.21	0.6(2)	^{69}Ga 3.9 mT	[4,5,8]
	B	1.8, 2.1, 2.2 eV N-B DAP	Multiple sites	2.01	axial		[5,6,8]
	Sc		1 A^0	1.97	2.49		[8,9]
	Sc		2 A$^-$	1.21 1.18 1.17	1.77 1.63 1.41		[5]

D DISTANT DONOR-ACCEPTOR PAIR RECOMBINATION

The observations of ODMR for donors and acceptors on specific emission bands confirm and enlarge upon the PL studies of distant DAP recombination. In all the cases discussed in the previous sections, distinct donor and acceptor resonances were detected on a specific emission band. Hence the exchange coupling is small and the donors and acceptors are distant. The data for the 6H polytype are most extensive. In material containing N and Al, a donor signal and three acceptor signals are detected on the 2.65 eV (blue) emission [3,4]. For samples doped with N and Ga, a donor and two acceptor signals were reported on the 2.2 eV (yellow) band [4]. With N- and B-doping, both the N-donor and deep acceptor signals were seen on bands from 2.2 eV (yellow) to 1.8 eV (red) [6]. In Sc-doped material, donor and deep-acceptor resonances were detected on luminescence at 2.2 eV (yellow) [5,8]. The ODMR reveals the symmetry of the site through the g-tensor and the localization of the wave function when hyperfine structure is resolved.

E   EXCITONS AT Ti IMPURITIES

Titanium impurities substitute for Si in SiC and produce a strong green luminescence in the wider bandgap polytypes. The PL and ODMR for Ti are particularly rich. The number of zero-phonon lines corresponds to the number of sites in a given polytype. The separation inherent in distant donor-acceptor pair recombination is absent in this case. The electron and hole overlap strongly in the neighbourhood of the impurity and are rightly considered as a bound exciton. This implies a very strong exchange coupling between the particles which leads to a spin-triplet state (S = 1). This leads to the spin Hamiltonian:

$$\mathcal{H} = \beta \mathbf{B} \mathbf{g} \mathbf{S} + D[S_z^2 - \tfrac{1}{3}(S)(S+1)] \qquad (3)$$

where D is the zero-field splitting. The anisotropies in g and D reflect the axial crystal structure.

The ODMR results have been reported by the Lehigh group [10-12]. ODMR has been studied for 4H, 6H and 15R polytypes. The results in 4H and 15R have been used to aid the site assignments for the ODMR in 6H. Hence, it is sufficient to summarize the results in this polytype.

Spectra with two distinct g-tensors are observed in the 6H polytype and labelled as Ti^A and Ti^C (see TABLE 3). Each of these spectra has further splittings, which have been shown to arise from a strong dependence of D on the mass of the Ti impurities for each isotope, with further splittings from the different Si isotopes. This dependence suggests a vibronic character for the excited state. For the Ti^C spectrum, there are additional splittings from hyperfine interactions for the magnetic Ti nuclei. The lifetimes of the two excitons are similar. However, the character of the ODMR for Ti^A and Ti^C differs due to a large difference in the spin-lattice relaxation. Furthermore, a third ODMR is expected since both the number of sites and the number of ZPLs is three. While no ODMR was detected from this site, Zeeman spectroscopy and level-crossing experiments give radically different spin Hamiltonian parameters (see TABLE 3) [12,13].

TABLE 3   Triplet ODMR from the ^{46}Ti impurity with ^{28}Si neighbours in 6H SiC.

Emission (ZPL)	Spectrum	Site	g-tensor $g\parallel$	g-tensor $g\perp$	Zero-field splitting (10^{-4} cm^{-1})	Ref
A_0 2.86 eV	Ti^A (ODMR)	Cubic (k_2)	1.935(0.5)	1.945(0.5)	+4936(2)	[10,11]
C_0 2.82 eV	Ti^C (ODMR)	Cubic (k_1)	1.957(0.5)	1.972(0.5)	-3644(2)	[10,12]
B_0 2.79 eV	(Zeeman)	Hexagonal (h)	1.95	0	-17,500(200)	[12,13]

The differences in the three excitons have been attributed to the hole's extreme sensitivity to the local environment [12]. This in turn may be evidence of a strong Jahn-Teller coupling.

The g-anisotropy for the B_0 exciton is similar to that for the shallow Al acceptor. However, the A_0 and C_0 excitons have a much smaller anisotropy with the parallel g-factor smaller than the perpendicular one. It appears that the sign of the crystal field reverses as the hole is localized. This behaviour is very similar to that observed for the acceptors as the binding energy increases (see TABLE 2).

F CROSS RELAXATION EXPERIMENTS AND PROCESS-INDUCED DEFECTS

The very rich ODMR observed for Ti in SiC led to another experiment called cross relaxation [11]. A change in emission occurs when the Zeeman splitting of two Ti-exciton states equals the splitting for a nearby, coupled defect. No microwaves are necessary but the sensitivity and resolution are very high when magnetic-field modulation is employed. In as-grown materials, cross relaxation reveals the N-donor in 4H and 6H SiC and the Al acceptor in 4H SiC (see TABLES 1 and 2). The linewidths are smaller than those observed by the ODMR of donor-acceptor pairs.

Both spin-½ and spin-1 centres have been observed following particle irradiation. Electron-irradiation of 6H SiC revealed new Ti-related triplets through ODMR and cross relaxation [11]. A spin-½ defect was also reported [11]. Similarly, electron-irradiation of the 3C polytype produced both an $S = ½$ centre, possibly the N-donor, and triplets [14]. Neutron-irradiation of both 4H and 6H polytypes produced luminescence-quenching donor resonances (see TABLE 1) [1].

Other defects were discovered after specific growth or annealing treatments. The donors in 3C SiC were first reported in a sample which had been annealed and quenched (see TABLE 1) [1]. A new spectrum was reported for 6H epilayers grown under Si-rich conditions [15]. The g-factor is near 2 and a hyperfine interaction is observed which is consistent with ^{29}Si. The spectrum was tentatively assigned to the Si_C antisite.

G CONCLUSION

The ODMR for the donor nitrogen in SiC is seen as a reduction in the photoluminescence signal but often the ODMR of donors is detected as an increased DAP signal in material doped with acceptors aluminium and gallium. Most results are for the 6H polytype but data in 3C and 4H are also found. In cubic material containing dislocations no ODMR of acceptors is observed due to degeneracy in the valence band edge. In hexagonal polytypes this degeneracy is lifted and studies of the ODMR of acceptors have been extensive. 6H material containing aluminium and nitrogen and gallium and nitrogen has been investigated and the hyperfine interactions give information on the depths of the centres. This is also found to be the case for 6H and 4H polytypes doped with boron and nitrogen. Material with Ti impurities has a strong green luminescence in polytypes with wide bandgaps. Electron irradiation reveals Ti-related ODMR triplets in the 6H polytype while neutron irradiation produces luminescence quenching donor resonances in 4H and 6H material.

REFERENCES

[1] N.G. Romanov, V.A. Vetrov, P.G. Baranov [*Sov. Phys.-Semicond. (USA)* vol.20 (1986) p.96-7]

[2] T.A. Kennedy, J.A. Freitas Jr., S.G. Bishop [*J. Appl. Phys. (USA)* vol.68 (1990) p.6170-3]

[3] Le Si Dang, K.M. Lee, G.D. Watkins, W.J. Choyke [*Phys. Rev. Lett. (USA)* vol.45 (1980) p.390-4]

[4] P.G. Baranov, V.A. Vetrov, N.G. Romanov, V.I. Sokolov [*Sov. Phys.-Solid State (USA)* vol.27 (1985) p.2085-6]

[5] P.G. Baranov, N.G. Romanov [*Mater. Sci. Forum (Switzerland)* vol.83-87 (1992) p.1207-12]

[6] N.G. Romanov, V.A. Vetrov, P.G. Baranov, E.N. Mokhov, V.G. Oding [*Sov. Tech. Phys. Lett. (USA)* vol.11 (1985) p.483-6]

[7] G.D. Watkins, Le Si Dang, K.M. Lee, W.J. Choyke [*J. Phys. Soc. Jpn. (Japan)* vol.49 (1980) p.619-22]

[8] P.G. Baranov, N.G. Romanov, V.A. Vetrov, V.G. Oding [*20th Int. Conf. Physics of Semiconductors* Eds E.M. Anastassakis, J.D. Joannopoulos (World Scientific, Singapore, New Jersey, London, Hong Kong, 1990) p.1855-8]

[9] P.G. Baranov, N.G. Romanov [*Magnetic Resonance and Related Phenomena*, Proc. 24th Congress Ampere, Pozan, 1988, p.85-99]

[10] K.M. Lee, Le Si Dang, G.D. Watkins, W.J. Choyke [*Solid State Commun. (USA)* vol.37 (1981) p.551-4]

[11] K.M. Lee, G.D. Watkins [*Phys. Rev. B (USA)* vol.26 (1982) p.26-34]

[12] K.M. Lee, Le Si Dang, G.D. Watkins, W.J. Choyke [*Phys. Rev. B (USA)* vol.32 (1985) p.2273-84]

[13] P.J. Dean, R.L. Hartman [*Phys. Rev. B (USA)* vol.5 (1972) p.4911-24]

[14] N.D. Wilsey, J.A. Freitas Jr., S.G. Bishop, P.B. Klein, M.A. Gipe [*Bull. Am. Phys. Soc. (USA)* vol.34 (1989) p.552]

[15] P.G. Baranov, N.G. Romanov [*Appl. Magn. Reson. (Austria)* vol.2 (1991) p.361-78]

CHAPTER 3

CARRIER PROPERTIES AND BAND STRUCTURE

3.1 Carrier mobilities and concentrations in SiC
3.2 Effective masses in SiC
3.3 Band structure of SiC: overview
3.4 Pressure effects on band structure of SiC

3.1 Carrier mobilities and concentrations in SiC

G.L. Harris, H.S. Henry and A. Jackson

January 1995

A   INTRODUCTION

Hall, and C-V, four-point probe and spreading resistance measurements to some extent provide a measure of the net impurity concentration of dopants in SiC ($[N_D - N_A]$ or $(N_A - N_D)$). In addition, Hall measurements provide a method for obtaining the mobility of the net carriers. These measurements have been applied to both n- and p-type SiC and its various polytypes. In this Datareview, we will report on the mobilities for most of the SiC polytypes under various growth conditions.

The mobility and carrier concentration in SiC are functions of the polytype, growth technique, starting substrate doping conditions and temperature. The information that follows is divided into polytype and the growth technique information is also provided. We do not claim to report on all the known measurements, but a representative number.

B   CARRIER MOBILITIES AND CONCENTRATIONS

The electron and hole mobilities in this section are at low field conditions where $\mu_e = q\tau/m_e^*$ and $\mu_p = q\tau/m_p^*$, where q is the electronic charge, τ is the average scattering time, m_e^* is the electron effective mass and m_p^* the hole effective mass. The principal scattering mechanisms controlling the scattering time in SiC are ionized impurity, acoustical phonon, piezoelectric, neutral impurity and polar optical phonon scattering. Under most conditions the ionized impurity and acoustical phonon scattering dominate for both 6H and 3C-SiC [1-4]. A detailed study of the scattering mechanisms in β-SiC is provided in [3]. The data since 1984 [1] indicates that in undoped material the Hall mobility has been increasing while the residual carrier concentration has been decreasing. This data has been fitted to the following curve:

$$\mu_n(n) = 4.82 \times 10^9 \times (n^{-0.424})$$

where n is the net electron carrier concentration as determined by Hall measurements. The range of the data is from approximately 8×10^{15} to 1×10^{19} cm^{-3}. This is a clear sign of improving crystal quality in 3C on Si. The same can be said about 6H material, both epitaxially and bulk grown [2]. For detailed temperature studies of carrier concentration and Hall mobility in 6H and 3C see [1,3,4].

The following tables list the carrier concentrations, Hall mobility, conditions of the growth technique and polytype. In some cases, C-V and other measurement techniques have been employed.

3.1 Carrier mobilities and concentrations in SiC

TABLE 1 Mobilities and carrier densities in β-SiC (3C-SiC).

Mobility (cm^2 V^{-1} s^{-1})	Carrier density (cm^{-3})	Sample information	Ref
510	3.0 x 10^{16}	Undoped, horizontal CVD on Si, 10 μm thick, n-type	[6]
464	1.3 x 10^{16}	Undoped, horizontal CVD on Si, 10 μm thick, n-type	
450	6.0 x 10^{15}	Undoped, horizontal CVD on Si, 10 μm thick, n-type	
322	2.4 x 10^{17}	N$_2$-doped, horizontal, > 10 μm thick	
280	3.8 x 10^{17}	N$_2$-doped, horizontal, > 10 μm thick	
186	8.2 x 10^{17}	N$_2$-doped, horizontal, > 10 μm thick	
270	2.5 x 10^{17}	Sample thickness 4.6 - 16.9 μm, undoped n-type, horizontal	[7]
310	1.6 x 10^{17}	Sample thickness 4.6 - 16.9 μm, undoped, horizontal	
305	1.6 x 10^{17}	Sample thickness 4.6 - 16.9 μm, undoped, horizontal	
245	1.6 x 10^{17}	Sample thickness 4.6 - 16.9 μm, undoped, horizontal	
890	1.0 x 10^{16}	All samples SiC platelets grown from carbon saturated silicon melts, 20 - 40 μm	[8]
760	7.5 x 10^{16}		
680	2.9 x 10^{16}		
410	1.2 x 10^{17}		
310	1.1 x 10^{17}		
14.4	8.1 x 10^{17}		
500	7 x 10^{16}	Undoped, horizontal CVD, n-type	[9]
40	3 x 10^{16}	Al-doped, horizontal CVD, p-type	
40	1 x 10^{17}	Al-doped, horizontal CVD, p-type, 6 - 50 μm thick	
600	5 x 10^{16}	Undoped, SiC on Si, on axis, n-type	[10]
800	1 x 10^{16}	Undoped, SiC on Si, on axis, n-type	
300	3 x 10^{17}	Undoped, SiC on Si, on axis, n-type	
200	8 x 10^{17}	Undoped, off axis, n-type	
190	4 x 10^{17}	Undoped, off axis, n-type	
100	5 x 10^{17}	Undoped, off axis, n-type	
90	5 x 10^{15}	Undoped β-SiC (111), as-grown on 6H-SiC, measured on substrate, 5 - 19.6 μm	
297	4 x 10^{16}	Undoped, horizontal CVD	[3]
93	3.95 x 10^{17}	Undoped, horizontal CVD	
138	1.87 x 10^{17}	Undoped, vertical CVD	
51	5.5 x 10^{16}	Undoped, vertical CVD, 7 - 15 μm	
800	8 x 10^{15}	Undoped, horizontal CVD, off axis	[1]
730	9.5 x 10^{15}	Undoped, horizontal CVD, off axis	
550	2.9 x 10^{16}	Undoped, horizontal CVD, off axis	
395	6.0 x 10^{16}	Undoped, horizontal CVD, off axis, thickness unknown	
20.1	3.943 x 10^{17}	TMA doped p-type on off axis with the substrate removed > 14 μm	[11]
21.1	3.32 x 10^{17}		
22.6	3.61 x 10^{17}	p-type with TMA doping source on off-axis Si substrate (measurements made with Si removed)	
27.6	3.44 x 10^{17}		
29.4	2.75 x 10^{17}		
$\rho(n) = 5.9 \times 10^{10} n^{-0.6703}$ $\mu(n) = 1.02 \times 10^8 n^{-0.326}$		Curve fitting on the data where n is net electron concentration SiC grown on Si in the range from $1 \times 10^{17} - 1.0 \times 10^{20}$ cm^{-3}	[4]

3.1 Carrier mobilities and concentrations in SiC

TABLE 2 Mobilities and carrier densities in 6H-SiC.

Mobility ($cm^2 V^{-1} s^{-1}$)	Carrier density (cm^{-3})	Sample information	Ref
350	$10^{15} - 10^{16}$	Homoepitaxially grown at 1500 °C using step controlled epitaxy, undoped n-type	[16]
247 - 253	8×10^{16}	Homoepitaxially grown, undoped 6H on 6H, n varied with Si/C ratio	[17]
60	4×10^{18}	n varied with Si/C ratio	
275	2.8×10^{17}	Homoepitaxial 6H grown $\perp$ c-axis on 6H substrates, n-type N_2-doped, Hall measurements from pn junctions	[18]
235	3.5×10^{17}		
265	5×10^{17}		
198	1.5×10^{18}		
60	1.6×10^{19}		
75	1.9×10^{19}		
70	2.3×10^{19}		
70	3.0×10^{19}		
25	3.9×10^{19}		
20	1.01×10^{20}		
101	1.2×10^{15}	p-type Al-doped	[18]
87	4.8×10^{15}		
83	2×10^{17}		
83	1.3×10^{18}		
43	1.3×10^{19}		
300	1.7×10^{17}	All samples are N_2-doped and grown by modified Lely technique	[12]
320	8×10^{16}		
320	8×10^{16}		
330	8×10^{16}	thickness not given	
280	1.3×10^{17}		
129	9.2×10^{17}	As-grown Acheson bulk	[3]
500	4×10^{16}	Sublimation, He ambient, n-type	[13]
38	1×10^{15}	Sublimation, Ar ambient, p-type	
39	6×10^{16}	Sublimation, Al ambient, p-type	
24	2.5×10^{16}	Sublimation, Al ambient, p-type	
38	2.5×10^{16}	Sublimation, Al ambient, p-type	
54	5×10^{13}	Sublimation, Ar ambient, p-type	
184	4.0×10^{17}	Sublimation grown, n-type	
264	1.46×10^{17}	Sublimation grown, n-type	
No Hall data provided	$\sim 10^{16}$	Undoped, $\rho \geq 2000 \Omega$ cm micropipes 10^2 to 10^3 cm^2, 50 - 75 μm diameter, sublimation bulk crystals	[14]
	$3 \times 10^{15} - 1.5 \times 10^{17}$	Epi-layers, a-axis, 6H depends on Si/C ratio	
	$8 \times 10^{14} - 8 \times 10^{16}$	Epi-layers, c-axis, 6H depends on Si/C ratio	
	$7 \times 10^{17} - 1.5 \times 10^{18}$	Epi-layers, prismatic low dependence on Si/C ratio	
No mobility measurements reported	5×10^{18}	p-type from TMA, generally $n_{prismatic} > n_{a-axis} > n_{c-axis}$ horizontal is on 6H-SiC	[15]

3.1 Carrier mobilities and concentrations in SiC

TABLE 2 continued

Mobility (cm² V⁻¹ s⁻¹)	Carrier density (cm⁻³)	Sample information	Ref
255	4.2×10^{16}	6H-bulk substrate physical vapour transport (sublimation), n-type samples were also analyzed by SIMS for nitrogen and boron	[19]
84	2.0×10^{18}		
80	2.7×10^{18}		
53	6.0×10^{18}		
34	1.9×10^{19}		
14	9.0×10^{19}		
200	3×10^{16}	6H-bulk substrate vertical reactor sublimation (physical vapour transport), n-type	[20]
200	5×10^{16}		
150	6×10^{17}		
140	8×10^{17}		
105	9×10^{17}		
75	2×10^{18}		
68	3×10^{18}		

TABLE 3 Mobilities and carrier densities in 4H-SiC.

Mobility (cm² V⁻¹ s⁻¹)	Carrier density (cm⁻³)	Sample information	Ref
450	1.1×10^{17}	n-type N₂-doped physical vapour transport c-axis, 3 - 4 μm thickness from pn junction Hall measurements on epi-layers	[19]
460	2×10^{17}		
463	2×10^{17}		
450	2.9×10^{17}		
350	2.5×10^{17}		
345	5×10^{17}		
205	1×10^{18}		
110	8×10^{18}		
50	8.5×10^{18}		
45	2.5×10^{19}		
45	3.1×10^{19}		
115	2×10^{15}	p-type Al-doped epi-layers	[18]
100	5.5×10^{16}		
99	1×10^{17}		
90	1×10^{18}		
60	4×10^{19}		
27	6×10^{16}	As-grown Acheson bulk 650 μm	[3]

3.1 Carrier mobilities and concentrations in SiC

TABLE 4 Mobilities and carrier densities in 15R and 27R SiC.

Mobility (cm^2 V^{-1} s^{-1})	Carrier density (cm^{-3})	Sample information	Ref
15R			
456		Physical vapour transport (sublimation) grown bulk crystal, n-type	[13]
500			
254	4.75 x 10^{17}	Physical vapour transport (sublimation) grown bulk crystal, n-type	
382	1.61 x 10^{17}		
1090 (110 - 140 K)	3 x 10^{16}	Original Lely growth process, n-type	[21]
1090 (110 - 140 K)	1.3 x 10^{16}		
	1 x 10^{18}	Physical vapour transport (sublimation) grown bulk crystal, n-type (from C-V data)	[22]
27R			
77	1.5 x 10^{18}	As-grown bulk modified Lely growth process	[3]

C CONCLUSION

Values for both the hole and electron mobilities and carrier densities in various SiC polytypes are listed. Ionized and neutral impurity, acoustic phonon, piezoelectric and polar optical phonon scattering mechanisms are all found in SiC. In general, mobilities have increased and carrier concentrations decreased with time, reflecting the improvement in crystal quality whether bulk or epitaxially-grown material is considered.

REFERENCES

[1]   S. Yoshida [*Proc. 1st High Temperature Electronics Meeting*, ISAS, (1991) p.22-41 (in Japanese)]

[2]   C. Carter [personal communication, Cree Research (1994)]

[3]   H. Haldane [M. Eng. Thesis, Howard University, May 1993]

[4]   J.A. Powell, L.G. Matus, M.A. Kuczmarski [*J. Electrochem. Soc. (USA)* vol.134 no.6 (1987) p.1558-65]

[5]   N.I. Buchan et al [*Trans. 2nd Int. High Temperature Electronics Conf.*, Phillips Laboratory, Sandia National Laboratories and Wright Laboratory, USA, 5 - 10 June 1994, vol.1, paper X-11]

[6]   A. Suzuki, A. Uemoto, M. Shigeta, K. Furukawa, S. Nakajima [*Appl. Phys. Lett. (USA)* vol.49 no.8 (1986) p.450-2]

[7]   B. Segall, S.A. Altterovitz, E.J. Haugland, L.G. Matus [*Appl. Phys. Lett. (USA)* vol.49 no.10 (1986) p.584-6]

[8]   W.E. Nelson, F.A. Halden, A. Rosengreen [*J. Appl. Phys. (USA)* vol.37 no.1 (1966) p.333-6]

[9] M. Yamanaka, H. Daiman, E. Sakuma, S. Misawa, S. Yoshida [*J. Appl. Phys. (USA)* vol.61 no.2 (1987) p.599-603]

[10] T. Tachibana, H.S. Kong, Y.C. Wang, R.F. Davis [*J. Appl. Phys. (USA)* vol.67 no.10 (1990) p.6375-81]

[11] C. Taylor [M. Eng. Thesis, Howard University, August 1994]

[12] W. Suttrop, G. Pensl, W.J. Chokye, R. Stein, S. Leibenzer [*J. Appl. Phys. (USA)* vol.72 (1992) p.3708]

[13] H.J. Van Daal [*Philips Res. Rep. (Netherlands)* Suppl.3 (1965) p.23-5]

[14] D.L. Barrett, R.B. Campbell [*J. Appl. Phys. (USA)* vol.38 (1967) p.53-5]

[15] H.M. Hobgood, J.P. McHugh, J. Greggi, R.H. Hopkins [*Inst. Phys. Conf. Ser. (UK)* no.137 (1994) p.7-12]

[16] A.A. Burk et al [*Inst. Phys. Conf. Ser. (UK)* no.137 (1994) p.29-32]

[17] H. Matsunami [*Inst. Phys. Conf. Ser. (UK)* no.137 (1994) p.45-50]

[18] D.J. Larkin, P.G. Neudeck, J.A. Powell, L.G. Matus [*Inst. Phys. Conf. Ser. (UK)* no.137 (1994) p.51-4]

[19] W.J. Schaffer, H.S. Kang, G.H. Neglay, J.W. Palmour [*Inst. Phys. Conf. Ser. (UK)* no.137 (1994) p.155-9]

[20] N.I. Buchan et al [*Trans. 2nd Int. High Temperature Electronics Conf.*, Phillips Laboratory, Sandia National Laboratories and Wright Laboratory, USA, 5 - 10 June 1994, vol.1, paper X-11]

[21] T. Nakata, K. Koga, Y. Matsushita, Y. Veda, T. Niina [*Amorphous and Crystalline SiC and Related Materials II* (Springer-Verlag, Berlin, vol.43 (1989) p.26-34]

[22] T. Troffer et al [*Inst. Phys. Conf. Ser. (UK)* no.137 (1994) p.173-6]

[23] N.I. Buchan, D.N. Henshall, W.S. Yoo, P.A. Mailloux, M.A. Tischler [*Inst. Phys. Conf. Ser. (UK)* no.137 (1994) p.113-6]

[24] H.J. van Daal [*Philips Res. Rep. (Netherlands)* Suppl.3 (1964) p.36-55]

3.2 Effective masses in SiC

S. Yoshida

August 1993

A INTRODUCTION

Electron and hole effective masses of various SiC polytypes have been determined by various methods, such as Hall measurements, Faraday rotation, Zeeman splitting of a photoluminescence line, electron cyclotron resonance, and infrared light reflection. There have also been several theoretical studies of the effective masses of 3C-SiC. The effective masses of electrons and holes thus obtained are listed in TABLE 1.

B HALL MEASUREMENTS

The density-of-states effective mass (m^*_d) of conduction electrons is obtained from the temperature dependence of carrier density, using a relation $N_c = 2g(2\pi m^*_d kT)^{3/2} h^{-3}$ and a charge equilibrium equation allowing for compensation, where N_c is the density-of-states of the conduction band, g the number of equivalent conduction band minima, k is Boltzmann's constant and h is Planck's constant. Using epitaxially grown n-type 6H-SiC, the electron effective mass was obtained by Wessels and Gatos [1]. A different approach was made by Violina et al [2] for n-type 6H-SiC and by Aivazova et al [3] for 3C-SiC using a temperature dependence of the electron concentration corresponding to the onset of degeneracy, where they put the number of conduction band minima at g=3.

Conductivity effective masses are derived from the temperature dependence of Hall mobility. Generally, values of the conductivity effective masses depend on types of carrier scattering. Applying the effective mass approximation for the impurity nitrogen, Lomakina and Vodakov [4] estimated the values of longitudinal effective mass m^*_l and transverse effective mass m^*_t for n-type 4H-, 6H-, 15R- and 33R-SiC using the values of the activation energy 33, 95, 47 and 51 meV, respectively. Suzuki et al [5] analysed the temperature dependence of Hall mobility of undoped and nitrogen-doped n-type 3C-SiC films using a theoretical model based on [6]. By comparing the theoretical results with experimental ones, they obtained the effective mass $m^*_e = 0.37 m_0$, where m_0 is the free-electron mass. The value is in fairly good agreement with the one obtained by the cyclotron measurements. The effective mass of the holes was evaluated by Van Daal et al [6] to be $1.0 m_0$ in p-type 6H-SiC.

C OPTICAL MEASUREMENTS

Dean et al [7] measured the Zeeman splitting of a luminescence line involving the 2p± donor state, obtaining the electron effective mass $m^*_t = (0.24 \pm 0.01) m_0$ and $m^*_t / m^*_l = 0.36 \pm 0.01$ for n-type cubic crystals. Measurements of infrared Faraday rotation due to free carriers were made by Ellis and Moss [8] at room temperature in a number of n-type hexagonal specimens belonging to the 6H and 15R polytypes of silicon carbide. One component of the total density-of-states effective mass was explicitly determined by this method. A value for the

3.2 Effective masses in SiC

TABLE 1 Electron and hole effective masses in 3C-, 4H-, 6H-, 15R- and 33R-SiC.

Electron effective mass			Hole effective mass			Method	Ref
m^*_e	m^*_t	m^*_l	m^*_h	m^*_{lh}	m^*_{hh}		
3C-SiC							
0.62						Hall effect	[3]
0.37						Hall effect	[5]
0.337[1]						Zeeman splitting	[7]
0.344[1]	0.24 ± 0.01	0.667 ± 0.02				ECR	[16]
0.347[1]	0.247 ± 0.011	0.667 ± 0.015		0.45		ECR	[17]
0.78	0.25 ± 0.01	0.67 ± 0.02[2]				I-V	[18]
		0.69				Theory	[19]
		0.7		0.13	0.275	Theory	[20]
	0.156	0.665		0.349 [100]	0.476	Theory	[21]
				0.186 [111]	1.2		
0.355			0.8			Theory	[25]
4H-SiC							
	0.24	0.19				Hall effect	[4]
	0.18	0.22				IR absorption, Hall	[14]
6H-SiC							
0.45[3]						Hall effect	[1]
0.48						Hall effect	[2]
	0.35	1.3				Hall effect	[4]
0.2			1.0			Hall effect	[6,23]
0.6			1.2			Hall effect	[22]
0.45[1]	0.25 ± 0.02	1.5 ± 0.2				Faraday rotation	[8,9]
	0.23 ± 0.03					IR reflection	[10]
	0.25					IR reflection	[11]
	0.25 ± 0.01	1.7 ± 0.2				IR reflection	[12]
	0.24 ± 0.01	0.34 ± 0.02				IR absorption, Hall	[13]
	0.26 ± 0.03					IR birefringence	[15]
15R-SiC							
0.28[1]	0.25	0.37				Hall effect	[4]
0.35[1]	0.28 ± 0.02	0.53				Faraday rotation	[8,9]
33R-SiC							
	0.49	0.49				Hall effect	[4]

(unit: electron mass m_0)

m^*_e (m^*_h): effective mass of electrons (holes)
m^*_{hh} (m^*_{lh}): effective mass of heavy (light) holes
m^*_t (m^*_l): transverse (longitudinal) effective mass of electrons

[1] The density of states effective mass m^*_d calculated by $m^*_d = (m^{*2}_t m^*_l)^{1/3}$
[2] Calculated from $m^*_t = (0.25 \pm 0.01) m_0$ and $m^*_2 = (0.41 \pm 0.01) m_0$, using the relation $m^*_2 = (m^*_t m^*_l)^{1/2}$
[3] For a detailed discussion see Allgaier [26].

6H-SiC polytype of $m^*_t = (0.25 \pm 0.02)\,m_0$ was given, and $m^*_t = (0.28 \pm 0.02)\,m_0$ for 15R-SiC by this method. The same authors [9] estimated a value of $m^*_l = (1.5 \pm 0.2)\,m_0$ for 6H-SiC, and $m^*_l = 0.53\,m_0$ for 15R-SiC evaluated from infrared absorption due to free-electron excitation. Similar values of the electron effective masses were given by a dispersion analysis of the infrared reflection spectra for heavily doped 6H-SiC [10-12]. From the infrared absorption and Hall measurements, Suttrop et al [13] obtained $m^*_t = (0.24 \pm 0.01)\,m_0$ and $m^*_l = (0.34 \pm 0.02)\,m_0$ for nitrogen-doped 6H-SiC by a fit of the effective-mass approximation formalism to the observed transition energies. Similarly, $m^*_t = 0.176\,m_0$ and $m^*_l = 0.224\,m_0$ were obtained for 4H-SiC [14]. Geidur et al [15] deduced $m^*_t = (0.26 \pm 0.03)\,m_0$ from the dispersion of birefringence of 6H-SiC.

D ELECTRON CYCLOTRON RESONANCE

More direct determination of electron effective masses was first performed by Kaplan et al [16] using electron cyclotron resonance (ECR) in n-type 3C-SiC epitaxially grown onto a silicon substrate. They obtained transverse effective mass $m^*_t = (0.247 \pm 0.011)\,m_0$, and longitudinal effective mass $m^*_l = (0.667 \pm 0.015)\,m_0$. The effective masses derived from cyclotron resonance agree, within experimental error, with the values obtained from Zeeman luminescence studies [7] of small 'bulk' crystals. An average effective mass of an electron, given by the equation $m_e^* = (m^{*2}_t m^*_l)^{1/3}$, is $0.344\,m_0$. Recently, similar ECR measurements were made by Kono et al [17].

E OTHER METHODS

Chaudhry [18] investigated the electrical transport properties of 3C-SiC/Si heterojunctions using current-voltage (I-V) and capacitance-voltage (C-V) characteristics, and found the density-of-states effective mass of electrons in the conduction band of 3C-SiC to be $0.78\,m_0$. This value is somewhat larger than Zeeman splitting, ECR and theoretical effective masses.

F THEORETICAL STUDIES

Recently, several theoretical studies have been carried out for the effective masses of electrons and holes in 3C-SiC. Talwar and Feng [19], using the energy-band structure calculation based on a tight-binding theory and the Green's-function framework, estimated $m^*_e = 0.69\,m_0$. This value is in good agreement with the cyclotron resonance [$m^*_e = (0.667 \pm 0.15)\,m_0$] and the Zeeman splitting [$m^*_e = (0.667 \pm 0.02)\,m_0$] data. Using a semi-empirical tight-binding approach, Li and Lin-Chung [20] obtained $m^*_l = 0.7\,m_0$ for 3C-SiC. Huang and Ching [21] calculated the band structure of the various semiconductors including 3C-SiC using an orthogonalised linear combination of atomic orbital method (LCAO), and obtained electron and hole effective masses. Results of all these theoretical studies are in good agreement with the experimental values measured by Zeeman splitting and electron cyclotron resonance.

G   CONCLUSION

In general, good agreement is found for the longitudinal and transverse effective masses, when measured by several techniques, within the various polytypes. There is a wider variability for the values of the overall electron effective mass. Theoretical studies yield values for electron effective masses which are in good agreement with measured values. Data on hole effective masses is scarce and much of what is available is from theoretical studies.

REFERENCES

[1]   B.W. Wessels, J.C. Gatos [*J. Phys. Chem. Solids (UK)* vol.38 (1977) p.345]
[2]   G.N. Violina, Yeh Liang-hsiu, G.F. Kholuyanov [*Sov. Phys.-Solid State (USA)* vol.5 (1964) p.2500]
[3]   L.S. Aivazova, S.N. Gorin, V.G. Sidyakin, I.M. Shvarts [*Sov. Phys.-Semicond. (USA)* vol.11 (1978) p.1069]
[4]   G.A. Lomakina, Yu.A. Vodakov [*Sov. Phys.-Solid State (USA)* vol.15 (1973) p.83]
[5]   A. Suzuki, A. Ogura, K. Furukawa, Y. Fujii, M. Shigeta, S. Nakajima [*J. Appl. Phys. (USA)* vol.64 (1988) p.2818]
[6]   H.J. van Daal, W.F. Knippenberg, J.D. Wasscher [*J. Phys. Chem. Solids (UK)* vol.24 (1963) p.109]
[7]   P.J. Dean, W.J. Choyke, L. Patrick [*J. Lumin. (Netherlands)* vol.15 (1977) p.299]
[8]   B. Ellis, T.S. Moss [*Proc. R. Soc. Lond. A (UK)* vol.299 (1967) p.383]
[9]   B. Ellis, T.S. Moss [*Proc. R. Soc. Lond. A (UK)* vol.299 (1967) p.393]
[10]   M.A. Il'in, A.A. Kukharskii, E.P. Rashevskaya, V.K. Subashiev [*Sov. Phys.-Solid State (USA)* vol.13 (1972) p.2078]
[11]   R. Helbig, C. Haberstroh, T. Lauterbach, S. Leibenzeder [*Abstract on Electrochem. Soc. Conf.* no.477 (1989) p.695]
[12]   A.V. Mel'nichuk, Yu.A. Pasechnik [*Sov. Phys.-Solid State (USA)* vol.34 (1992) p.227-]
[13]   W. Suttrop, G. Pensl, W.J. Choyke, R. Stein, S. Leibenzeder [*J. Appl. Phys. (USA)* vol.72 (1992) p.3708]
[14]   W. Götz et al [*J. Appl. Phys. (USA)* vol.72 (1993) p.3332]
[15]   S.A. Geidur, V.T. Prokopenko, A.D. Yas'kov [*Sov. Phys.-Solid State (USA)* vol.20 (1978) p.1654]
[16]   R. Kaplan, R.J. Wagner, H.J. Kim, R.F. Davis [*Solid State Commun. (USA)* vol.55 (1985) p.67]
[17]   J. Kono et al [*Physica B (Netherlands)* vol.184 (1993) p.178]
[18]   M.I. Chaudhry [*IEEE Electron Device Lett. (USA)* vol.12 (1991) p.670]
[19]   D.N. Talwar, Z.C. Feng [*Phys. Rev. (USA)* vol.44 (1991) p.3191]
[20]   Y. Li, P.J. Lin-Chung [*Phys. Rev. B (USA)* vol.36 (1987) p.1130]
[21]   M.Z. Huang, W.Y. Ching [*J. Phys. Chem. Solids (UK)* vol.46 (1985) p.977]
[22]   J.A. Lely, F.A. Kroger [*Halbleiter und Phosphore* (Vieweg, Braunschweig, 1958) p.525]
[23]   H.J. van Daal [*Philips Res. Rep. (Netherlands)* suppl.3 (1965)]
[24]   Yu.M. Altaiskii, I.M. Zaritskii, V.Ya. Zevin, A.A. Konchits [*Sov. Phys.-Solid State (USA)* vol.12 (1971) p.2453]
[25]   I.B. Ermolovich, V.F. Peklun [*Ukr. Fiz. Zh. (Ukraine)* vol.21 (1976) p.73]

[26]   R.S. Allgaier [*J. Phys. Chem. Solids (UK)* vol.40 (1979) p.327]

3.3 Band structure of SiC: overview

S. Yoshida

August 1993

A INTRODUCTION

SiC is the only IV-IV compound to form stable and long-range ordered structures (polytypes). Over 100 different such polytypes have been observed. These polytypes are semiconductors with a varying band structure, such as bandgap and the location of the conduction band minimum in the k-space. The energy band structures of SiC in the zinc blende (3C-) and the wurzite structures (2H-SiC) have been calculated theoretically by many authors since the first report in 1956 by Kobayashi [1]. However, little is known about the band structures of the other SiC polytypes. The absence of band structure calculations for the polytypes other than 3C- and 2H-SiC is due to the relatively large number of atoms per primitive unit cell, or a large supercell, which exceedingly complicates the calculations.

From the theoretical calculations as well as the optical measurements, it is known that all the polytypes have the valence band maximum at the zone centre (Γ-point). However, the location of the conduction band minimum in k-space depends on the polytype, for example, X-point for 3C- and K-point for 2H-SiC. All the polytypes, studied thus far, have indirect bandgaps, which increase monotonically with the hexagonality of the polytypes h, from $E_g = 2.417$ eV for 3C-SiC (h=0) to $E_g = 3.33$ eV for 2H-SiC (h=1). TABLE 1 shows the values of the hexagonality, the indirect bandgap, and its temperature coefficient for various typical polytypes [2-5].

TABLE 1 Hexagonalities, observed minimum indirect* and direct bandgaps at 4 K and their temperature dependences for various typical SiC polytypes [2-5].

Polytype	Jagodzinski notation	Hexagonality h	Minimum bandgap eV (Indirect)	Minimum bandgap eV (Direct)	$dE_{g.ind}/dT$
3C	k	0.00	2.39[1]	5.3	-5.8×10^{-4}
8H	hkkk	0.25	2.728		
21R	hkkhkkk	0.29	2.853		
6H	hkk	0.33	3.023		-3.3×10^{-4}
33R	hkkhkkhkkhk	0.36	3.01		
15R	hkhkk	0.40	2.986		
4H	hk	0.50	3.263		
2H	h	1.00	3.33	4.39	

* exciton energy gap E_{gx}
[1] bandgap $E_g = E_{gx} + E_{ex} = 2.417$ eV

B BAND STRUCTURE OF 3C-SiC

The cubic modification (3C-SiC) is the only SiC with zinc blende structure. The band structure of 3C-SiC has been calculated theoretically by Kobayashi [1], and estimated from optical absorption and reflection [6], luminescence [7], soft X-ray emission [8] and X-ray photoelectron [9] spectra.

Kobayashi [9], Bassani and Yoshimine [10], and Herman et al [11] calculated the band structure by using the orthogonalised plane wave (OPW) method, and showed the conduction band minimum locates at the X-point and the valence band maximum at the Γ-point in the k-space, resulting in an indirect bandgap. On the other hand, Haeringen and Junginger [13,14], Hemstreet and Fong [14-16] and Nemoshkalenko et al [17,18] calculated the band structure by the empirical pseudopotential method (EPM) and non-local EPM (NEPM), adjusting the atomic form factors at some band edges. They found the values of the various principal bandgaps thus calculated are in good agreement with those obtained from the optical measurements. Lubinsky et al [19] calculated in the first-principles Hartree-Fock-Slater model, or a Bloch linear-combination-of-atomic-orbitals (LCAO) band formulation, and showed the major components for each zone centre level are Γ_{1v}:C(2s), Γ_{15v}:C(2p), Γ_{1c}:Si(3s), Γ_{15c}:Si(3p), and Γ_{12c}:Si(3d), where subscripts v and c denote valence band and conduction band, respectively. Recently, band structure calculations based on a tight-binding theory have been carried out by Huang and Ching [21], Li and Lin-Chung [22], and Talwar and Feng [23]. Craig and Smith [24] performed the calculation using the periodic large unit cell-modified intermediate neglect of differential overlap (LUC-MINDO) method based on self-consistent field molecular orbital treatment.

The calculated values of the principal energy gaps reported are tabulated in TABLE 2 as well as those obtained from optical measurements [17,25]. All the results indicate the minimum indirect bandgap of 2.417 eV is assigned as X_{1c}-Γ_{15v}. However, some specific transitions remain unclear. For example, Junginger and Haeringen [13], Li and Lin-Chung [22], and Hemstreet and Fong [14-16] asserted that the threshold for direct transition occurs at the Γ point (Γ_{15v}-Γ_{1c} ~ 6.0 eV), while Bassani and Yoshimine [10], Herman et al [11], and Lubinsky et al [19] presented a calculated threshold for direct transition at the X-point. Li and Lin-Chung [22] chose the 6 eV direct transition at the Γ-point as one of the inputs of the semi-empirical band calculation. Bimberg et al [26] calculated the valence band spin-orbit splitting to be 10.2 meV, which is in good agreement with that obtained from wavelength modulated absorption measurements [27,28]. This value is much smaller than those of usual semiconductors, which brings about strong coupling between the light hole valence band and the spin-orbit split off band, resulting in their highly non-parabolic dispersion.

Hoechst and Tang [29] investigated the band structure with synchrotron radiation by angular resolved valence-band photoemission spectroscopy, and found good agreement in the band structure mapped along Γ-X with theoretical calculations by Lubinsky et al [19].

Wang et al [30], Li and Lin-Chung [22], and Talwar and Feng [23] have calculated the formation energy and the bound electronic states of native defects. Li and Lin-Chung [22] showed that isolated silicon and carbon vacancies and a divacancy complex of Si and C sites induce gap states, while no defect-induced state is found in the energy gap for either the isolated Si or C antisite defects or for the pair of antisite defects in SiC. Kohyama et al [31],

3.3 Band structure of SiC: overview

and Lambrecht and Segall [32] obtained the atomic and electronic structure of the {122} grain boundary using the self-consistent tight-binding (SCTB) method, and the (110) antiphase domain boundary using the linear muffin-tin-orbital (LMTO) band structure method. Wenchang et al [33] studied the electronic structure of ideal and the (2x1) reconstructed (100) surfaces with the extended Hückel band calculation.

TABLE 2 Calculated and experimental values of the principal energy gaps of 3C-SiC.

Γ_{15v}-Γ_{1c} eV	Γ_{15v}-Γ_{15c} eV	X_{5v}-X_{1c} eV	X_{1c}-X_{3c} eV	L_{3v}-L_{1c} eV	Γ_{15v}-L_{1c} eV	Γ_{15v}-X_{1c} eV	L_{3v}-X_{1c} eV	Γ_{15v}-K_{1c} eV	Ref
6.8	8.6	5.8**	3.2	9.9	6.8	2.7*	6.0		[10]
5.9	7.8	5.3**	2.6	6.4	5.5	2.3*	3.1		[11]
5.14**	10.83	5.27	3.24	6.75	5.93	2.4*	3.26		[13]
5.92**	6.49	6.36	3.08	6.02	4.38	2.35*	3.9		[16]
	6.8	5.60	1.1	6.4	4.9	2.3*	3.8		[17,18]
6.5	7.2	5.3**	3.7	8.0	6.5	2.4*	3.1	4.3	[19]
	9.02	6.16	2.38	10.39	8.77	2.39*	4.01		[21]
6.0**	7.0	6.44	3.1	5.62	3.59	2.4*	4.43		[22]
6.0	7.0		3.29			2.4*	4.4	3.03	[23]
6.0			3.1		4.2	2.417*k	3.55	3.0	[16]

* minimum indirect bandgap, ** minimum direct bandgap.
k [27].

C   BAND STRUCTURE OF 2H-SiC

Another simple crystal structure among SiC polytypes is wurzite 2H-SiC, which is known to have the largest bandgap. There are four atoms per unit cell, two anions and two cations in the wurzite structure. The unit cell is of hexagonal close-packed structure, i.e. the hexagonality h is unity. The band structure calculations for 2H-SiC have been reported by several authors. Herman et al [11], Haerigen and Junginger [12], and Hemstreet and Fong [14-16] calculated by using EPM and showed that the conduction band minimum locates at the K-point and the valence band maximum at the Γ-point in the k-space, resulting in an indirect bandgap. The calculation of the band structure by the empirical tight-binding (ETB) method has been performed by Tuncay and Tomak [34]. They showed the lowest lying valence bands derived mainly from C(2s) and Si(3s) orbitals, the upper valence band exhibits C(2p) and Si(3s) hybridisation, and lower conduction bands are of C(s) and Si(s) character, while uppermost conduction bands show C(p) and Si(p) character. Gavrilenko and co-workers [35,36] calculated three kinds of SiC polytypes including 2H-SiC by the first-principles self-consistent linear muffin-tin orbital (LMTO) atomic sphere approximation (ASA) method. However, the calculated value for the lowest forbidden gap, 2.76 eV, is much less than the experimental value. They pointed out that the reason for this is the use of the local-density approximation in LMTO theory.

The calculated values of the principal energy gaps reported are tabulated in TABLE 3 as well as those obtained from the optical measurements [3]. Herman et al [11] assigned the minimum indirect bandgap as the Γ_{6v}-K_{2c} transition. The value of this indirect Γ-K gap in the case of the EPM calculation has been adjusted to 3.35 eV. The symmetry of the top of the valence band (Γ_6) obtained by Gavrilenko and co-workers [35,36] agrees with Herman et al and Hemstreet and Fong, but in the calculation by Junginger and Haeringen [13] a Γ_1 symmetry has been obtained for the highest valence band. Hemstreet and Fong obtained a value for the crystal field splitting near the valence band edge (Γ_{6v}-Γ_{1v}) of about 0.45 eV, while 0.12 eV has been obtained from the calculation using the supercell technique and a first-principles pseudopotential approach by Qteish et al [37]. Qteish et al also discussed the band lineup between 3C- and 2H-SiC and obtained the valence-band offset of 0.13 eV.

TABLE 3 Calculated and experimental values of the principal energy gaps of 2H-SiC.

	Γ_{6v}-K_{2c} eV	Γ_{1v}-K_{2c} eV	Γ_{6v}-M_{1c} eV	Γ_{1v}-M_{1c} eV	Γ_{6v}-L_{1c} eV	Γ_{1v}-L_{1c} eV	Γ_{6v}-Γ_{1c} eV	Γ_{1v}-Γ_{1c} eV
Herman[a]	3.3*	3.7	4.0	4.4	3.3	3.7	6.0**	6.4
Junginger[b]	4.02	3.35**	4.42	3.75	4.49	3.82	5.09	4.46**
Hemstreet[c]	3.3*	3.77	3.4	3.87	3.81	4.28	4.39**	4.86
Tuncay[d]							5.87**	
Gavrilenko[e]	2.67*	3.96	3.29	4.58			6.07**	7.36
Experiment[f]	3.33							

* minimum indirect bandgap, ** minimum direct bandgap.
[a] [11], [b] [13], [c] [16], [d] [34], [e] [35], [f] [3].

Lee and Joannopoulos [38] calculated the ideal and relaxed electronic structures and optimum relaxed geometries of 2H-SiC ($10\bar{1}0$) and ($11\bar{2}0$) surfaces as well as those of the 3C-SiC (110) surface by using the transfer matrix method and discussed the reduction of the total energy due to the relaxation and bona fide surface bands.

D   BAND STRUCTURE OF 4H- AND 6H-SiC

As mentioned in the introduction, little is known about the band structures of the polytypes other than 3C- and 2H-SiC, resulting from the relatively large number of atoms per primitive unit cell of these polytypes, which exceedingly complicates the calculation. Junginger and Haeringen [13] calculated the band structures of 4H- and 6H-SiC as well as 3C- and 2H-SiC by using EPM. However, they obtained the energy levels for only some symmetry points, i.e. Γ, K, H, M, and L-points. For 4H-SiC, an indirect Γ-M gap of 2.8 eV has been found, to be compared with the experimentally obtained energy gap of 3.26 eV. For 6H-SiC, on the other hand, an indirect Γ-M gap of 2.4 eV is found, to be compared with the experimental value of 3.0 eV. In this respect, it should be remarked that the minimum of the conduction band does not necessarily lie at the M-point. Recently, Gavrilenko and co-workers [35,36] performed the calculation of the band structures of 4H- and 6H-SiC as well as 2H-SiC by the first-principles self-consistent LMTO-ASA method. Electron energy levels and densities of states are obtained in the range ±15 eV around the top of the valence band. They obtained the lowest energy gaps in 4H- and 6H-SiC as 2.89 and 2.92 eV, respectively, which are much closer to the experimental values compared with those obtained by EPM. However, the trend of the changes of the bandgap (slight increase) calculated going from 2H- to 6H-SiC is

3.3 Band structure of SiC: overview

opposite to the experimental one. They also discussed the subband structure of the conduction band in 4H- and 6H-SiC caused by the superstructure along the c-axis. Good agreement between calculated and experimental values of intersubband gaps in the conduction band of SiC indicates that the orbital basis used in their calculation is sufficient to obtain a realistic band structure of SiC.

The calculated values of the principal energy gaps reported are tabulated in TABLES 4 and 5 as well as those obtained from the optical measurements [5]. The lowest direct transitions in 6H-SiC have been reported to be near 4.6 eV [39] and 5.5 eV [40]. These values agree well with the value obtained by Gavrilenko and co-workers [35,36]. According to their calculations, the direct optical gaps in 4H- and 6H-SiC with energy lower than 6 eV occur only near the M-point in the k-space.

TABLE 4 Calculated and experimental values of the principal energy gaps of 4H-SiC.

	M_{4c}-Γ_{1v} eV	M_{1c}-Γ_{1v} eV	M_{4c}-Γ_{6v} eV	M_{1c}-Γ_{6v} eV	L_{1c}-Γ_{1v} eV	M_{4c}-Γ_{5v} eV
Junginger[a] Gavrilenko[b]	2.8 2.89	3.15	3.22	3.57	3.58	3.7
Experiment[c]	3.263					

[a] [13], [b] [35], [c] [2].

TABLE 5 Calculated and experimental values of the principal energy gaps of 6H-SiC.

	M_{4c}-Γ_{1v} eV	L_{1c}-Γ_{1v} eV	M_{4c}-Γ_{6v} eV	L_{1c}-Γ_{6v} eV	M_{1c}-Γ_{1v} eV	M_{4c}-Γ_{5v} eV
Junginger[a] Gavrilenko[b]	2.45 2.92	2.51	2.67	2.73	2.9	2.91
Experiment[c]	3.023					

[a] [13], [b] [35], [c] [2].

Dang et al [41] measured the optical magnetic resonance at the fundamental absorption edge of 6H-SiC and discussed the structure of the valence band top. The splitting of the valence band by the hexagonal crystal field is larger than the spin-orbit splitting, which is about 7 meV for free excitons. Therefore, the uppermost valence band state is the doublet Γ_9 for the C_{6v} crystal group of the 6H polytype and $\Gamma_5 + \Gamma_6$ for the C_{3v} point group of the lattice sites, which are degenerate because of time-reversal symmetry. Humphreys et al [25] derived the spin-orbit splittings of 6H and 15R to be as small as 7 meV from the wavelength modulated absorption measurements. Choyke and Patrick [42] estimated the splitting as 4.8 meV from the photoluminescence measurements.

E CONCLUSION

All SiC polytypes have the valence band maximum at the zone centre but the conduction band minimum is polytype dependent. The indirect bandgaps increase with polytype hexagonality from 2.39 eV for 3C-SiC to 3.33 eV for 2H-SiC. In 3C-SiC, isolated silicon and carbon vacancies and a divacancy complex of carbon and silicon sites produce states in the forbidden gap but antisite defects do not. Theoretical values for the energy gap in 2H-SiC do not correspond to the experimental values. The valence band offset between 3C- and 2H-SiC is found to be 0.13 eV. Relatively little work has been carried out on the other polytypes. Theory and experiment are also at variance for the energy gap of 4H- and 6H-SiC polytypes.

REFERENCES

[1] S. Kobayashi [*J. Phys. Soc. Jpn. (Japan)* vol.11 (1956) p.175]
[2] W. Choyke, D.R. Hamilton, L. Patrick [*Phys. Rev. A (USA)* vol.133 (1964) p.1163]
[3] L. Patrick, D.R. Hamilton, W.J. Choyke [*Phys. Rev. (USA)* vol.143 (1966) p.526]
[4] R. Dalven [*J. Phys. Chem. Solids (UK)* vol.26 (1965) p.439]
[5] W.J. Choyke, L. Patrick [in *A High Temperature Semiconductor*, Eds J.R. O'Connor, J. Smiltens (Pergamon, Oxford, 1960) p.306]
[6] L. Patrick, W.J. Choyke [*Phys. Rev. (USA)* vol.186 (1969) p.775]
[7] W.J. Choyke, L. Patrick [*Phys. Rev. (USA)* vol.187 (1969) p.1041]
[8] G. Wiech [in *Soft X-ray Band Spectra*, Ed. D.J. Fabian (Academic, New York, 1968)]
[9] S. Kobayashi [*J. Phys. Soc. Jpn. (Japan)* vol.13 (1958) p.261]
[10] F. Bassani, M. Yoshimine [*Phys. Rev. (USA)* vol.130 (1963) p.20]
[11] F. Herman, J.P. van Dyke, R.L. Kortum [*Mater. Res. Bull. (USA)* vol.4 (1969) p.S167]
[12] W. van Haeringen, H.G. Junginger [*Solid State Commun. (USA)* vol.7 (1969) p.1135-]
[13] H.G. Junginger, W. van Haeringen [*Phys. Status Solidi (Germany)* vol.37 (1970) p.709]
[14] L.A. Hemstreet Jr., C.Y. Fong [*Solid State Commun. (USA)* vol.9 (1971) p.643]
[15] L.A. Hemstreet Jr., C.Y. Fong [*Phys. Rev. B (USA)* vol.6 (1972) p.1464]
[16] L.A. Hemstreet Jr., C.Y. Fong [in *Silicon Carbide-1973*, Eds R.C. Marshall, J.W. Faust Jr., C.E. Ryan (University of South Carolina, Columbia, 1973) p.284]
[17] V.V. Nemoshkalenko, V.G. Aleshin, M.T. Panchenko, A.I. Senkevich [*Sov. Phys.-Solid State (USA)* vol.15 (1974) p.2318]
[18] V.V. Nemoshkalenko, V.G. Aleshin, M.T. Panchenko, A.I. Senkevich [*Sov. Phys.-Solid State (USA)* vol.19 (1974) p.38]
[19] A.R. Lubinsky, D.E. Elliis, G.P. Painter [*Phys. Rev. B (USA)* vol.11 (1975) p.1537]
[20] V.G. Aleshin, Yu.N. Kucherenko [*Solid State Commun. (USA)* vol.19 (1976) p.903]
[21] M.Z. Huang, W.Y. Ching [*J. Phys. Chem. Solids (UK)* vol.46 (1985) p.977]
[22] Y. Li, P.J. Lin-Chung [*Phys. Rev. B (USA)* vol.36 (1987) p.1130]
[23] D.N. Talwar, Z.C. Feng [*Phys. Rev. B (USA)* vol.44 (1991) p.3191]
[24] B.I. Craig, P.V. Smith [*Phys. Status Solidi B (Germany)* vol.154 (1989) p.K127]
[25] R.G. Humphreys, D. Bimberg, W.I. Choyke [*J. Phys. Soc. Jpn. (Japan)* vol.49 suppl.A519 (1980)]

[26] D. Bimberg, M. Altarelli, N.O. Lipari [*Solid State Commun. (USA)* vol.40 (1981) p.437]
[27] R.G. Humphreys, D. Bimberg, W.J. Choyke [*Solid State Commun. (USA)* vol.39 (1981) p.163]
[28] D.S. Nedzvetskii, B.V. Novikov, N.N. Prokofeva, M.B. Teitman [*Sov. Phys.-Semicond. (USA)* vol.2 (1968) p.914]
[29] H. Hoechst, M. Tang [*J. Vac. Sci. Technol. A (USA)* vol.5 (1987) p.1640]
[30] C. Wang, J. Bernholc, R.F. Davis [*Phys. Rev. B (USA)* vol.36 (1987) p.1130]
[31] M. Kohyama, S. Kose, M. Kinoshita, R. Yamamoto [*J. Phys., Condens. Matter (UK)* vol.2 (1990) p.7809]
[32] W.R.L. Lambrecht, B. Segall [*Phys. Rev. B (USA)* vol.41 (1990) p.2943]
[33] Lu Wenchang, Y. Weidong, Z. Kaiming [*J. Phys., Condens. Matter (UK)* vol.3 (1991) p.9079]
[34] C. Tuncay, M. Tomak [*Phys. Status Solidi B (Germany)* vol.127 (1985) p.709]
[35] V.G. Gavrilenko, A.V. Postnikov, N.I. Klyui, V.G. Litovchenko [*Phys. Status Solidi B (Germany)* vol.162 (1990) p.477]
[36] V.G. Gavrilenko, S.I. Frolov [*Proc. SPIE (USA)* vol.1361 (1990) p.171]
[37] A. Qteish, V. Heine, R.J. Needs [*Phys. Rev. B (USA)* vol.45 (1992) p.6534]
[38] D.H. Lee, J.D. Joanopoulos [*J. Vac. Sci. Technol. (USA)* vol.21 (1982) p.351]
[39] W. Choyke, L. Patrick [*Phys. Rev. (USA)* vol.172 (1968) p.769]
[40] B.E. Wheelen [*Solid State Commun. (USA)* vol.4 (1966) p.173]
[41] Le Si Dang, K.M. Lee, G.D. Watkins [*Phys. Rev. Lett. (USA)* vol.45 (1980) p.390]
[42] W.J. Choyke, L. Patrick [*Phys. Rev. (USA)* vol.127 (1962) p.1868]

3.4 Pressure effects on band structure of SiC

I. Nashiyama

August 1993

A   INTRODUCTION

The pressure dependences of the physical properties, including bandgaps [1] and phonon frequencies [2,3], have been measured mainly for 3C-SiC. In recent first-principles theoretical calculations [4-6], the explored ground-state properties of 3C-SiC, i.e. equilibrium lattice constant, bulk modulus and its pressure derivative, frequencies of selected phonons, Grüneisen parameters and deformation potentials, were shown to be in good agreement with the experiments. Chang and Cohen [7] predicted using ab initio pseudopotential calculations within the framework of the local density approximation (LDA) that the cubic phase of SiC transforms into the more ionic rocksalt structure as pressure is applied above 60 GPa. Kohyama et al [8] also discussed the nature of the bonding and the phase stability by use of the self-consistent tight-binding model and showed the possibility of a pressure-induced phase transition to the β-Sn structure.

B   PRESSURE EFFECTS ON BAND STRUCTURE

By the empirical Paul's rule [9], the pressure coefficient of the indirect gaps between Γ_v and X_c (or Δ_c) are about -15 meV GPa^{-1} for diamond and zinc blende structure semiconductors, e.g. -14.1 meV GPa^{-1} for Si and -13.4 meV GPa^{-1} for GaAs. However, the pressure derivative for diamond was reported to be +5 meV GPa^{-1} [10], which is attributed to the absence of d-orbitals in the low-lying valence states of the carbon atoms. As SiC is considered to be an intermediate substance between Si and diamond, the pressure derivatives of the properties are expected to be different from those of Si and diamond. Kobayashi et al [1] measured the pressure dependence of absorption spectra in 3C-SiC up to 14 GPa and obtained the pressure derivative of the fundamental gaps between Γ_{15v} and X_{1c} as -1.9 meV GPa^{-1}. This value is the smallest coefficient in the absolute value for the Γ_v-X_c gaps in all tetrahedrally coordinated semiconductors and insulators ever investigated. They considered this to arise from the partial contribution of d-orbitals of Si at the X-point in the conduction band. Cheong et al [11] reanalysed the experimental results by Kobayashi et al [1] precisely, and obtained dE/dP = -3.4 meV GPa^{-1} and found strong sublinear behaviour for the pressures of 1 to 1.5 GPa.

The ab-initio total energy-pseudopotential calculations based on the LDA have underestimated the values of bandgaps by about 30 - 50 % compared with experiments, e.g. the indirect bandgap of 3C-SiC was estimated as 1.21 eV [11], which is 50 % of the measured value of 2.42 eV. However, the pressure variations of the bandgaps in semiconductors have been correctly described [12]. Cheong et al [11] calculated the linear (dE/dP) and sublinear (d^2E/dP^2) pressure coefficients for cubic SiC at symmetry points, and showed the dE/dP value is in good agreement with the experimental value reported by Kobayashi et al [1], though the sublinear coefficient is much smaller than the measured value. Other bandgaps (Γ_{15v}-Γ_{1c} and Γ_{15v}-L_{1c}) are shown to increase with pressure and have pressure coefficients larger by an order

3.4 Pressure effects on band structure of SiC

of magnitude than those for the fundamental gap. For the direct Γ_{15v}-Γ_{1c} gap, sublinear behaviour is found to be more significant above 20 GPa and its second-order pressure coefficient is much larger compared to that of the fundamental gap.

Van Camp et al [13] also calculated the first and second pressure derivatives by the ab-initio norm-conserving non-local pseudopotential method of the energy differences between the Γ-, X-, and L-states of the valence and lowest conduction band and the top of the valence band in Γ_{15} for 3C-SiC. The first-order pressure coefficients for the direct (Γ_{15v}-Γ_{1c}) and indirect (Γ_{15v}-X_{1c}) bandgaps are 61.7 meV GPa^{-1} and -1.1 meV GPa^{-1}, respectively. The signs of these values are the same for all semiconductors by Paul's rule. However, the magnitudes are quite different from those of other semiconductors.

TABLE 1 gives the first and second pressure derivatives for 3C-SiC relative to the top valence (Γ_{15v}) band obtained by the calculations [11,13] as well as those obtained by optical measurements [1].

TABLE 1 Calculated and experimental values of the band energies (E in eV), first (dE/dP in meV GPa^{-1}) and second (d^{2}E/dP^2 in meV GPa^{-2}) pressure derivatives of the energy for 3C-SiC at symmetry points with respect to the top of the valence band Γ_{15v}.

	Band energy (eV)			dE/dP (meV GPa^{-1})			d^{2}E/dP^2 (meV GPa^{-2})		
	Cheong[a]	Camp[b]	Expt.	Cheong[a]	Camp[b]	Expt.[c]	Cheong[a]	Camp[b]	Expt.[c]
Γ_{15v}	0.0	0.0	0.0						
Γ_{1v}		-15.56			-36.6			1.318	
Γ_{1c}	6.27	5.46	6.0	55.0	61.7		-2.97 x 10^{-3}	-0.724	1.06 x 10^{-3}
Γ_{15c}					35.8			0.06	
X_{1v}		-10.66			-12.0			0.984	
X_{3v}		-7.96			-11.2			0.98	
X_{5v}		-3.28			-5.1			1.022	
X_{1c}	1.21	0.91	2.42	-3.3	-1.1	-3.4	8.03 x 10^{-5}	0.674	
X_{3c}					19.8			0.394	
X_{5c}					62.8			0.26	
L_{1v}		-12.09			-17.9			1.05	
L_{1v}		-8.67			-24.4			1.022	
L_{3v}		-1.12			1.3			0.63	
L_{1c}	5.32		4.2	39.5	47.6		-1.58 x 10^{-3}	-0.242	
L_{3c}					22.3			0.218	
L_{1c}					29.4			-0.238	

[a] [11], [b] [13], [c] [1].

Choyke et al [14] examined the energy bandgap shifts under a generalised axial stress, which includes the case of hydrostatic pressure and uniaxial and biaxial stress theoretically, and discussed the observed photoluminescence line shifts in CVD grown 3C-SiC films due to the internal stress in the films on Si substrates. The effect of strain or stress on energy bands is described by the strain-orbit Hamiltonian and deformation potential coefficients. They discussed the influence of the small spin-orbit splitting compared with the splitting of the heavy-hole and light-hole band by the stress. Epitaxial growth of thin films on a thick substrate may lead to the introduction of layer stress and strains due to the differences of lattice constants and thermal expansion coefficients. In the case of 3C-SiC on Si, the thermal expansion coefficient of 3C-SiC is slightly larger than that of Si. The lattice constant of 3C-SiC at room temperature is 4.359 Å and is smaller than that of Si (5.430 Å at RT). Consequently, as an epitaxial 3C-SiC film is grown on a Si substrate, we have a tensile biaxial stress inside the SiC film. The bandgap change due to a tensile biaxial stress is given as:

$$\Delta E_g = P[2a(S_{11}+2S_{12}) + b(S_{11}-S_{12})]$$

where P is the pressure, S_{ij} are the elastic compliances and a and b are the deformation potentials. As $a<0$, $b<0$ and $P>0$ for a tensile stress, we have $\Delta E_g < 0$. Their [14] photoluminescence measurements confirm this conclusion and they obtained the value of $dE/dP \sim -10$ meV GPa^{-1}.

C   CONCLUSION

For the 3C-SiC polytype, an increase in pressure has been predicted to produce both a rocksalt, above 60 GPa, and a β-Sn transition from the original zinc blende structure. The pressure derivative of the fundamental bandgap has been measured as -1.9 meV GPa^{-1}, the smallest value in any tetrahedrally coordinated semiconductor or insulator, although re-analysis of these results gave a value of -3.4 meV GPa^{-1}. Theoretical studies confirm this latter value but the second pressure derivative is predicted to be much smaller than the measured value. Epitaxial growth of 3C-SiC on a silicon substrate leads to biaxial stress in the film and a consequent change in the bandgap. Photoluminescence measurements confirm this to be the case.

REFERENCES

[1] M. Kobayashi, M. Yamanaka, M. Shinohara [*J. Phys. Soc. Jpn. (Japan)* vol.58 (1989) p.2673]
[2] S.S. Mitra, O. Brafman, W.B. Daniels, R.K. Crawford [*Phys. Rev. (USA)* vol.186 (1969) p.942]
[3] D. Olego, M. Cardona, P. Vogl [*Phys. Rev. B (USA)* vol.25 (1982) p.3878]
[4] N. Churcher, K. Kunc, V. Heine [*Solid State Commun. (USA)* vol.56 (1985) p.177; *J. Phys. C, Solid State Phys. (UK)* vol.19 (1986) p.4413]
[5] P.J.H. Dentneer, W. van Haeringen [*Phys. Rev. B (USA)* vol.33 (1986) p.2831]
[6] W.R.L. Lambrecht, B. Segall, M. Methfessel, M. van Schilfgaarde [*Phys. Rev. (USA)* vol.44 (1991) p.3685]
[7] K.J. Chang, M.L. Cohen [*Phys. Rev. B (USA)* vol.35 (1987) p.8196]
[8] M. Kohyama, S. Kose, M. Kinoshita, R. Yamamoto [*J. Phys., Condens. Matter (UK)* vol.2 (1990) p.7791]
[9] W. Paul [*J. Appl. Phys. (USA)* vol.32 (1961) p.2028]
[10] S. Fahy, K.J. Chang, S.G. Louie, M.L. Cohen [*Phys. Rev. B (USA)* vol.35 (1987) p.5856]
[11] B.H. Cheong, K.J. Chang, M.L. Cohen [*Phys. Rev. B (USA)* vol.44 (1991) p.1053]

[12] K.J. Chang, S. Froyen, M.L. Cohen [*Solid State Commun. (USA)* vol.50 (1984) p.105]

[13] P.E. van Camp, V.E. van Doren, J.T. Devreese [*Phys. Status Solidi B (Germany)* vol.146 (1988) p.573]

[14] W.J. Choyke, Z.C. Feng, J.A. Powell [*J. Appl. Phys. (USA)* vol.64 (1988) p.3163]

CHAPTER 4

ENERGY LEVELS

4.1 Energy levels of impurities in SiC
4.2 Deep levels in SiC

4.1 Energy levels of impurities in SiC

I. Nashiyama

August 1993

A INTRODUCTION

This Datareview lists the energy levels of impurities from groups II, III, V and VI in several of the SiC polytypes. The most widely studied impurity is nitrogen as this is believed to cause the n-type character of high purity material. Boron, aluminium and gallium have also received a great deal of attention as acceptors in SiC. Reports of beryllium-doping and oxygen-doping have also been made. The standard techniques of Hall measurements and photoluminescence have mainly been used in these studies.

B MULTIPLE DONOR AND ACCEPTOR LEVELS

Impurity atoms in SiC substitute on either the silicon or carbon sublattice. Nitrogen as well as other donor impurities, such as phosphorus, occupy the carbon sites [1-3]. Boron may substitute on the carbon sublattice or may occupy either the carbon or the silicon site (or both sites), in order to minimise the total free energy of the system. Aluminium atoms substitute only on the silicon sublattice [4]. The SiC polytype does not affect the site preference of these impurities, though energy levels of impurity atoms differ in the different polytypes. In addition, owing to the long unit cells of various SiC polytypes except 3C- and 2H-SiC, many inequivalent lattice sites exist, which are divided into two species: a cubic-like atomic configuration of the first- and second-neighbour atoms and a hexagonal-like atomic configuration. Stacking sequences in the c-axis direction and numbers of cubic-like and hexagonal-like sites in 2H-, 3C-, 4H-, 6H- and 15R-SiC are given in TABLE 1 [5]. These inequivalent sites cause site-dependent impurity levels.

TABLE 1 SiC polytypes and numbers of inequivalent sites.

Polytype	Stacking sequence	Number of inequivalent sites	
		hexagonal-like	cubic-like
2H	AB	1	0
3C	ABC	0	1
4H	ABAC	1	1
6H	ABCACB	1	2
15R	ABCACBCABACABCB	2	3

The ionisation energies of the electronically active impurities have been determined primarily by photoluminescence techniques and Hall measurements. Ionisation energy levels of such impurities as nitrogen and some of the group III elements (aluminium, gallium, boron) in 3C-, 4H-, 6H- and 15R-SiC polytypes are compiled in TABLE 2. Nitrogen gives relatively shallow donor levels. In contrast, other p-type dopants have deep-level acceptor states.

4.1 Energy levels of impurities in SiC

TABLE 2 Impurity levels in 3C-, 4H-, 6H-, 15R- and 33R-SiC (unit: meV).

Nitrogen	Aluminium	Gallium	Boron	Method	Ref
3C-SiC					
54.5 ± 0.3	260 ± 15			PL	[13-15]
	248 + E_x			PL	[17]
53.6 ± 0.5				PL	[18]
56.5	254	343	735	PL	[19]
55			735	PL	[23]
	230			PL	[30]
118	179 + E_x			PL	[36]
		331 + E_x		PL	[43]
54				PL	[51]
48				Hall	[21]
	160			Hall	[49]
4H-SiC					
80, 130	155 + E_x			PL	[16]
	168 + E_x	249 + E_x	628 + E_x	PL	[17]
h:66, k:124	191	267	647	PL	[19]
	160 + E_x			PL	[38]
		250		PL	[40]
h:52.1, k:91.8				IR	[20]
h:45, k:100				Hall	[20]
40	180			Hall	[39]
6H-SiC					
170, 200, 230				PL	[7]
h:100, k:155	h:239, k:249	h:317, k:333	h:698, k:723	PL	[19,46]
100, 150	165 + E_x			PL	[16]
	231 + E_x			PL	[17]
180, 210, 240				TL	[28]
	280, 390, 490			PL/EL	[31]
			430, 540, 650	EL	[32]
180				IR	[35]
h:81, k:138, 142				IR	[53]
150			390	Hall	[22]
	270			Hall	[27]
81 - 95				Hall	[34]
50 - 70				Hall	[39]
66				Hall	[44]
h:85, k:125				Hall	[53]
	190		300	DLTS	[52]
15R-SiC					
140, 160, 160, 200				PL	[8]
h:64, k:112	h:206, 221 k:223, 230, 236	h:282, 300 k:305, 311, 320	h:666, k:700	PL	[19]
40 - 52				Hall	[34]
40				Hall	[39]
33R-SiC					
150 - 230				PL	[12]

k: cubic-like site, h: hexagonal-like site. PL: photoluminescence, Hall: Hall effect, IR: infrared absorption, EL: electroluminescence, TL: thermoluminescence, DLTS: deep level transient spectroscopy.
E_x (exciton energy) = 13.5 meV (3C) [33], 20 meV (4H) [41], 78 meV (6H) [42], 40 meV (15R) [19].

C   NITROGEN AND GROUP III ELEMENT IMPURITIES

High-purity SiC is reported as n-type in character. This fact has been normally explained by residual nitrogen. Donor-type native defects are also possible sources of residual carriers [6]. Hamilton et al [7] studied the photoluminescence spectra of nitrogen-doped 6H-SiC crystals and showed that nitrogen replaces carbon and forms three donor levels with ionisation energies 0.17, 0.20 and 0.23 eV, which correspond to the three inequivalent sites occupied by nitrogen atoms in the lattice. Site-dependent impurity levels of nitrogen atoms in 15R-SiC were found to be 0.14, 0.16, 0.16 and 0.20 eV by Patrick et al [8]. From studies of photoluminescence, Choyke and co-workers [7-12] established that the number of zero-phonon spectral lines due to excitons bound to neutral nitrogen donors agrees with the number of inequivalent sites in the various SiC polytypes.

Nitrogen donor levels and aluminium acceptor levels were usually determined from photoluminescence of donor-acceptor pairs. Long and co-workers [13-15] determined the ionisation energies of a donor (nitrogen) $E_D = 54.5 \pm 0.3$ meV and an acceptor (aluminium) $E_A = 260 \pm 15$ meV, from the donor-acceptor pairs and free to bound luminescence spectra of 3C-SiC (Al, N). Similarly, Hagen et al [16] investigated the low temperature luminescence spectra of 6H- and 4H-SiC doped with nitrogen and aluminium. From the luminescence spectra and the temperature dependence of the emission intensity, Suzuki et al [17] determined that the ionisation energies of Al, Ga and B in 4H-SiC were 168 meV + E_x, 249 meV + E_x, and 628 meV + E_x, respectively, where E_x is the exciton binding energy. In 6H- and 3C-SiC, they found that the ionisation energies of the aluminium acceptor were 231 meV + E_x and 248 meV + E_x, respectively. The ionisation energy of the nitrogen donor in 3C-SiC $E_D = 53.6 \pm 0.5$ meV was derived by Dean et al [18] from the measurements of the Zeeman splitting of photoluminescence lines. Site-dependent impurity levels caused by the inequivalent sites in 4H-, 6H- and 15R-SiC were extensively studied by Ikeda et al [19] for nitrogen donors and aluminium, gallium and boron acceptors. Recently, Götz et al [20] performed Hall-effect and infrared-absorption measurements on n-type 4H-SiC, obtaining the ionisation energies of the $1s(A_1)$ ground state of the nitrogen donor $E_D(k) = 91.8$ meV at the k site and $E_D(h) = 52.1$ meV at the h site, where k stands for the cubic-like site and h for the hexagonal-like site.

Segall et al [21] performed Hall measurements on n-type 3C-SiC epitaxial films and made detailed analyses of the temperature-dependent carrier concentrations. They found that the donor ionisation energy E_D depends on the donor concentration N_D with the relation $E_D(N_D) = 48 - 2.6 \times 10^{-5} N_D^{1/3}$ meV. A similar dependence was observed by Lomakina et al [22] for n- and p-type 6H-SiC.

The optical activation energy of boron is about 0.7 eV [3,23], in contrast to the thermal activation energy of 0.39 eV [22] calculated from the temperature dependence of the Hall mobility. This difference was discussed by Veinger et al [24]. They found, from Hall and ESR measurements, that the deeper level is an activator for the high-temperature luminescence but not seen in the ESR spectra and that the shallow level is a paramagnetic state.

D OTHER IMPURITIES

The beryllium atom is apparently an acceptor in pure or n-type material [9]. In heavily-doped p-type SiC, however, it seems to act as a compensating donor impurity with a 2^+ charge. Two acceptor levels, 0.42 and 0.60 eV, of beryllium in 6H-SiC were determined from the temperature dependence of the Hall effect (the hole density) in p-type 6H-SiC prepared by the diffusion of Be [25]. Padlasov et al [26] found that doping of 3C-SiC with phosphorus during crystal-growth gives rise to a donor centre, which has an ionisation energy 95 ± 3 meV, and doping with oxygen gives rise to an acceptor centre with an ionisation energy 180 ± 5 meV.

E CONCLUSION

Donors, such as nitrogen and phosphorus, occupy the carbon sites in SiC while aluminium, an acceptor, occupies the silicon sublattice and boron can substitute on either site. The site preference for a given impurity is not polytype dependent but its energy level is dependent on the particular polytype. In 3C-SiC nitrogen gives shallow donor levels but p-type dopants have deep level acceptor states, e.g. Al at ~250 meV. Nitrogen is thought to be the cause of n-type behaviour in high purity material but native defect donors are also possible. The donor ionisation energy in 3C-SiC epitaxial films follows the normal dependence on donor concentration. Beryllium is an acceptor in pure or n-type material but can act as a compensating impurity in p-type material. Phosphorus is a donor and oxygen is an acceptor in 3C-SiC, as expected.

REFERENCES

[1]   H.H. Woodbury, G.W. Ludwig [*Phys. Rev. (USA)* vol.124 (1961) p.1083]
[2]   G.E.G. Hardeman [*J. Phys. Chem. Solids (UK)* vol.24 (1963) p.1223]
[3]   S.H. Hagen, A.W.C. van Kemenade [*Phys. Status Solidi A (Germany)* vol.33 (1976) p.97]
[4]   Y. Tajima, W.D. Kingery [*Commun. Am. Ceram. Soc. (USA)* vol.65 (1982) p.C27]
[5]   L. Patrick [*Phys. Rev. (USA)* vol.127 (1962) p.1878]
[6]   H. Okumura et al [*Jpn. J. Appl. Phys. (Japan)* vol.27 (1988) p.1712]
[7]   D.R. Hamilton, W.J. Choyke, L. Patrick [*Phys. Rev. (USA)* vol.131 (1963) p.127]
[8]   L. Patrick, D.R. Hamilton, W.J. Choyke [*Phys. Rev. (USA)* vol.132 (1963) p.2023]
[9]   W.J. Choyke, L. Patrick [*Phys. Rev. (USA)* vol.127 (1962) p.1868]
[10] W.J. Choyke, D.R. Hamilton, L. Patrick [*Phys. Rev. A (USA)* vol.133 (1964) p.1163-]
[11] W.J. Choyke, D.R. Hamilton, L. Patrick [*Phys. Rev. A (USA)* vol.137 (1965) p.1515-]
[12] W.J. Choyke, D.R. Hamilton, L. Patrick [*Phys. Rev. A (USA)* vol.139 (1965) p.1262-]
[13] N.N. Long, D.S. Nedzvetskii, N.K. Prokofeva, M.B. Riefman [*Opt. Spectrosc. (USA)* vol.29 (1970) p.388]
[14] N.N. Long, D.S. Nedzvetskii, N.K. Prokofeva, M.B. Riefman [*Opt. Spectrosc. (USA)* vol.30 (1971) p.165]

[15] N.N. Long, D.S. Nedzvetskii [*Opt. Spectrosc. (USA)* vol.35 (1973) p.645]
[16] S.H. Hagen, A.W.C. van Kemenade, J.A.W. van der Does de Bye [*J. Lumin. (Netherlands)* vol.8 (1973) p.18]
[17] A. Suzuki, H. Matsunami, T. Tanaka [*J. Electrochem. Soc. (USA)* vol.124 (1977) p.241]
[18] P.J. Dean, W.J. Choyke, L. Patrick [*J. Lumin. (Netherlands)* vol.15 (1977) p.299]
[19] M. Ikeda, H. Matsunami, T. Tanaka [*Phys. Rev. B (USA)* vol.22 (1980) p.2842]
[20] W. Götz et al [*J. Appl. Phys. (USA)* vol.72 (1993) p.3332]
[21] B. Segall, S.A. Alterovitz, E.J. Haugland, L.G. Matus [*Appl. Phys. Lett. (USA)* vol.49 (1986) p.584]
[22] G.A. Lomakina [*Sov. Phys.-Solid State (USA)* vol.7 (1965) p.475]
[23] H. Kuwabara, S. Yamada [*Phys. Status Solidi A (Germany)* vol.30 (1975) p.739]
[24] A.I. Veinger et al [*Sov. Tech. Phys. Lett. (USA)* vol.6 (1980) p.566]
[25] Yu.P. Maslakovets, E.N. Mokhov, Yu.A. Vodakov, G.A. Lomakina [*Sov. Phys.-Solid State (USA)* vol.10 (1968) p.634]
[26] S.A. Podlasov, V.G. Sidyakin [*Sov. Phys.-Semicond. (USA)* vol.20 (1986) p.462]
[27] H.J. van Daal, W.F. Knippenberg, J.D. Wasscher [*J. Phys. Chem. Solids (UK)* vol.24 (1963) p.109]
[28] I.S. Gorban, A.F. Gumenyuk, Tu.M. Suleimanov [*Sov. Phys.-Solid State (USA)* vol.8 (1967) p.2746]
[29] A.A. Kal'nin, V.V. Pasynkov, Yu.M. Tairov, D.A. Yas'kov [*Sov. Phys.-Solid State (USA)* vol.8 (1967) p.2381]
[30] G. Zanmarchi [*J. Phys. Chem. Solids (UK)* vol.29 (1968) p.1727]
[31] Yu.S. Krasnov, T.G. Kmita, I.V. Ryzhikov, V.I. Pavlichenko, O.T. Sereev, Yu.M. Suleimanov [*Sov. Phys.-Solid State (USA)* vol.10 (1968) p.905]
[32] V.I. Pavlichenko, I.V. Ryzhikov [*Sov. Phys.-Solid State (USA)* vol.10 (1969) p.2977-]
[33] D.S. Nedzvetskii, B.V. Novikov, A.K. Prokofeva, M.B. Reifman [*Sov. Phys.-Semicond. (USA)* vol.2 (1969) p.914]
[34] S.H. Hagen, C.J. Kapteyns [*Philips Res. Rep. (Netherlands)* vol.25 (1970) p.1]
[35] I.M. Purtseladze, L.G. Khavtasi [*Sov. Phys.-Solid State (USA)* vol.12 (1970) p.1007]
[36] W.J. Choyke, L. Patrick [*Phys. Rev. B (USA)* vol.2 (1970) p.4959]
[37] A. Suzuki, H. Matsunami, T. Tanaka [*Jpn. J. Appl. Phys. (Japan)* vol.12 (1973) p.1083]
[38] H. Matsunami, A. Suzuki, T. Tanaka [in *Conf. on Silicon Carbide* (abstracts), Miami Beach, Florida, USA, 17 - 20 Sept. 1973 (University of South Carolina, 1973) p.618]
[39] H. Kang, R.B. Hilborn Jr. [in *Conf. on Silicon Carbide* (abstracts), Miami Beach, Florida, USA, 17 - 20 Sept. 1973 (University of South Carolina, 1973) p.493]
[40] A. Suzuki, H. Matsunami, T. Tanaka [*Jpn. J. Appl. Phys. (Japan)* vol.14 (1975) p.891]
[41] G.B. Dubrovskii, V.I. Sankin [*Sov. Phys.-Solid State (USA)* vol.17 (1975) p.1847]
[42] V.I. Sankin [*Sov. Phys.-Solid State (USA)* vol.17 (1975) p.1191]
[43] H. Kuwabara, K. Yamanaka, S. Yamada [*Phys. Status Solidi A (Germany)* vol.37 (1976) p.K157]
[44] A. Suzuki, M. Ikeda, N. Nagao, H. Matsunami, T. Tanaka [*J. Appl. Phys. (USA)* vol.47 (1976) p.4546]
[45] L.S. Aivazova, S.N. Gorin, V.G. Sidyakin, I.M. Shvarts [*Sov. Phys.-Semicond. (USA)* vol.11 (1977) p.1069]

[46] M. Ikeda, H. Matsunami, T. Tanaka [*J. Lumin. (Netherlands)* vol.20 (1979) p.111]
[47] H. Kuwabara, S. Yamada, Y. Uchida [*Phys. Status Solidi A (Germany)* vol.57 (1980) p.K45]
[48] M. Ikeda, H. Matsunami [*Phys. Status Solidi A (Germany)* vol.58 (1980) p.657]
[49] M. Yamanaka, H. Daimon, E. Sakuma, S. Misawa, S. Yoshida [*J. Appl. Phys. (USA)* vol.61 (1987) p.599]
[50] A. Suzuki, A. Ogura, K. Furukawa, Y. Fujii, M. Shigeta, S. Nakajima [*J. Appl. Phys. (USA)* vol.64 (1988) p.2818]
[51] J.A. Freitas Jr., S.G. Bishop, P.E.R. Nordquist Jr., M.L. Gipe [*Appl. Phys. Lett. (USA)* vol.52 (1988) p.1695]
[52] W. Suttrop, G. Pensl, P. Lanig [*Fall Meeting of the Electrochem. Soc.* (abstracts), Hollywood, Florida, USA, 15 - 20 Oct. 1989 (Electrochem. Soc., Pennington, New Jersey, 1989) abstr. no.494, p.716]
[53] W. Suttrop, G. Pensl, W.J. Choyke, R. Stein, S. Leibenzeder [*J. Appl. Phys. (USA)* vol.72 (1992) p.3708]

4.2 Deep levels in SiC

M.G. Spencer

September 1994

A   INTRODUCTION

In developing a material for use in device applications, it has been found that deep energy levels, in the forbidden energy gap, play an important role. Deep levels can act as carrier recombination or trapping centres and affect the performance of electronic and opto-electronic devices. Deep levels have been a subject of investigation for over thirty years and several excellent reviews are available: Grimmeiss [1], Neumar and Kosai [2], Milnes [3] and on the capacitance measurement techniques Lang [4].

Deep levels can be described by the Shockley-Read-Hall recombination statistics [5]. However, for a large number of deep states, the capture cross section for one type of carrier is many times larger than that for the other carrier. The state, therefore, interacts principally with only one of the band edges and can be characterised as either an electron or a hole trap. Capacitance techniques, such as DLTS (Deep Level Transient Spectroscopy), are particularly convenient for the determination of trap type and concentration. If additional experimental information is present to allow charge state determination, then the states can be characterised as deep acceptors or donors.

Silicon carbide presents some interesting opportunities for the study of deep levels. First, SiC is a material which exhibits over 170 different polytypes, thus making for a rich variety of possible defects. Second, almost all of the impurities in SiC have ionisation energies of greater than several kT. If the deep levels are defined as energy states, which are several kT from either the valence or conduction band edges, then most impurity or defect states in SiC can be thought of as deep levels. However, in the classifying of deep levels, a distinction is often made as to whether or not the state is hydrogenic like or non-hydrogenic like. Since the experimental data on SiC impurities is incomplete, we will not attempt to classify energy levels using this criterion.

In this Datareview, we concentrate on deep levels measured by capacitance and admittance techniques; those measured by other techniques are detailed in Datareview 4.1. For completeness, trap parameters for major defects and impurities obtained from all techniques are listed. Capacitance techniques have proven useful for the characterisation of deep states in semiconductor devices. In particular, states which are non-radiative can be analysed by this technique. If the state under study is one which principally determines the conductivity of the crystal, the techniques of admittance spectroscopy are used. The set-up for doing capacitance and admittance spectroscopy on SiC is identical to that used for other semiconductors with the exception of the necessity to operate the system at higher temperatures in order to access potentially deeper levels in the energy gap. The data are summarised in TABLE 1.

4.2 Deep levels in SiC

TABLE 1 Ionisation energies of deep levels in SiC.

Impurity/ Centre	Sym[a,b,c,d]	6H-SiC eV	Ref	3C-SiC eV	Ref	4H-SiC eV	Ref	Comment
N	E_D^T	0.085, 0.125	[25]			0.033, 0.09	[21,22]	note e
	E_D^O	0.081, 0.137, 0.142	[25]	0.053	[26]			
Al	E_A^T	0.27	[23]			0.27	[23,17]	
	E_A^O	0.257		0.271	[26]			
B	E_A^T	0.39	[24]					See D centre
	E_A^O	0.73	[21,30]	0.73	[28,29]	$0.628 + E_x$	[29]	
Be	E_A^T	0.4, 0.6	[18]					
Ga	E_A^T	0.29	[19]					
	E_A^O	0.35	[19]			$0.249 + E_x$	[29]	
Sc	E_A^T	0.24, 0.55	[20,13]					
Ti			[26]					note f
V			[27]					note g
D Centre	E_A^T	$E_v + 0.58$	[8-10]					See B
i Centre		$E_v + 0.52$	[11]			0.57	[17]	
Z_1	Elec T	$E_c - 0.62$	[6,7]					
Z_2	Elec T	$E_c - 0.64$	[6,7]					
E_1	Elec T	$E_c - 0.33$	[7]					note h
E_2	Elec T	$E_c - 0.4$	[7]					
E_3	Elec T	$E_c - 0.570$	[7]					
E_4	Elec T	$E_c - 0.570$	[7]					
ML_1	Hole T			$E_v + 0.4$	[16]			note i
SCE_1	Elec T			$E_c - 0.34$	[14]			
SCE_2	Elec T			$E_c - 0.68$	[14]			
R	Elec T	$E_c - 1.27$	[11,12]					
S	Elec T	$E_c - 0.35$	[11,12]					
N_{II}	Elec T			$E_c - 0.49$	[15]			note j

a E_D^T or E_A^T are acceptor or donor ionisation energies determined by thermal experiments such as Hall effect or admittance spectroscopy.

b E_A^O or E_D^O are acceptor or donor ionisation energies determined by optical techniques. E_x is the exciton binding energy.

c   Elec T is a general designation of a level which principally captures electrons rather than holes; when unoccupied the level may be donor like or acceptor like.

d   Hole T is a general designation of a level which principally captures holes rather than electrons; when unoccupied the level may be donor like or acceptor like.

e   6H-SiC has three inequivalent lattice sites for donors, one hexagonal like and two cubic like. Therefore there are three binding energies for the 6H donor.

f	Titanium is an isoelectronic centre which can exhibit a bright line in photoluminescence spectra.
g	The vanadium centre was studied by electron spin resonance and infrared luminescence.
h	The levels E1-E4 were introduced by electron irradiation.
i	This level was measured by microwave absorption techniques.
j	This level exhibited a broad peak roughly centred at the indicated energy. The level was introduced by neutron irradiation.

B DEEP LEVELS IN 6H-SiC

B1 Deep Levels in Lely or Modified Lely 6H-SiC

Bulk SiC is grown by the Lely method or the modified Lely method. In the case of the former, platelet crystals from $1\,cm^2$ to $4\,cm^2$ in size are produced. In the case of the latter, it is possible to obtain material with a diameter of two inches or greater. Defects in 'as-grown bulk material' have been investigated by Zhang et al [6] using DLTS techniques as well as IR absorption in bulk SiC. In this work, two defect levels are reported. The energy states of these defects (denoted Z_1 and Z_2) are closely spaced at energies of $0.620\,eV$ and $0.640\,eV$ from the conduction band. The defect concentration for both levels was reported as approximately $10^{16}\,cm^{-3}$. Since the ionisation energies of the two states did not depend on electric field, it was concluded that both states were acceptor like. It is suggested that these are self defects (such as carbon vacancies). Both Z defects were shown to be thermally stable with respect to annealing up to temperatures of $1700\,°C$. These defects are not seen in material grown by epitaxial techniques, but can be seen after these films have been irradiated with either electrons or protons.

B2 Deep Levels in Epitaxial 6H-SiC

As can be seen from the earlier section the literature on SiC is small. However, several publications exist on studies of epitaxial material. There are three epitaxial systems for the growth of SiC, sublimation epitaxy (SE), liquid phase epitaxy (LPE) or chemical phase deposition (CVD). Most of the defect studies have been done on material grown by sublimation epitaxy or liquid phase epitaxy. Of all the deep centres studied by capacitance techniques the boron centre has received the most attention. It was observed that epitaxial layers grown by LPE [8] or SE [9] of 6H-SiC doped in-situ with boron or implanted with boron exhibit at least two levels when probed by capacitance techniques. One of these two levels is the shallow boron level which has been determined to have an ionisation energy of about $0.39\,eV$ from the valence band edge and the second level has been termed the D centre which is located $0.58\,eV$ from the valence band edge as measured by DLTS. The D centre is a deep acceptor and seems to occur only in combination with the boron atom (implantation studies with other impurities have failed to produce the D centre). The concentration of the D centre tracks that of boron concentration and the presence of the D centre is thought to be responsible for the intense $2.42\,eV$ luminescence line seen in 6H material under UV excitation [8,10]. In addition to the D centre, at least two other centres have been reported. In CFLPE (container free liquid phase epitaxy), these centres have been denoted R and S and are located at energies of $0.35\,eV$ and $1.27\,eV$ respectively below the conduction band [11]. As opposed to the D centre the R and S states are electron traps. The R and S levels have been seen in both p-n junctions and Schottky barrier structures [12]. Finally, there is a report of the determination of the scandium acceptor level by DLTS techniques. This impurity produces a deep state in 6H-SiC at an energy level of 0.52 eV from the conduction band edge [13].

B3 Radiation Studies in 6H-SiC

There have been a few reports on radiation damage in SiC [7]. In this area, the effects of ions and electrons have been considered. If irradiation is performed, six deep states are produced in 6H-SiC. These states have been denoted E1-E4, Z_1 and Z_2. After thermal annealing, only the two Z states remain. It should be noted that these are the same Z states observed in as-grown bulk material. It should also be noted that the defects reported are rather shallow in energy and there are no reports of semi-insulating material produced by radiation damage.

C DEEP STATES IN 3C-SiC

There are only a handful of deep level studies in 3C material. This is due to lack of device quality material in this polytype. Currently it is possible to grow only epitaxial films of 3C material. However, recent progress has been made toward high quality epitaxial and bulk 3C material and this situation may change. All of the reported 3C studies have been performed on material that was grown on silicon substrates. It is well known that material grown in this manner contains over 1×10^7 dislocations and/or stacking faults cm^{-2}. Zhou et al [14] have done DLTS studies on as-grown SiC films. They found two levels in the as-grown films with energies of 0.34 eV and 0.68 eV respectively. There is some evidence that the latter level is related to oxygen treatment. The concentrations of these states, as measured by DLTS techniques, was approximately $2 - 5 \times 10^{16}$ cm^{-3}. The aforementioned study was performed using Schottky barriers on n-type material and hence only electron traps could be detected. Furthermore, the samples were 4 - 6 μm in thickness. Nagesh and Farmer [15], working with 14 - 16 μm samples, could not detect traps in as-grown material. Okumura et al [16] measured the energy level of deep states in 3C material using a combination of photo-conductivity and microwave absorption and from their results they inferred the presence of a hole trap with activation energy of 0.48 eV. This hole trap was determined to be a mobility killer for CVD films on Si and thought to be a structural defect. Nagesh and Farmer [15] have done DLTS studies on neutron irradiated 3C-SiC. These authors found a broad peak produced by neutron irradiation. The peak had a thermal activation energy of 0.49 eV from the conduction band. The introduction rate for the aforementioned defect was 1.0 cm^{-1} (defects cm^{-3} MeV^{-1}, equivalent neutron cm^{-2}). About 90% of these defects could be removed by annealing at 350 °C.

D DEEP STATES IN 4H-SiC

The author could find only one study of 4H-SiC [17]. That study was performed on epitaxial 4H films which were doped by ion implantation with Al. It was found that these films exhibited two levels. The first, thought to be associated with Al, was at an energy of 0.26 eV with respect to the valence band edge and the second denoted a centre located at 0.57 eV with respect to the valence band edge.

E CONCLUSION

Most impurity and defect states in SiC can be considered as deep levels. Both capacitance and admittance spectroscopy provide data on these deep levels which can act as donor or acceptor traps. Bulk 6H-SiC contains intrinsic defects which are thermally stable, up to 1700°C. In epitaxial films of 6H-SiC a deep acceptor level is seen in boron-implanted samples but not when other impurities are implanted. Other centres, acting as electron traps, are also seen in p-n junction and Schottky barrier structures. Irradiation of 6H-SiC produces 6 deep levels, reducing to 2 after annealing. Only limited studies have been carried out on the 3C-SiC polytype, in the form of epitaxial films on silicon substrates. No levels were seen in thick films but electron traps were seen in thin n-type films and a hole trap (structural defect) was found to be a mobility killer. Neutron irradiation produces defects most of which can be removed by annealing. Two levels were found in Al-implanted 4H-SiC.

REFERENCES

[1] H.J. Grimmeiss [*Annu. Rev. Mater. Sci. (USA)* vol.7 (1977) p.341]
[2] G.F. Neumark, K.K. Kosai [*Semicond. Semimet. (USA)* vol.19 (1983) p.1-74]
[3] A.G. Milnes [*Deep Impurities in Semiconductors* (Wiley Interscience, New York, 1973)]
[4] G.L. Miller, D.V. Lang, L.C. Kimmerling [*Annu. Rev. Mater. Sci. (USA)* vol.7 (1977) p.377]
[5] W. Shockley, W.T. Read Jr. [*Phys. Rev. (USA)* vol.87 (1952) p.835]
[6] H. Zhang, G. Pensl, A. Dornen, S. Leibenzeder [*Ext. Abstr. Electrochem. Soc. (USA)* vol.89-2 (1989) p.699]
[7] H. Zhang, G. Pensl, P. Glasow [*Ext. Abstr. Electrochem. Soc. (USA)* vol.89-2 (1989) p.714]
[8] W. Suttrop, G. Pensl, P. Lanig [*Appl. Phys. A (USA)* vol.51 (1991) p.231]
[9] M.M. Anikin, A.A. Lebedev, A.L. Syrkin, A.V. Suvorov [*Sov. Phys.-Semicond. (USA)* vol.19 (1985) p.69]
[10] M.M. Anikin, N.I. Kuznetsov, A.A. Lebedev, A.M. Strel'chuk, A.L. Sykin [*Sov. Phys.-Semicond. (USA)* vol.24 no.8 (1990) p.869]
[11] M.M. Anikin, A.S. Zubrilov, A.A. Lebedev, A.P. Strel'chuk, A.E. Cherenkov [*Sov. Phys.-Semicond. (USA)* vol.25 no.3 (1991) p.289]
[12] M.M. Anikin et al [*Sov. Phys.-Semicond. (USA)* vol.25 no.2 (1991) p.199]
[13] V.S. Ballandovich [*Sov. Phys.-Semicond. (USA)* vol.25 no.2 (1991) p.175]
[14] P. Zhou, M.G. Spencer, G.L. Harris, K. Fekade [*Appl. Phys. Lett. (USA)* vol.50 no.11 (1987) p.1384]
[15] V. Nagesh, J.W. Farmer, R. Davis, H.S. Kong [*Appl. Phys. Lett. (USA)* vol.50 no.17 (1987) p.27]
[16] K. Okumura, K. Endo, S. Misawa, E. Sakuma, S. Yoshida [*Springer Proc. Phys. (Germany)* vol.43 (1989) p.94]

[17] M.M. Anikin, A.A. Lebedev, A.L. Syrkin, A.V. Suvorov [*Sov. Phys.-Semicond. (USA)* vol.20 no.12 (1986) p.1357]
[18] Yu.P. Maslokovets, E.N. Mokhov, Yu.A. Vodakov, G.A. Lomakina [*Sov. Phys.-Solid State (USA)* vol.10 (1968) p.634]
[19] Yu.A. Vodakov et al [*Phys. Status Solidi A (Germany)* vol.35 (1976) p.37]
[20] Yu.M. Tairov, I.I. Khlebneinkov, V.F. Tsvetkov [*Phys. Status Solidi (Germany)* vol.25 (1974) p.349]
[21] G.A. Lomakina, G.F. Kholuyanov, R.G. Verenehikova, E.H. Kokhov, Yu.A. Vodakov [*Fiz. Tekh. Poluprovodn. (Russia)* vol.6 (1972) p.1133]
[22] S.H. Hagen, A.W.C. van Kemenade, G.A.W. van der Does de Bye [*J. Lumin. (Netherlands)* vol.8 (1973) p.18]
[23] G.A. Lomakina [in *Silicon Carbide 1973* Eds R.C. Marshall, J.W. Faust, C.E. Ryan (University of South Carolina, Columbia, South Carolina, 1974) p.520]
[24] G.A. Lomakina [*Sov. Phys. (USA)* vol.7 (1965) p.475]
[25] W. Suttrop, G. Pensl, W.J. Choyke, R. Stein, S. Lebenzeder [*J. Appl. Phys. (USA)* vol.72 no.8 (1992) p.3708]
[26] W.J. Choyke [*Optical and Electronic Properties of SiC* Ed. R. Freer, NATO ASI Series 'The Physics and Chemistry of Carbides, Nitrides and Borides', Manchester, UK (1989)]
[27] J. Schneider et al [*Appl. Phys. Lett. (USA)* vol.56 (1990) p.1184]
[28] S. Yamada, H. Kuwabera [in *Silicon Carbide 1973* Eds R.C. Marshall, J.W. Faust, C.E. Ryan (University of South Carolina, Columbia, South Carolina, 1974) p.305]
[29] A. Suzuki, H. Matsunami, T. Tanaka [*J. Electrochem. Soc. (USA)* vol.124 no.2 (1977) p.241]
[30] D.V. Dem'yanchik, G.A. Karatynina, E.A. Konorova [*Sov. Phys.-Semicond. (USA)* vol.9 (1975) p.900]

CHAPTER 5

SURFACE STRUCTURE, METALLIZATION AND OXIDATION

5.1 Surface structure and metallization of SiC
5.2 Oxidation of SiC

5.1 Surface structure and metallization of SiC

SURFACE STRUCTURE, METALLIZATION AND OXIDATION

R. Kaplan and V.M. Bermudez

1. INTRODUCTION

5.1 Surface structure and metallization of SiC

R. Kaplan and V.M. Bermudez

May 1993[1]

A INTRODUCTION

Recent progress in the experimental investigation of SiC surfaces has reflected the increasing availability of high quality single crystals and the development of methods for obtaining clean ordered surfaces in ultra-high vacuum (UHV). During the last few years a number of distinct surface phases have been identified and characterized. The origins and properties of these surface phases increasingly have been subjected to theoretical investigation. While the picture is very far from complete, at the present time a quantitative understanding of several intrinsic SiC surface reconstructions is within reach, and other reconstructions have received detailed experimental examination. In addition, there has been considerable work on the related subject of metallization of well-characterized surfaces. The aim of this Datareview is to organize and describe the current status of research on the atomic scale structure and composition of crystalline SiC surfaces and the evolution and properties of such surfaces following metallization. As discussed elsewhere in this volume, SiC exhibits well over 100 structural polytypes. Virtually all of the reported research on crystalline surfaces has involved β (cubic) or α(6H) (hexagonal) polytypes, since these are the most readily available in a form suitable for characterization. Investigation of the properties of concern - surface structure, composition and reactivity on an atomic scale - requires a UHV environment and a variety of electron, optical and ion spectroscopies and other techniques.

A brief enumeration of the relevant UHV characterization techniques follows this introduction. Methods of surface preparation are next considered, after which progress in the determination of the structure and composition of stable surface phases of β and 6H crystals is reviewed. The final section describes work on the SiC interface with 15 different metals, both as-prepared and after thermal annealing.

B CHARACTERIZATION TECHNIQUES

The most often used technique for studying SiC surface structure has been low energy electron diffraction (LEED). (See TABLE 1, which lists references to published work on SiC surfaces organized according to the experimental technique used.) The utility of LEED in SiC investigations is enhanced by the material's high Debye temperature, which leads to good pattern contrast, most importantly at elevated sample temperature and high primary energy. While most of the LEED investigations have involved only determination of two-dimensional (i.e., lateral) atomic periodicities, during the last year several three-dimensional structure determinations based on multiple scattering analysis of diffraction intensities have been published. Structure studies utilizing scanning tunnelling microscopy (STM) have appeared recently. Thus far, the SiC surfaces imaged by STM show much greater disorder than

---

[1] (Editor's note: at the time of going to press, June 1995, the area of knowledge reviewed here had not changed substantially since the cut-off date, May 1993.)

corresponding Si surfaces, and the spatial resolution obtained for the latter has not yet been achieved for SiC. Total energy calculations for SiC surface structure determination have been performed for several surface orientations and configurations. Medium energy ion scattering (MEIS) has also provided important information on surface structure. For determination of surface composition, contamination and chemical bonding, the techniques of choice have been Auger electron spectroscopy (AES), and X-ray and ultraviolet photoemission spectroscopy (XPS and UPS, respectively). In AES the Si(LVV) and C(KLL) transitions are particularly useful in that the lineshapes and positions are sensitive to the chemical bonding of the atoms. In XPS the C 1s and Si 2p core levels are readily probed using 1253.6 eV Mg Kα radiation, while the Si 2p core level and SiC valence band can be observed with greater surface sensitivity with the 151.4 eV Mζ line from a Zr anode. HeI (21.2 eV) and HeII (40.8 eV) radiation in UPS also provides surface sensitivity for valence band studies. Additional information on composition, bonding and electronic structure has been provided by electron energy loss spectroscopy (EELS), soft X-ray spectroscopy (SXS), and ellipsometry. Chemical deposition studies, incorporating some of the foregoing techniques together with in-situ reflection high energy electron diffraction (RHEED), have provided further insights.

C   SURFACE PREPARATION

As early as 1975, Von Bommel et al [1] showed that atomically clean ordered surfaces of 6H-SiC could be obtained by annealing in UHV at 800 °C, but surface precipitation of graphite was observed even at this relatively low temperature, presumably associated with Si loss. Dayan [2,3] subsequently reported similar results for smooth high quality films of (100) β-SiC grown on Si substrates. Surface oxide desorbed at either 1010 or 1080 °C, yielding (3x2) or c(2x2) structures respectively, as determined by LEED. The different behaviours among the (100) films are thought to reflect different surface compositions determined by the manner in which film deposition was terminated. Annealing at higher temperatures caused further structural changes, which could not be reversed by additional annealing, since they are caused by the depletion of surface Si. The (100) films exhibited surface graphite after brief annealing at 1285 °C. Subsequent work to be discussed has confirmed that, while annealing in UHV can yield oxide-free surfaces, this preparation method suffers from three distinct disadvantages. First, the results obtained are sample dependent. Second, the surface phase transformations can proceed only in the direction of successively Si-poorer compositions. Third, the most Si-poor phase is found to be accompanied by graphite precipitation and considerable disorder. The phenomenon of annealing-induced surface graphitization has been studied by a number of workers [1,4-10].

In addition to annealing, one of the traditional methods for obtaining clean surfaces in UHV is ion bombardment. Since the bombarded surfaces are badly damaged, annealing is required to restore surface order. This method therefore suffers from most of the problems encountered when SiC surfaces are prepared by annealing alone. Early work by Kaplan [11] showed that LEED patterns could be observed from (100) surfaces bombarded with 700 eV Kr ions at 650 °C. The C(KLL) AES lineshape showed that short range order was present on the clean surface. Absence of long range order, however, was evidenced by high background intensity, broad diffraction beams, and lack of reconstruction in the LEED patterns. It may be concluded that ion bombardment is inappropriate both for SiC surface structure studies and for investigation of chemical reactions, metallization and any other surface process where broken bonds, diffusion etc. play an important role. Details of the consequences of ion bombardment and subsequent annealing of SiC have been reported by several workers [7-9,11].

TABLE 1 Surface-science techniques applied to β- and α(6H)-SiC. (The numbers indicate the appropriate literature references.)

Technique	SiC polytype and orientation		
	β-(100)	β-(111)	6H-(0001)
AES	[2,3,7,8,11,12,14-17, 42,57,58,61-65]	[15]	[1,5,9,14,15,49-51,59,60]
EELS	[3,11,14-18,44,65]	[15]	[9,15,59]
Ellipsometry			[4,68]
LEED	[2,3,8,11,14-17,26,28,29,42]	[15]	[1,14,15,50,51,66]
MEIS	[26]		
RHEED	[20,27]		
STM	[39-41]	[40,53,54]	
SXS	[70]		
UPS	[16,69]		[67]
XPS	[10,16,18,21,22,44,71]		[5,9,50,59,67,71]
TPD, ESD	[17,19,24]		[72]
Theory	[25,34-38,43,45-47,73,74]	[23,34,37,43]	[55,56]

AES	Auger Electron Spectroscopy
EELS	Electron Energy Loss Spectroscopy
LEED	Low Energy Electron Diffraction
MEIS	Medium Energy Ion Scattering
RHEED	Reflection High Energy Electron Diffraction
STM	Scanning Tunnelling Microscopy
SXS	Soft X-ray Spectroscopy
UPS	Ultraviolet Photoemission Spectroscopy
XPS	X-ray Photoemission Spectroscopy
TPD	Temperature Programmed Desorption
ESD	Electron Stimulated Desorption

Exposure to high energy laser radiation sometimes is successful in removing contamination and leaving a clean ordered surface. This approach has been investigated by Pehrsson and Kaplan [12] for β-SiC (100) using the 193 nm line from an excimer laser. Multiple pulses of energy fluence between 400 and 600 mJ cm^{-2} first removed surface oxide and then partially regenerated long range order as judged from the observed weak, high background LEED patterns. Further exposure generated a surface graphitic carbon layer. This approach consequently does not appear promising for surface structure studies. Excimer laser radiation has also been used to anneal Ga-implanted 6H-SiC [13]. Recrystallization of the amorphous implanted region and redistribution of the Ga were observed, but surface structure was not examined since the annealing was done in air.

In order to avoid the shortcomings of the surface preparation methods previously described, techniques based on the chemical reduction of surface oxide at modest temperatures have been investigated. Kaplan and Parrill [14] reported disappearance of virtually all the oxide from both 6H and β samples after several minutes exposure at 970 °C to a Ga flux in the range 0.1 - 1.0 equivalent monolayers sec^{-1}. The remaining oxide could be removed only by further exposure at higher temperature and was thought to be trapped at near-surface grain boundaries inaccessible to the Ga beam. A similar approach but using a Si beam instead of Ga has proven the most successful of all surface preparation methods tried to date. With this method, Kaplan [15] and Parrill and Chung [16] were able to obtain clean surfaces without heating above 1000 °C. The arriving Si atoms convert surface oxide to SiO which is volatile in this temperature range, thus removing the oxide without otherwise disturbing the surface. Contaminant C is similarly converted to SiC, thereby becoming part of the crystal structure. By means of further Si exposure or annealing, the whole range of SiC surface structures from Si-terminated to C-terminated phases could be obtained and studied. It was subsequently shown by Bermudez and Kaplan [17] that the most homogeneous and highly ordered C-terminated (100) structure could be achieved by thermal decomposition of ethylene, C_2H_4, on an initially Si-terminated surface maintained at 800 - 1100 °C. This work is typical of a number of UHV [18-22] and theoretical [23-25] studies of SiC growth and of Si and C deposition from gaseous sources which yield insight into the origins of surface structure and composition.

D   SURFACE STRUCTURE AND COMPOSITION

The preponderance of research to date has been on (100) surfaces of β-SiC in the form of films deposited on Si substrates. Work on this orientation will be reviewed first, followed by a description of results obtained on β(111) and 6H(0001) (which appear to be indistinguishable in LEED) and on 6H(000$\bar{1}$) surfaces.

D1   β-SiC (100)

The existence of a number of distinct (100) surface phases each exhibiting a unique composition and structural unit cell is now well established. Studies of the evolution of structure with changing composition have utilized Si deposition [15,16,20,26,27], C deposition [17,20,28], and Si depletion via annealing [2,3,14-16,26] to alter the surface Si-to-C ratio. In order of decreasing Si content, the observed phases are (3x2), (5x2), (2x1) and c(4x2) (this pair have the same composition), c(2x2) and (1x1). The latter is probably a partially disordered phase in which reconstruction is inhibited.

Elimination of surface dangling bonds by the formation of 'dimer pairs' - adjacent atoms displaced towards each other to increase overlap of the unpaired electrons between them - is the single most important distortion employed by SiC surfaces to lower their energy. This distortion causes a doubling of the unit cell dimension parallel to the dimer axis. Different arrangements of the surface dimers can then lead to a variety of surface structures. The situation is further complicated by the presence of distinguishable surface domains. A consequence of covalent bonding in the zinc blende structure is that the direction of dangling bonds projected in the (100) surface rotates 90 degrees with each successive atom layer. Since nominally flat, real (100) surfaces will contain areas connected by single layer steps, a mosaic of domains characterized by two orthogonal dangling bond directions will result. This gives rise to LEED patterns which are superpositions of diffraction from orthogonally reconstructed unit cells unless means are used to suppress one domain type. In the following, unless otherwise noted, the LEED patterns are representative of such two-domain surfaces.

D1.1 (2x1) and c(4x2) phases

The (2x1) and c(4x2) phases are now commonly accepted as characteristic of the ideal (100) surface terminated in a single complete layer of Si atoms. The Si atoms are coupled to form arrays of dimer pairs. Supporting this conclusion are measurements of AES lineshapes [15,16], XPS shifts and lineshapes [16], MEIS shifts and intensities [26], RHEED growth studies [27], and analysis of LEED patterns [15,16] and intensities [29]. The latter structural analysis by automated tensor LEED concluded that the dimers are buckled (i.e., their axes are tilted relative to the surface) and that the dimer atoms exhibit a small charge transfer. This property allows us to think of the dimers as vectors, whose relative orientations help determine the size and shape of the unit cell. The various possibilities have been investigated for Si (100) surfaces by means of total energy calculations [30,31]. It was found that at sufficiently low temperature 'antiferromagnetic' ordering of the buckled dimers yields a c(4x2) unit cell. Thermal disordering renders the dimers uncorrelated, thus returning the reconstruction to (2x1). Such order-disorder transitions have been experimentally observed for Si [32] and Ge [33] at 150 and 220 K, respectively. For SiC the c(4x2) persists at least to 873 K [15]. This increased stability may reflect a stronger dimer-dimer interaction due to smaller spacing of atoms on SiC as compared to Si surfaces. Occurrence of the SiC c(4x2) reconstruction is found to be inhibited by surface roughness and contamination [15,16]. Theoretical investigation of the SiC (100) surface has been confined thus far to the (2x1) reconstruction, with no consideration of the dimer interactions leading to the c(4x2). Using the periodic MINDO (modified intermediate neglect of differential overlap) method Craig and Smith [34,35] found that the dimers are indeed buckled, with charge transfer occurring from the inner to the outer atom of the pair. This is consistent with the partially ionic character of SiC. Due to their relatively greater electronegativity the C atoms are negatively charged, i.e., they display increased valence electron density. Thus, the uppermost C atoms attract the positively charged inner Si dimer pair atoms, and repel the outer negatively charged dimer atoms. Other calculations [36-38] for the SiC (100) surface also predicted (2x1) dimer reconstructions but yielded symmetric rather than buckled dimers in disagreement with experimental observation of the c(4x2) reconstruction. Unlike the case for Si, STM [39-41] studies of (100) SiC have thus far failed to resolve fine details of the surface structure. Thus, at present, the only experimental evidence for buckling of the dimers is the observation of the c(4x2) reconstruction.

Studies of the (2x1) and c(4x2) phases using AES, XPS and UPS have generally been consistent with expectations for a Si-terminated surface and serve primarily as a baseline for comparison of spectra from other reconstructions [15,16]. Caution is required when interpreting small shifts or lineshape changes, which are known to arise due to slight O or H contamination or to the presence of more than one surface phase. Similar considerations apply to EELS results, for which most of the observed peaks remain unassigned. Exceptions are the bulk plasmon excitation at about 21.5 eV, and a sharp peak just below 2 eV presumably associated with surface states.

D1.2 c(2x2) phase

By analogy with the ideal Si-terminated SiC (100) surface there exists the corresponding ideal surface terminated in a complete monolayer of C atoms. According to recent experimental and theoretical work, these C atoms also form pairs, but the pairs are organized into a staggered array thus yielding a c(2x2) reconstruction instead of (2x1). The nature of the C-C and C-Si bonding in the c(2x2) phase is still a subject of controversy. Earlier AES studies [3,14-16] established that the surface Si/C ratio decreased strongly during the transformation from (2x1) to c(2x2), consistent with formation of a C-terminated structure. The appearance of surface C-C bonding and changes in the surface state density accompanying c(2x2) formation have been deduced from valence band photoemission [16] and EELS [15,16] spectra. Convincing evidence supporting the conclusion that the c(2x2) structure is indeed terminated in a single C monolayer has also been provided by studies of medium energy ion scattering [26], of the room-temperature reaction of Ti with c(2x2) surfaces [42] and of C deposition on Si-terminated (2x1) surfaces [17,27]. STM studies [39,40] of samples prepared by annealing at 1100 °C have shown regions with c(2x2) as well as other geometries. However, the surfaces were partially disordered and inhomogeneous due to the annealing, and no correlation with other properties could be made. It should be emphasized that interpretation of experiments on c(2x2) surfaces obtained by annealing is likely to be complicated by the presence of other reconstructions, surface graphite and considerable disorder. A much more homogeneous, highly ordered c(2x2) phase can be obtained by chemical vapour deposition of a C monolayer on the easily-prepared Si-terminated surface, as previously described [17]. When using this approach it is necessary to consider the possible presence of adsorbed H following dissociation of the C precursor; however, such H can easily be desorbed by a relatively mild anneal.

Some theoretical studies [34,35,37,38,43] have concluded that the C atoms on C-terminated (100) surfaces form a regular (non-staggered) array of symmetric dimers, thus behaving similarly to Si atoms on Si-terminated surfaces. However, it has recently become evident that the situation may be more complicated with regard both to the bonding of C atoms within pairs and to the bonding of these pairs to the underlying Si layer. The suggestion that this surface comprises a staggered array of C dimers was first made by Hara et al [26]. Based on LEED, AES, EELS and electron-stimulated H$^+$ desorption results, Bermudez and Kaplan [17] proposed a model for the c(2x2) structure in which sp^3 single-bonded C-C pairs form a staggered array at four-fold hollow sites of the underlying Si atom layer. The assumed C-C separation is 1.54 Å. Each C bridges two nearest-neighbour Si atoms. Only C dangling bonds are present, and these are active sites for the thermally reversible adsorption of H. This surface is known [44] to be relatively inert to room temperature adsorption of O_2, consistent with its lack of Si dangling bonds. The C pairs or 'dimers' are analogous to the Si dimers on Si-terminated surfaces. An alternative model, based on tensor-LEED structural analysis, has been proposed by Powers et al [28]. In this model the surface is terminated in a staggered array of C pairs or 'C₂ groups' on Si bridge sites. Second layer Si atoms are weakly dimerized for the H-free (annealed) surface, and the optimized C pair spacing is 1.31 Å, implying that the C₂ groups are double bonded and sp² hybridized. The total energy calculation of Craig and Smith [45] for the C-terminated c(2x2) structure strongly favours the staggered dimer model described above, except that the predicted C-C bond length is 1.368 Å, implying a double bond in disagreement with the result of [17]. On the other hand, related calculations by Badziag [46,47] support the Si bridge site model, including weak dimerization of second layer Si atoms. However, the predicted C-C bond length for the H-free surface is now 1.20 Å (as compared with the value 1.31 Å derived from tensor LEED [28]), implying a C triple bond for the C₂ group. Clearly further work is needed to establish the structure of the C-terminated c(2x2) surface.

D1.3 (3x2) and (5x2) phases

The (3x2) phase is obtained from the (2x1) or c(4x2) by addition of less than a monolayer of Si to the already Si-terminated surface. Its characteristic LEED pattern, first reported by Dayan [3] and later by Kaplan for single domain samples [15], contains sharp one-third order and streaked half order beams. The latter become less streaked with repeated cycles of desorption and re-deposition of the additional Si. XPS, AES and EELS measurements [15,16] have demonstrated the existence of Si-Si bonds on the (3x2) surface. Examination of single-domain diffraction patterns [15,27] confirmed that the direction of tripled periodicity of the reconstructed unit cell is perpendicular to the dimer axis, while the doubled periodicity occurs along the dimer axis direction. These features are contained in the (3x2) structure originally proposed by Dayan [3] based on the defect model of Pandey [48]. Every third row of Si adatoms atop the Si-terminated surface is missing. The energy is lowered by dimerization of the adatoms and π-bonding of second layer Si atoms adjacent to the vacancy rows. The latter allow the Si adlayer to be accomodated on the underlying SiC lattice with its 20% smaller lattice spacing. Disorder in the orientation of the Si adlayer buckled dimers could explain [15] the observed streaking of the half-order diffraction beams. This model, however, appears incompatible with recent observations [26,27] of the sequential appearance of (5x2) and (3x2) reconstructions as the Si adatom coverage increased. Instead it has been suggested [26,27] that there exists a class of Nx2 reconstructions, consisting of dimer rows N lattice spacings apart, yielding 1/N monolayer coverage. While STM experiments on β-SiC (100) [39,40] have shown regions of (3x2) reconstruction, their resolution was insufficient to yield detailed structural information.

D1.4 (1x1) phase

Several surface preparation methods are known to yield (1x1) LEED patterns. Annealing at temperatures approaching 1300 °C leads to (1x1) diffraction with broadened beams and high background intensity, indicative of partial disordering. As detailed in Section C, such surfaces also exhibit graphitic C bonding; the combined presence of disorder and graphitic 'contamination' presumably inhibits reconstruction. Hara et al [26] reported observing (1x1) LEED after annealing a Si-rich surface sufficiently to yield termination by a half monolayer of C, i.e., intermediate between the surface compositions of the (2x1) and c(2x2) phases. Further details regarding this structure were not given. It may be surmised that the C half monolayer was disordered. Finally, Matsunami et al [20] observed a (1x1) RHEED pattern during gas-source molecular beam epitaxial film growth when the sample was exposed to acetylene, C_2H_2, at about 1000°C. Without further information it is only possible to speculate that various C precursor molecules experience different adsorption-dissociation-bonding paths on the SiC surface which may lead to different, possibly metastable, C atom arrangements.

D2 α-SiC (0001) and (000$\bar{1}$); β-SiC (111)

Experimental studies of these surface orientations are few in number and somewhat mutually contradictory. The crystal structure can be envisioned as composed of bilayers parallel to the surface, with Si and C in opposing planes each forming three bonds directed into the bilayer, and one pointing out normal to the layer. One consequence of this structure is that a sample with plane-parallel sides will tend to have one Si-terminated and one C-terminated surface. A second consequence is that a single complete pseudomorphic adlayer cannot exist on either surface, since it would possess three dangling bonds per atom. It is possible that the characterization methods considered in this Datareview cannot distinguish between Si-terminated (0001) and (111) surfaces, or between C-terminated (000$\bar{1}$) and ($\bar{1}\bar{1}\bar{1}$) surfaces. This follows from the observation that the atomic arrangements are the same for the two structures within the sampling depth of the electrons or ions used in the measurements; only at greater depths does the stacking of the atom bilayers differ. Experimental results to date show no differences between the corresponding α- and β-SiC orientations.

The earliest work on these orientations is that of Van Bommel [1] using platelets of α-SiC. Annealing at 600°C was reported to give clean surfaces. Surprisingly, for both the Si- and C-terminated surfaces the AES intensity ratio is C(KLL)/Si(LVV) = 0.50. (The convention seems to have developed of citing the ratio C/Si for this orientation, and Si/C for the (100) case.) The Tottori University group [49-51] has used the annealing temperature dependence of the C(KLL)/Si(LVV) ratio to indicate the presence of a clean 'stoichiometric' surface. A minimum at 1000°C was interpreted as showing that the Si oxide had been desorbed but that graphitization at higher temperatures had not yet commenced. This procedure yielded C(KLL)/Si(LVV) values of 0.26 and 0.38 for Si and C faces, respectively [50]; although, in an earlier publication [49] somewhat different values, 0.32 and 0.54, were given. The result C(KLL)/Si(LVV) = 0.63 was observed [51] for a C-terminated sample misoriented 5 degrees towards (11$\bar{2}$0). In a different approach to obtaining clean surfaces, Kaplan and Parrill [14] reduced and desorbed the surface oxide by exposure to a Ga flux at 900°C with the result C(KLL)/Si(LVV) = 0.29 for the Si face and 1.3 for the C face. It is evident from the scatter in the cited values that the surfaces for this orientation remain ill-defined. The problem is mirrored by the corresponding disagreement regarding diffraction results. For the Si-terminated surface, previously cited references variously report LEED patterns having (3x3) and ($\sqrt{3}$x$\sqrt{3}$)R30° geometry, with additional structures due to oriented graphite overlayers following high temperature annealing. For the C-terminated surface, the reported LEED patterns are mostly (1x1), with mention also of (2x2) and (3x3) and of ($\sqrt{3}$x$\sqrt{3}$)R30° after further annealing. The misoriented C-face sample referred to above displayed (1x1) LEED.

Some insight into this situation is provided by the only work reported so far on the β-SiC (111) orientation [15]. The sample was a film deposited in registry on a (111) Si substrate. Measurement of AES intensities indicated that the surface was Si-terminated. The surface was cleaned and its composition reversibly altered by a suitable combination of exposure to a Si flux and annealing, as previously described for (100) samples. After exposure to a Si flux at 850 °C, a sharp uniform (3x3) LEED pattern was observed with C(KLL)/Si(LVV) = 0.17. Details of the Si(LVV) and the EELS lineshapes were suggestive of an additional Si bilayer residing on the Si-terminated surface. A defect model for this structure based on the dimer-adatom-stacking fault (DAS) model [52] was proposed. Annealing at 950 °C caused the LEED pattern to change from (3x3) to ($\sqrt{3} \times \sqrt{3}$)R30° and C(KLL)/Si(LVV) to increase from 0.17 to 0.29. Whether this reflects loss of part or all of the excess Si bilayer could not be determined. By analogy with adsorption of group III or IV atoms on Si (111), the observed $\sqrt{3}$ LEED pattern might be due to 1/3 of a monolayer of Si adatoms each bonding to three Si atoms of the SiC surface. With additional annealing the LEED pattern became (1x1), but the data were suggestive of surface disorder and the onset of segregation of surface graphite. This sequence of structure transformations is similar to that exhibited by the β-SiC (100) surface. None of the structures discussed in this section has yet to be spatially resolved in STM experiments, which have focused on ordered graphite surface overlayers [40,53,54]. The only calculation to date for these orientations has been carried out by Badziag [55,56] for the ($\sqrt{3} \times \sqrt{3}$)R30° phase of α-SiC (000$\bar{1}$). The model favoured by Badziag contains Si adatoms with C atoms substituting for Si in the second layer of the unreconstructed surface. The latter sites are the H_5 locations directly below the adatoms. There has not yet been any theoretical investigation of the most frequently observed reconstructions - (1x1) for the C termination and (3x3) and ($\sqrt{3} \times \sqrt{3}$)R30° for the Si-terminated surface.

E INTRODUCTION TO METAL-SILICON CARBIDE INTERFACES

An understanding of the chemical and physical structure of the metal-SiC interface is important in providing the framework for a systematic approach to contact properties. The published works in this area may be divided into two categories, based on the nature of the SiC substrate. In one group are those in which the initial SiC surface is well-characterized in stoichiometry and well-ordered in crystallographic structure. This type of experiment usually involves a final surface preparation in UHV. In the second group are those studies involving so-called 'practical' surfaces in which the substrate is prepared either by ex-situ chemical treatment or by in-situ ion bombardment prior to metallization. As discussed above, such surfaces are usually disordered and/or contaminated (e.g., by oxygen or free carbon) and exhibit a non-stoichiometric surface C/Si ratio. For those metals for which both types of substrates have been studied, the results will be seen to be quite different in many respects. Each class is relevant in its own right, the former providing information on an ideal system under controlled conditions and the latter being more closely related to actual contact fabrication.

This section will focus almost exclusively on experimental and theoretical work dealing with the microscopic chemical and physical properties of metal-SiC interfaces. By this is meant those processes occurring during deposition of the first few monolayers of metal and during subsequent thermal and chemical treatment. Excluded are most studies of so-called 'solid-state' reactions, involving macroscopically thick layers of metal and SiC, which deal mainly with the mechanical and crystallographic properties of the interface. Also omitted are those works dealing only with electrical properties and providing little or no other interface characterization. These are discussed in other Chapters.

5.1 Surface structure and metallization of SiC

The presentation is arranged alphabetically according to metal and chronologically within each metal grouping. Since this Datareview is in the nature of an annotated bibliography, the various results in the literature are merely summarized succinctly and uncritically. At the beginning of each section one will find statements such as 'Metal X forms a silicide and a carbide' or 'Metal X forms a silicide but not a carbide'. This is intended only as a rough guide to the relevant metal-Si and metal-C chemistries which are often, in fact, quite complex. Reviews of the composition and properties of metal silicides have been given by Murarka [75,76] and by Schlesinger [77]. Similar information for metal carbides is given by Kosolapova [78]. Finally, TABLE 2 summarizes the range of experimental results covered in this section.

TABLE 2 Summary of experimental studies of metal/SiC interfaces[a].

Metal	Silicide	Carbide	'Practical'	'UHV'
Aluminium	No	Yes	α [79-81] β [8,84]	β [82]
Chromium	Yes	Yes		β [8]
Copper	Yes	No		α [49,51]
Gold	No	No	β [87]	β [88,89]
Iron	Yes	Yes	β [91,92]	β [93,94]
Manganese	Yes	Yes	α [95]	
Molybdenum	Yes	Yes	α [97] β [96,97]	
Nickel	Yes	Yes	α [98(?)[b],99,100]	β [101]
Palladium	Yes	No	α [98(?)[b],102,103] β [103]	
Platinum	Yes	No		α [104] β [104]
Silver	No	No	β [105]	
Tantalum	Yes	Yes	β [96]	
Tin	Yes	No	β [106]	
Titanium	Yes	Yes	α [107,110(?)[b],111] β [108,109]	β [42]
Tungsten	Yes	Yes	β [96]	

(a) The 'silicide' and 'carbide' columns, respectively, indicate whether a given metal forms bulk silicides or carbides which are stable at room temperature. The different phases have not necessarily all been observed at the corresponding metal/SiC interface. 'Practical' and 'UHV' refer to the SiC surface preparation (see text). The designations 'α' and 'β' refer, respectively, to hexagonal α-SiC and cubic β-SiC surfaces, and the numbers are the corresponding literature references.

(b) For [98], the SiC lattice type was not specified. For [110], polycrystalline SiC whiskers were used.

F   RESULTS FOR SPECIFIC METAL-SILICON CARBIDE INTERFACES

F1   Aluminium

Al forms a carbide but not a silicide. Bermudez [79] reported AES and EELS data for Al deposited on Ar$^+$-ion bombarded α-SiC near 25 °C. Metal islands are formed which, upon annealing at ~600 °C, cluster in C-rich areas of the surface. At higher temperature, reaction with free C occurs to form Al$_4$C$_3$. Porte [80] performed core-level and valence-band XPS on similar samples and found that features associated with metallic Al disappear after annealing at about 600 °C as the Al intermixes with the substrate and reacts to form the carbide. Bellina et al [8] deposited Al on ion-bombarded β-SiC but did not observe any clear indication of carbide formation after subsequent annealing. Yasuda et al [81] deposited Al on chemically etched and cleaned Si- and C-terminated α-SiC. Both surfaces show O contamination, with a higher level on the latter. Auger sputter profiling shows that O remains at the interface but, after a 900 °C anneal, diffuses toward the Al surface. I-V and C-V measurements were also reported which indicate that contacts formed on the Si-terminated surface are more thermally-stable than those on the C-face. Bermudez [82] reported XPS, EELS and LEED results for Al on well-characterized SiC(100)-c(4x2) and -(2x1) surfaces. In this case the interface is relatively sharp, with essentially no Al carbide formation. Apparently the Al$_4$C$_3$ observed after annealing Al on 'practical' SiC surfaces arises from reaction with free C or with broken Si-C bonds rather than with the carbidic C in the SiC lattice. The Al layer growth mode is of the 'Stranski-Krastonov' type, with the first few monolayers growing uniformly and subsequent deposition forming islands. In the final stages of thermal desorption of the deposited layer, ordering of the remaining Al is observed in LEED. Li et al [83] performed quantum chemical cluster calculations dealing with the adhesion of Al to various SiC surfaces. Depending on the SiC surface orientation and on the Al adsorption site, the bond strength with Al can be as much as 50 % higher than the Al-Al bond strength. Maruyama and Ohuchi [84] performed XPS studies of the effects of H$_2$O on the Al-SiC interface. The initial surfaces were β-SiC cleaned by Ar$^+$-ion bombardment at 800 °C. Annealing the Al/SiC sample at 600 °C results in carbide formation and the release of free Si, and both processes are greatly enhanced for samples exposed to H$_2$O vapour before annealing. It was suggested that reaction with H$_2$O weakens Si-C bonds. Wenchang et al [85] used extended Hückel theory to treat the structure of the Al/SiC(100) interface. The interface is found to be sharp, with Al adsorption at bridging sites favoured for both the Si- and C-terminated surfaces. The Al-C interaction is found to be stronger than that between Al and Si and to become more so with increasing Al coverage.

F2   Chromium

Cr forms a silicide and a carbide. Bellina et al [8], using LEED and AES, studied Cr on a β-SiC(100)-c(2x2) surface and on a surface disordered by ion bombardment. On the former, Cr desorbs during annealing with no evidence of reaction. On the damaged surface, Cr carbide is observed following anneals in the 600-750 °C range. Higher temperature leads to desorption of the Cr carbide, leaving an apparently clean and well-ordered SiC(100)-(2x1) surface.

F3 Copper

Cu forms a silicide but not a carbide. Hatanaka et al [49] reported AES, LEED and sputter-profiling results for Cu on clean and well-ordered Si- and C-terminated α-SiC(0001) surfaces. Before Cu deposition the samples exhibited, respectively, ($\sqrt{3}$ x $\sqrt{3}$)R30° and (3x3) LEED patterns. Nishimori et al [51] performed similar experiments for α-SiC(0001) surfaces cut 5° off-axis toward the [11$\bar{2}$0] direction. The samples were chemically cleaned then annealed at 1000 °C in UHV. Before Cu deposition, the off-axis samples were clean and exhibited (1x1) LEED patterns. All three types of surfaces [49,51] give the same results for Cu deposition. No interfacial reaction occurs near 25 °C. After a 250 °C anneal, a thin Cu silicide film appears at the surface of the thick (~ 120 Å) layer of metal and at the SiC interface. After the 250 °C anneal a (1x1) LEED pattern is observed which (surprisingly) persists throughout the sputter profiling, even to the point where Cu is no longer detectable in AES. Annealing at 1000 °C then restores the clean-surface LEED pattern. After a 500 °C anneal a thick layer of intermixed Cu and SiC is seen with no indication of Cu silicide. Anderson and Ravimohan [86] performed molecular orbital calculations for the adsorption of Cu on the Si- and C-terminated (0001) surfaces of α-SiC. The Cu-Si bond is found to be stronger than the Cu-C bond.

F4 Gold

Au forms neither a silicide nor a carbide. Mizokawa et al [87] reported AES and XPS sputter-profiling results for Au on chemically-cleaned β-SiC. The interface is sharp compared to that of Au/Si. Hatanaka et al [49] and Nishimori et al [51] studied Au on various α-SiC surfaces as described above in connection with their results for Cu. Considerable intermixing of Au and Si resulted from a 200 °C anneal. Parrill and Chung [88,89] performed an extensive series of AES, EELS, LEED, UPS and XPS for ordered and well-characterized SiC(100)-(3x2), -(2x1) and -c(2x2) surfaces. On a Si-terminated surface an interfacial alloy forms near 25 °C, with a thickness dependent on the surface Si content. Some Si also out-diffuses to the Au surface. For Si-deficient surfaces, Au forms clusters at 25 °C as on a graphite substrate. Interfacial C blocks the Au-Si interaction. The structure of the Au-SiC interface is thus a strong function of the composition of the outermost few monolayers of the initial substrate. Wenchang et al [90] used extended Hückel theory to treat the adsorption of Au on β-SiC(111) surfaces. In general agreement with the experimental results of Parrill and Chung [88,89], the Au-Si interaction is found to be stronger than that between Au and C. Intermixing between Au and β-SiC is found to be possible on a Si- but not on a C-terminated surface. The calculations also predict the existence of surface states on both the clean and Au-adsorbed surfaces.

F5 Iron

Fe forms a silicide and a carbide. Kaplan et al [91] used AES and EELS to study Fe films grown by Fe(CO)$_5$ decomposition. The substrate was β-SiC cleaned by Kr$^+$-ion bombardment at elevated temperature which gives a disordered but relatively stoichiometric surface. Intermixing at the interface occurs at temperatures as low as 190 °C, with higher temperatures causing extensive intermixing and formation of both Fe silicide and Fe carbide. Geib et al

[92] reported AES and SEM results for Fe films sputter-deposited on chemically-cleaned β-SiC. An Fe carbide layer formed at the interface, due to C contamination incurred during Fe deposition, and prevented reaction with the SiC at temperatures up to 700 °C. At 800 °C a 'violent reaction' occurs resulting in a rough surface layer comprised of mixed Fe silicide and Fe carbide. Mizokawa et al [93] performed AES experiments for thin Fe films vapour-deposited on (100)-c(2x2) surfaces. Near 25 °C the first monolayer grows as a uniform, disordered film. With increasing Fe thickness, the interface develops into a layer of Fe silicide adjacent to the SiC with another region of mixed Fe carbide and graphite between the silicide and the unreacted metal. Annealing at ~700 °C causes decomposition of the Fe carbide, rapid in-diffusion of Fe and accumulation of graphitic C at the surface. Above ~1000 °C Fe completely dissolves, leaving a mixture of SiC and free Si and C at the surface. Mizokawa et al [94] have also done a similar study for much thicker Fe films. Annealing at 250 °C gives a mixture of Fe silicide and Fe carbide at the interface. Carbon also diffuses along grain boundaries in the film, forming Fe carbide at the surface. Annealing at 540 °C causes accumulation of Fe carbide and free Si at the interface and also segregation of Si to the surface where a thin layer of mixed Fe silicide, Fe carbide and free C is formed. The Fe carbide at the interface appears to act as a diffusion barrier inhibiting more complete reaction of the Fe film. Grain boundaries in the thick Fe film are also found to be important in providing reactive surfaces and diffusion channels.

F6 Manganese

Mn forms a silicide and a carbide. Okajima and Miyazaki [95] used AES, XPS and X-ray diffraction (XRD) to study Mn vapour-deposited on samples formed from hot-pressed α-SiC powder. Annealing at 450 °C causes out-diffusion of C which reacts to form Mn carbide. The Si remaining at the interfaces intermixes with the overlayer at about 550 °C to form Mn silicide.

F7 Molybdenum

Mo forms a silicide and a carbide. Geib et al [96] reported Auger sputter-profiling results for Mo sputter-deposited on chemically-cleaned β-SiC. Thermodynamic considerations indicate that reaction of Mo with SiC should not be favourable. Other than a thin reacted layer at the immediate interface, there is no indication of reactive intermixing even after an 800 °C anneal. Hara et al [97] reported AES, Rutherford backscattering (RBS), XRD and transmission electron microscopy (TEM) measurements on Mo layers formed either by plasma deposition or by evaporation on chemically-cleaned α- and β-SiC. Prolonged (1 hr) anneals at 1200 °C give an outer layer of Mo carbide on top of a layer of Mo silicide.

F8 Nickel

Ni forms a silicide and a carbide. Pai et al [98] performed MeV ion backscattering measurements for Ni vapour-deposited on chemically cleaned SiC substrates of unspecified lattice type. A Ni-rich silicide phase is observed after a 500 °C anneal which, at higher temperature, tends toward Ni_2Si. By contrast, $NiSi_2$ is the stable phase formed for Ni on bulk

Si. Ohdomari et al [99] performed similar experiments, together with AES and XRD, for Ni on chemically-cleaned Si- and C-terminated faces of α-SiC(0001). Annealing at 600 °C in forming gas (a mixture of H_2 and N_2) gives a Ni_2Si layer with dissolved, unreacted C. Höchst et al [100] used core-level and valence-band XPS to study Ni deposited on β-SiC cleaned by ion bombardment. Interfacial reaction occurs at 25 °C to give a Si-rich silicide, believed to be NiSi. For a Ni thickness of more than a few monolayers, the surface film exhibits XPS features associated with bulk, metallic Ni. The Ni layer is stable at temperatures up to 600 °C, above which a mixture of SiC and Ni_2Si is formed. Slijkerman et al [101] performed LEED and RBS measurements for Ni vapour-deposited on clean and well-ordered (100)-c(2x2) surfaces. No intermixing is observed at 25 °C. Annealing at 300 to 600 °C gives a disordered layer of Ni_2Si. The free C released as a result of silicide formation does not react with the Ni but diffuses to the surface of the silicide layer.

F9 Palladium

Pd forms a silicide but not a carbide. Bermudez [102] reported AES and EELS data for Pd on ion-bombarded α-SiC. The different stable silicides (Pd_4Si, Pd_3Si and Pd_2Si) can be distinguished by the Si $L_{2,3}VV$ lineshapes. Whereas Pd_2Si is formed by reaction with bulk Si, Pd deposited on ion-bombarded α-SiC near 25 °C forms a mixture of Pd_4Si and Pd_3Si. Annealing increases the Si content somewhat, leading to a mixed Pd_3Si+Pd_2Si phase. Sputter profiling indicates a build-up of unreacted C at the interface which presumably acts as a diffusion barrier. Pai et al [98] performed MeV ion backscattering measurements for Pd vapour-deposited on chemically cleaned SiC substrates of unspecified lattice type. Only Pd_3Si was seen after a 500 °C anneal, which gradually converted to Pd_2Si with annealing at 900 °C. Nakamura et al [103] have examined Pd on α- and β-SiC whiskers using atom-probe field ion microscopy and observed formation of a silicide, of unspecified composition, after a 600 °C anneal.

F10 Platinum

Pt forms a silicide but not a carbide. Anderson and Ravimohan [86] performed molecular orbital calculations for the adsorption of Pt on the Si- and C-terminated (0001) surfaces of α-SiC. The Pt-Si bond is found to be stronger than the Pt-C bond. Bermudez and Kaplan [104] reported AES and LEED results for Pt on ordered and well-characterized α-SiC(0001), β-SiC(100) and β-SiC(111) surfaces. On (100)-(2x1), deposition of a few monolayers of Pt near 25 °C leads to a uniform layer with some indication of intermixing. Annealing at ≤ 1000 °C causes aggregation of the Pt into islands which, at higher temperature, react to form a mixture of Pt_2Si and PtSi and free, graphitic C. Vapour deposition of Si on this surface, followed by annealing, consumes the free C and regenerates SiC. A (2x2) LEED pattern also results which is believed to be due to formation of an ordered PtSi layer. The importance of impurity O in determining the structure of the interface was also examined. On the (111)-(3x3) surface, which is terminated in a Si bilayer, Pt behaves very differently. Pt silicide is seen even before annealing, but no free C is observed, suggesting that, under these conditions, Pt does not react with the SiC lattice itself. With further annealing, part of the Pt Auger intensity is lost, presumably due to Pt in-diffusion, leaving behind a thin disordered Pt silicide surface layer on an apparently intact SiC substrate. On the Si-terminated

(0001)-(√3 x √3)R30° surface, Pt behaves in much the same manner as on the (100) and (111) surfaces of Si. Annealing a thin layer leads to the in-diffusion of almost all the Pt and the appearance of a (2√3 x 2√3)R30° superstructure in LEED. One conclusion of this work is that interfacial chemistry can be controlled by adjusting the stoichiometry and structure of the initial surface.

F11   Silver

Ag does not form a silicide or a carbide. Niles et al [105] studied Ag on ion-bombarded β-SiC using core-level and valence-band photoemission. Three-dimensional island growth is observed during deposition near 25 °C. The interface is non-reactive and exhibits ohmic behaviour. Annealing in the 400-600 °C range leads to coalescence of the Ag, with bare SiC between clusters, and higher temperatures lead to evaporation of essentially all the metal.

F12   Tantalum

Ta forms a silicide and a carbide. Geib et al [96] reported Auger sputter-profiling results for Ta sputter-deposited on chemically-cleaned β-SiC. As expected from thermodynamic considerations, Ta reacts during annealing, forming a macroscopically rough layer of intermixed Ta carbide and Ta silicide.

F13   Tin

Sn forms a silicide but not a carbide. Niles et al [106] have used core-level and valence-band XPS to study Sn on ion-bombarded β-SiC. The first monolayer grows as a uniform layer of α-Sn, while subsequent layers form clusters of β-Sn. Annealing in the 400-1000 °C range leads to out-diffusion of Si into the Sn layer with formation of Sn silicide. The reacted surface layer withstands anneals at as high as 1000 °C without desorption of Sn.

F14   Titanium

Ti forms a silicide and a carbide. As an historical point, the first published Auger study of a metal-SiC interface was that of Chamberlain [107] for Ti on ion-bombarded α-SiC. Thick Ti films (~350 nm) were deposited by electron-beam evaporation and sputter-profiled after different anneals. Carbon appears to mix more readily with the Ti than does Si, leading to Ti carbide formation with no clear indication of Ti silicide. Taubenblatt and Helms [108] deposited Ti on thin SiC films grown in-situ by exposing (100) Si wafers to C_2H_4. The interest was in determining the effect of a thin SiC contamination layer on the Ti/Si interaction. Near 25 °C, TiC is formed at the interface with no indication of silicide. Annealing in the 300-500 °C range gives a very rough interface. Si penetrates the mixed SiC and TiC interface at defect sites and reacts rapidly with Ti. Up to ~500 °C, TiC grows at the interface. At higher temperature, TiC decomposes resulting in more silicide and a regeneration of SiC at the interface. The thermodynamics of interface formation was also discussed. Bellina and Zeller [109] used LEED and AES to study Ti films deposited on ion-bombarded β-SiC surfaces which were either Si- or C-rich, depending on subsequent thermal treatment. Ti reacts with SiC, forming TiC and free Si which diffuses to the TiC surface. Free C at the SiC surface readily forms TiC which impedes subsequent Ti-SiC reaction. Annealing Ti/SiC at 750°C gives a (1x3) LEED pattern, believed to result from formation of an ordered Ti silicide. Hasegawa et al [42] reported AES and LEED data for Ti on SiC(100)-c(2x2). Near 250°C, Ti bonds with the C in the c(2x2) termination layer without breaking C-Si backbonds. Annealing at 900°C causes SiC decomposition and growth of TiC islands. Tan et al [110] used AES and XPS to study Ti on the surfaces of SiC-coated boron fibres. In general agreement with the results of Bellina and Zeller [109], fibres with a high coverage of surface C react to form TiC but little Ti silicide, whereas those with less excess C give an intermixed layer of Ti silicide and carbide. Spellman et al [111] reported LEED, TEM, RBS and XPS results for the epitaxial growth of Ti on chemically-cleaned α-SiC(0001). A (1x1) LEED pattern, similar to that for the substrate, is found for the deposited films. Strain resulting from the 4% lattice mismatch is partly relieved through creation of dislocations. Anderson and Ravimohan [86] performed molecular orbital calculations for the adsorption of Ti on the Si- and C-terminated (0001) surfaces of α-SiC. The Ti-C bond is found to be stronger than the Ti-Si bond. Mehandru and Anderson [112] performed similar calculations for Ti adsorption on both ideally-terminated and relaxed β-SiC(100) and (111) surfaces. On the unrelaxed (100) surfaces, strong Ti-Si and Ti-C interactions are found due to the presence of dangling Si and C bonds. Dimer pairing on the SiC surface reduces the bonding interactions with Ti.

F15 Tungsten

W forms a silicide and a carbide. Geib et al [96] reported Auger sputter-profiling results for W sputter-deposited on chemically-cleaned β-SiC. Thermodynamic considerations indicate that reaction of W with SiC should not be favourable. Other than a thin reacted layer at the immediate interface, there is no indication of reactive intermixing even after an 800°C anneal.

G CONCLUSION

This Datareview has described the known surface phases which exist on both α- and β-SiC. Surface treatments by annealing in UHV, by ion bombardment and by laser irradiation are not suitable to prepare SiC surfaces for further study. Chemical reduction of surface oxides is the preferred route to surface preparation, particularly using a Si flux at temperatures <1000°C. A distinction is drawn between 'ideal' surfaces prepared in UHV and 'practical' ones where substrates are chemically treated or ion bombarded prior to metallisation. Processes occurring during deposition of the first few monolayers of metal and subsequent treatments are discussed in terms of chemical and physical properties. A total of 15 metal-SiC combinations are reviewed and discussed in terms of silicide and carbide formation.

REFERENCES

[1] A.J. Van Bommel, J.E. Crombeen, A. Van Tooren [*Surf. Sci. (Netherlands)* vol.48 no.2 (1975) p.463-72]

[2] M. Dayan [*J. Vac. Sci. Technol. A (USA)* vol.3 no.2 (1985) p.361-6]

[3] M. Dayan [*J. Vac. Sci. Technol. A (USA)* vol.4 no.1 (1986) p.38-45]

[4] F. Meyer, G.J. Loyen [*Acta Electron. (France)* vol.18 no.1 (1975) p.33-8]

[5] K. Miyoshi, D. Buckley [*Appl. Surf. Sci. (Netherlands)* vol.10 no.3 (1982) p.357-76]

[6] S. Adachi, M. Mohri, T. Yamashina [*Surf. Sci. (Netherlands)* vol.161 no.2/3 (1985) p.479-90]

[7] J.J. Bellina Jr., M.V. Zeller [*Appl. Surf. Sci. (Netherlands)* vol.25 no.4 (1986) p.380-90]

[8] J.J. Bellina Jr., J. Ferrante, M.V. Zeller [*J. Vac. Sci. Technol. A (USA)* vol.4 no.3 (1986) p.1692-5]

[9] L. Muehlhoff, W.J. Choyke, M.J. Bozack, J.T. Yates Jr. [*J. Appl. Phys. (USA)* vol.60 no.8 (1986) p.2842-53]

[10] D.R. Wheeler, S.V. Pepper [*Surf. Interface Anal. (UK)* vol.10 no.2/3 (1987) p.153-62]

[11] R. Kaplan [*J. Appl. Phys. (USA)* vol.56 no.6 (1984) p.1636-41]

[12] P.E. Pehrsson, R. Kaplan [*J. Mater. Res. (USA)* vol.4 no.6 (1989) p.1480-90]

[13] S.Y. Chou, Y. Chang, K.H. Weiner, T.W. Sigmon, J.D. Parsons [*Appl. Phys. Lett. (USA)* vol.56 no.6 (1990) p.530-2]

[14] R. Kaplan, T.M. Parrill [*Surf. Sci. (Netherlands)* vol.165 no.2/3 (1986) p.L45-52]

[15] R. Kaplan [*Surf. Sci. (Netherlands)* vol.215 no.1/2 (1989) p.111-34]

[16] T.M. Parrill, Y.W. Chung [*Surf. Sci. (Netherlands)* vol.243 no.1-3 (1991) p.96-112]

[17] V.M. Bermudez, R. Kaplan [*Phys. Rev. B (USA)* vol.44 no.20 (1991) p.11149-58]

[18] F. Bozso, J.T. Yates Jr., W.J. Choyke, L. Muehlhoff [*J. Appl. Phys. (USA)* vol.57 no.8 (1985) p.2771-8]

[19] M.J. Bozack, L. Muehlhoff, J.N. Russell Jr., W.J. Choyke, J.T. Yates Jr. [*J. Vac. Sci. Technol. A (USA)* vol.5 no.1 (1987) p.1-8]

[20] H. Matsunami, M. Nakayama, T. Yoshinobu, H. Shiomi, T. Fuyuki [*Springer Proc. Phys. (Germany)* vol.43 (1989) p.157-61]

[21] C.D. Stinespring, J.C. Wormhoudt [*J. Appl. Phys. (USA)* vol.65 no.4 (1989) p.1733-42]

[22] P.A. Taylor, M. Bozack, W.J. Choyke, J.T. Yates Jr. [*J. Appl. Phys. (USA)* vol.65 no.3 (1989) p.1099-105]

[23] T. Takai, T. Halicioğlu, W.A. Tiller [*Surf. Sci. (Netherlands)* vol.164 no.2/3 (1985) p.327-40]

[24] Y. Ohshita [*J. Cryst. Growth (Netherlands)* vol.110 no.3 (1991) p.516-22]

[25] B.I. Craig, P.V. Smith [*Surf. Sci. (Netherlands)* - in press]

[26] S. Hara et al [*Surf. Sci. (Netherlands)* vol.231 no.3 (1990) p.L196-200]

[27] T. Yoshinobu, I. Izumikawa, H. Mitsui, T. Fuyuki [*Appl. Phys. Lett. (USA)* vol.59 no.22 (1991) p.2844-6]

[28] J.M. Powers, A. Wander, P.J. Rous, M.A. Van Hove, G.A. Somorjai [*Phys. Rev. B (USA)* vol.44 no.20 (1991) p.11159-66]

[29] J.M. Powers, A. Wander, M.A. Van Hove, G.A. Somorjai [*Surf. Sci. (Netherlands)* vol.260 no.1-3 (1992) p.L7-10]

[30] J. Ihm, D.H. Lee, J.D. Joannopoulos, J.J. Xiong [*Phys. Rev. Lett. (USA)* vol.51 no.20 (1983) p.1872-5]

[31] A. Saxena, E.T. Gawlinski, J.D. Gunton [*Surf. Sci. (Netherlands)* vol.160 no.2 (1985) p.618-40]

[32] T. Tabata, T. Aruga, Y. Murata [*Surf. Sci. (Netherlands)* vol.179 no.1 (1987) p.L63-70]

[33] S.D. Kevan [*Phys. Rev. B (USA)* vol.32 no.4 (1985) p.2344-50]

[34] B.I. Craig, P.V. Smith [*Surf. Sci. (Netherlands)* vol.233 no.3 (1990) p.255-60]

[35] B.I. Craig, P.V. Smith [*Springer Ser. Surf. Sci. (Germany)* vol.24 (1991) p.545-9]

[36] J.N. Carter [*Solid State Commun. (USA)* vol.72 no.7 (1989) p.671-4]

[37] S.P. Mehandru, A.B. Anderson [*Phys. Rev. B (USA)* vol.42 no.14 (1990) p.9040-9]

[38] L. Wenchang, Y. Weidong, Z. Kaiming [*J. Phys. Condens. Matter (UK)* vol.3 no.46 (1991) p.9079-86]

[39] C.-S. Chang, N.-J. Zheng, I.S.T. Tsong, Y.-C. Wang, R.F. Davis [*J. Am. Ceram. Soc. (USA)* vol.73 no.11 (1990) p.3264-8]

[40] C.-S. Chang, N.-J. Zheng, I.S.T. Tsong, Y.C. Wang, R.F. Davis [*J. Vac. Sci. Technol. B (USA)* vol.9 no.2 (1991) p.681-4]

[41] A.J. Steckl, S.A. Mogren, M.W. Roth, J.P. Li [*Appl. Phys. Lett. (USA)* vol.60 no.12 (1992) p.1495-7]

[42] S. Hasegawa, S. Nakamura, N. Kawamoto, H. Kishibe, Y. Mizokawa [*Surf. Sci. (Netherlands)* vol.206 no.1/2 (1988) p.L851-6]

[43] T. Takai, T. Halicioğlu, W.A. Tiller [*Surf. Sci. (Netherlands)* vol.164 no.2/3 (1985) p.341-52]

[44] V.M. Bermudez [*J. Appl. Phys. (USA)* vol.66 no.12 (1989) p.6084-92]

[45] B.I. Craig, P.V. Smith [*Surf. Sci. (Netherlands)* vol.256 no.1/2 (1991) p.L609-12]

[46] P. Badziag [*Phys. Rev. B (USA)* vol.44 no.20 (1991) p.11143-8]

[47] P. Badziag [*Surf. Sci. (Netherlands)* vol.269/270 (1992) p.1152-6]

[48] K.C. Pandey [*Proc. 17th Int. Conf. on the Phys. of Semicond.* Eds D.J. Chadi, W.A. Harrison (Springer, N.Y., USA, 1985) p.55-8]

[49] H. Hatanaka, H. Tokutaka, K. Nishimori, N. Ishihara [*J. Vac. Soc. Jpn. (Japan)* vol.30 no.5 (1987) p.265-9]

[50] S. Nakanishi, H. Tokutaka, K. Nishimori, S. Kishida, N. Ishihara [*Appl. Surf. Sci. (Netherlands)* vol.41/42 no.1-4 (1989) p.44-8]

[51] K. Nishimori, H. Tokutaka, S. Nakanishi, S. Kishida, N. Ishihara [*Jpn. J. Appl. Phys. (Japan)* vol.28 no.8 (1989) p.L1345-8]

[52] K. Takayanagi, Y. Tanishiro, S. Takahashi, M. Takahashi [*Surf. Sci. (Netherlands)* vol.164 no.2/3 (1985) p.367-92]

[53] C.S. Chang, I.S.T. Tsong, Y.C. Wang, R.F.Davis [*Surf. Sci. (Netherlands)* vol.256 no.3 (1991) p.354-60]

[54] M.-H. Tsai, C.S. Chang, J.D. Dow, I.S.T. Tsong [*Phys. Rev. B (USA)* vol.45 no.3 (1992) p.1327-32]

[55] P. Badziag [*Surf. Sci. (Netherlands)* vol.236 no.1/2 (1990) p.48-52]

[56] P. Badziag, M.A. Van Hove, G.A. Somorjai [*Springer Ser. Surf. Sci. (Germany)* vol.24 (1991) p.550-4]

[57] Y. Mizokawa, T. Miyasato, S. Nakamura, K.M. Geib, C.W. Wilmsen [*Surf. Sci. (Netherlands)* vol.182 no.3 (1987) p.431-8]

[58] P. Morgen, K.L. Seaward, T.W. Barbee Jr. [*J. Vac. Sci. Technol. A (USA)* vol.3 no.6 (1985) p.2108-15]

[59] F. Bozso, L. Muehloff, M. Trenary, W.J. Choyke, J.T. Yates Jr. [*J. Vac. Sci. Technol. A (USA)* vol.2 no.3 (1984) p.1271-4]

[60] B. Jorgensen, P. Morgen [*Surf. Interface Anal. (UK)* vol.16 no.1-12 (1990) p.199-202]

[61] C. Jardin, H. Khalaf, M. Ghamnia [*J. Microsc. Spectrosc. Electron. (France)* vol.13 no.1 (1988) p.65-74]

[62] J.T. Grant, T.W. Haas [*Phys. Lett. (Netherlands)* vol.33A no.6 (1970) p.386-7]

[63] J.E. Rowe, S.B. Christman [*Solid State Commun. (USA)* vol.13 no.3 (1973) p.315-8]

[64] R. Weissmann, W. Schnellhammer, R. Koschatzky, K. Muller [*Appl. Phys. (Germany)* vol.14 no.3 (1977) p.283-7]

[65] A.J. Nelson et al [*J. Vac. Sci. Technol. A (USA)* vol.8 no.3 (1990) p.1538-43]

[66] K. Nishimori, H. Tokutaka, Z. Okinaga, S. Kishida, N. Ishihara [*Appl. Surf. Sci. (Netherlands)* vol.41/42 no.1-4 (1989) p.49-52]

[67] S.V. Didziulis, J.R. Lince, P.D. Fleischauer, J.A. Yarmoff [*Inorg. Chem. (USA)* vol.30 no.4 (1991) p.672-8]

[68] Yu.M. Tairov, V.F. Tsvetkov, Yu. Laukhe [*Sov. Phys.-Solid State (USA)* vol.19 no.12 (1977) p.2134-5]

[69] H. Hoechst, M. Tang, B.C. Johnson, J.M. Meese, G.W. Zajac, T.H. Fleisch [*J. Vac. Sci. Technol. A (USA)* vol.5 no.4 (1987) p.1640-3]

[70] M. Iwami, M. Kusaka, M. Hirai, H. Nakamura, K. Shibahara, H. Matsunami [*Surf. Sci. (Netherlands)* vol.199 no.3 (1988) p.467-75]

[71] T.M. Parrill, V.M. Bermudez [*Solid State Commun. (USA)* vol.63 no.3 (1987) p.231-5]

[72] V.M. Bermudez, T.M. Parrill, R. Kaplan [*Surf. Sci. (Netherlands)* vol.173 no.1 (1986) p.234-44]

[73] B.I. Craig, P.V. Smith [*Physica B (Netherlands)* vol.170 no.1-4 (1991) p.518-22]

[74] B.I. Craig, P.V. Smith [*Solid State Commun. (USA)* vol.81 no.7 (1992) p.623-6]

[75] S.P. Murarka [*J. Vac. Sci. Technol. A (USA)* vol.17 no.4 (1980) p.775-92]

[76] S.P. Murarka [*Annu. Rev. Mater. Sci. (USA)* vol.13 (1983) p.117-37]

[77] M.E. Schlesinger [*Chem. Rev. (USA)* vol.90 no.4 (1990) p.607-28]

[78] T.Ya. Kosolapova [*Carbides - Properties, Production and Applications* (Plenum, New York, 1971)]

[79] V.M. Bermudez [*Appl. Phys. Lett. (USA)* vol.42 no.1 (1983) p.70-2]

[80] L. Porte [*J. Appl. Phys. (USA)* vol.60 no.2 (1986) p.635-8]

[81] K. Yasuda, T. Hayakawa, M. Saji [*IEEE Trans. Electron Devices (USA)* vol.ED-14 no.9 (1987) p.2002-8]

[82] V.M. Bermudez [*J. Appl. Phys. (USA)* vol.63 no.10 (1988) p.4951-9]

[83] S. Li, R.J. Arsenault, P. Jena [*J. Appl. Phys. (USA)* vol.64 no.11 (1988) p.6246-53]

[84] B. Maruyama, F.S. Ohuchi [*J. Mater. Res. (USA)* vol.6 no.6 (1991) p.1131-4]

[85] L. Wenchang, Z. Kaiming, X. Xide [*Phys. Rev. B (USA)* vol.45 no.19 (1992) p.11048-53]

[86] A.B. Anderson, Ch. Ravimohan [*Phys. Rev. B (USA)* vol.38 no.2 (1988) p.974-7]

[87] Y. Mizokawa, K.M. Geib, C.W. Wilmsen [*J. Vac. Sci. Technol. A (USA)* vol.4 no.3 (1986) p.1696-1700]

[88] T.M. Parrill, Y.W. Chung [*J. Vac. Sci. Technol. A (USA)* vol.6 no.3 (1988) p.1589-92]

[89] T.M. Parrill, Y.W. Chung [*Surf. Sci. (Netherlands)* vol.271 no.3 (1992) p.395-406]

[90] L. Wenchang, Y. Ling, Z. Kaiming [*Chin. Phys. Lett. (China)* vol.9 no.2 (1992) p.101-4]

[91] R. Kaplan, P.H. Klein, A. Addamiano [*J. Appl. Phys. (USA)* vol.58 no.1 (1985) p.321-6]

[92] K.M. Geib, C.W. Wilmsen, J.E. Mahan, M.C. Bost [*J. Appl. Phys. (USA)* vol.61 no.12 (1987) p.5299-302]

[93] Y. Mizokawa, S. Nakanishi, S. Miyase [*Jpn. J. Appl. Phys. (Japan)* vol.28 no.12 (1989) p.2570-5]

[94] Y. Mizokawa, S. Nakanishi, S. Miyase [*Jpn. J. Appl. Phys. (Japan)* vol.28 no.12 (1989) p.2576-80]

[95] Y. Okajima, K. Miyazaki [*Jpn. J. Appl. Phys. (Japan)* vol.24 no.8 (1985) p.940-3]

[96] K.M. Geib, C. Wilson, R.G. Long, C.W. Wilmsen [*J. Appl. Phys. (USA)* vol.68 no.6 (1990) p.2796]

[97] S. Hara et al [*Jpn. J. Appl. Phys. (Japan)* vol.29 no.3 (1990) p.L394-7]

[98] C.S. Pai, C.M. Hanson, S.S. Lau [*J. Appl. Phys. (USA)* vol.57 no.2 (1985) p.618-9]

[99] I. Ohdomari, S. Sha, H. Aochi, T. Chikyow, S. Suzuki [*J. Appl. Phys. (USA)* vol.62 no.9 (1987) p.3747-50]

[100] H. Höchst, D.W. Niles, G.W. Zajac, T.H. Fleisch, B.C. Johnson, J.M. Meese [*J. Vac. Sci. Technol. B (USA)* vol.6 no.4 (1988) p.1320-5]

[101] W.F.J. Slijkerman, A.E.M.J. Fischer, J.F. van der Veen, I. Ohdomari, S. Yoshida, S. Misawa [*J. Appl. Phys. (USA)* vol.66 no.2 (1989) p.666-73]

[102] V.M. Bermudez [*Appl. Surf. Sci. (Netherlands)* vol.17 no.1 (1983) p.12-22]

[103] S. Nakamura, Y. Hasegawa, T. Hashizume, T. Sakurai [*J. Phys. Colloq. (France)* vol.47 no.11 (1986) p.C7-309-14]

[104] V.M. Bermudez, R. Kaplan [*J. Mater. Res. (USA)* vol.5 no.12 (1990) p.2882-93]

[105] D.W. Niles, H. Höchst, G.W. Zajac, T.H. Fleisch, B.C. Johnson, J.W. Meese [*J. Vac. Sci. Technol. A (USA)* vol.6 no.3 (1988) p.1584-8]

[106] D.W. Niles, H. Höchst, G.W. Zajac, T.H. Fleisch, B.C. Johnson, J.M. Meese [*J. Appl. Phys. (USA)* vol.65 no.2 (1989) p.662-7]

[107] M.B. Chamberlain [*Thin Solid Films (Switzerland)* vol.72 no.3 (1980) p.305-11]

[108] M.A. Taubenblatt, C.R. Helms [*J. Appl. Phys. (USA)* vol.59 no.6 (1986) p.1992-7]

[109] J.J. Bellina Jr., M.V. Zeller [*Mater. Res. Soc. Symp. Proc. (USA)* vol.97 no.1 (1987) p.265-70]

[110] B.J. Tan, L. Hwan, S.L. Suib, F.S. Galasso [*Chem. Mater. (USA)* vol.3 no.2 (1991) p.368-78]

[111] L.M. Spellman et al [*Mater. Res. Soc. Symp. Proc. (USA)* vol.221 no.1 (1991) p.99-104]

[112] S.P. Mehandru, A.B. Anderson [*Surf. Sci. (Netherlands)* vol.245 no.1/2 (1991) p.333-44]

5.2 Oxidation of SiC

J.J. Kopanski

October 1992[1]

A   INTRODUCTION

Thermal oxidation of the two most common forms of single-crystal silicon carbide with potential for semiconductor electronics applications is discussed: 3C-SiC formed by heteroepitaxial growth by chemical vapour deposition on silicon, and 6H-SiC wafers grown in bulk by vacuum sublimation or the Lely method. SiC is also an important ceramic and abrasive that exists in many different forms. Its oxidation has been studied under a wide variety of conditions. Thermal oxidation of SiC for semiconductor electronic applications is discussed in the following section. Insulating layers on SiC, other than thermal oxide, are discussed in Section C, and the electrical properties of the thermal oxide and metal-oxide-semiconductor capacitors formed on SiC are discussed in Section D.

B   THERMAL OXIDATION

For semiconductor electronic applications, thermal oxides on SiC are employed as a masking material for ion implantation and dry etching, as a gate insulator for field-effect devices, and as a surface passivation. Oxidation can also be used to etch the surface of SiC, as well as for polarity determination and for the delineation of defects and boundaries in SiC [1]. The slow oxidation rate of deposited SiC has been used for local oxidation inhibition of silicon [2].

Each of SiC's crystalline polytypes has a distinct oxidation rate under the same oxidation conditions [1,3,4]. For the various SiC polytypes, the oxidation rate on the (0001) Si faces increases with the decrease in the percentage of hexagonality of the SiC polytype, while the growth rate on the $(000\bar{1})$ C faces does not depend dramatically on polytype [3]. The dramatic difference in oxidation rates between opposite faces of the polar SiC crystal has long been known. Intermediate faces have intermediate oxidation rates [3]. As with other semiconductors, conduction type, dopant density, surface roughness and crystalline quality should also be expected to have an effect on the oxidation rate [5-7]. Selective oxidation at antiphase boundaries has been reported for wet oxidation of 3C-SiC heteroepitaxial layers, but not for dry oxidation [8-10].

The oxidation rates of the cubic (3C) [1,3,9,11-15] and hexagonal (6H or 4H) [3,4,16-23] polytypes have been studied. SiC has been oxidised with procedures very similar to those used in Si integrated circuit fabrication. Dry and wet oxidation have been employed in the typical open-ended, resistance-heated quartz tubes, at temperatures ranging from 850 to 1250°C. Since, in general, SiC oxidises at a much slower rate than Si, temperatures in excess of 1000°C are more practical.

---

[1] (Editor's note: at the time of going to press, June 1995, the area of knowledge reviewed here had not changed substantially since the cut-off date, October 1992.)

5.2 Oxidation of SiC

The composition of wet thermal oxide on SiC has been established to be close to stoichiometric SiO_2 from Auger electron spectroscopy sputter depth profiles and from the refractive index of the oxide determined by ellipsometry [4,12,14,19,21,24-27]. Dry oxide on 3C-SiC has been found to contain much more silicon than stoichiometric SiO_2 [27]. The carbon content of the SiC thermal oxide layer, away from the interface, determined by Auger spectroscopy, has been reported as: at below detection limits for wet oxide on 6H-SiC [25]; at below detection limits [24,26] and also at 2% for dry oxide on 3C-SiC [27]; and at 14% for wet oxide on 3C-SiC [27]. The oxide on SiC grown below 1200°C is amorphous, but above 1200°C the oxide grown is increasingly crystalline [15,18,20,22-24,28].

Several studies [1,10,12,14,16,19] have interpreted SiC thermal oxidation rates with respect to the Deal and Grove phenomenological model developed for Si [29]. However, the oxidation of the Si face of 6H-SiC does not conform to this model under the oxidation conditions examined [16,19]. According to the Deal and Grove model, the oxide thickness, X, as a function of oxidation time, t, follows a mixed linear and parabolic relationship:

$$X^2 + AX = B(t + \tau) \tag{1}$$

where τ takes into account any oxide which may form before the general relationship becomes valid (τ has been determined to be zero for wet and dry oxidation of SiC). For short oxidation times and thin oxide layers,

$$X \approx (B/A)t \tag{2}$$

B/A is called the linear rate constant. Once a thick oxide forms, the oxidation relationship reduces to a parabolic oxidation law,

$$X^2 = Bt \tag{3}$$

where B is called the parabolic rate constant. Both rate constants have a simple exponential dependence on temperature, T, described by an activation energy, E_a, of the form:

$$\sim Ce^{-E_a/kT} \tag{4}$$

where C is a pre-exponential constant and k is Boltzmann's constant. The activation energies of the linear and parabolic rate constants obtained from various studies are summarised in TABLE 1. The oxidation rate of SiC is dependent on the partial pressure of oxygen, and the activation energy of the parabolic rate constant has been found to be inversely proportional to the amount of O_2 present [10,15,22]. Unlike the oxidation of Si, the wet oxidation of 3C-SiC is slower than dry oxidation at temperatures below 1050°C and near atmospheric pressure.

5.2 Oxidation of SiC

TABLE 1 Activation energies of the linear and parabolic rate constants for the wet and dry oxidation of silicon carbide.

Material	Oxidation conditions	Linear RC (E_a kcal mole^{-1})	Parabolic RC (E_a kcal mole^{-1})	Ref
C face 6H-SiC $N_d \approx 2 \times 10^{17}$ cm^{-3}	8.3 cm^3 s^{-1} O$_2$ 970 - 1245 °C	85	47	[16]
(000$\bar{1}$)C 6H-SiC Lely method $N_d \approx 4 \times 10^{17}$ cm^{-3}	1.7 cm^3 s^{-1} O$_2$ bubbled through 95 °C water 850 - 1100 °C	26	48	[19]
CVD 3C-SiC $N_d \approx 2 \times 10^{17}$ - 2×10^{19} cm^{-3}	11.7 cm^3 s^{-1} O$_2$ 16.7 cm^3 s^{-1} H$_2$ 1000 - 1250 °C	74	50	[12]
CVD 3C-SiC $N_d \approx 2 \times 10^{15}$ - 5×10^{16} cm^{-3}	2.5 cm^3 s^{-1} O$_2$ 1000 - 1200 °C	58	34	[9]
"	1 cm^3 s^{-1} O$_2$ bubbled through 98 °C water 1000 - 1200 °C	60	127	[9]
"	1 cm^3 s^{-1} Ar bubbled through 98 °C water 1000 - 1200 °C	67	153	[9]

RC: Rate constant

There has been much speculation about the oxidation mechanisms of SiC (see for example [28]). Because of the variety of polytypes, degree of crystallinity, and varying concentrations of impurities of the SiC used in oxidation studies, as well as the effects of oxygen partial pressure and oxide crystallisation, no one model obviously explains the myriad SiC oxidation rate data. Because the measured activation energy of the parabolic rate constant is larger than the activation energy for the diffusion of either H$_2$O or O$_2$ in SiO$_2$, it often has been suggested that the wet oxidation mechanism of SiC is limited by inward diffusion of O$_2$ and the outward diffusion of CO [12,19,24]. Other recent studies have identified the diffusion of molecular and ionic oxygen as the rate controlling mechanisms [15,18,20,22,23]. The difference in composition between wet and dry oxide grown on 3C-SiC at 1100 °C has been used as evidence that wet and dry oxidation proceed by different mechanisms for 3C-SiC [27].

Oxidation of semiconductor SiC can result in redistribution of the dopants in the SiC crystal [30]. Boron and aluminium were found to be depleted from the surface of 6H-SiC during oxidation; nitrogen and phosphorus were found to pile up at the surface of 6H-SiC.

C OTHER INSULATING LAYERS ON SiC

A variety of insulating layers, other than thermal oxide, have been deposited on SiC and are summarised in TABLE 2. It has been speculated that certain deposited insulators may prove to have better dielectric strength and longer useful lifetime at temperatures greater than 300 °C than the thermally grown oxide layer on SiC. Deposited insulators are also assured of containing no residual carbon. Silicon nitride (Si_3N_4) and aluminium oxide (Al_2O_3) have been employed as sodium barriers and in memory device applications for silicon integrated circuits. The studies to date have been of insulators deposited on 3C-SiC grown on silicon. Insulators studied include: chemical vapour deposited (CVD) silicon dioxide [5,6], CVD silicon nitride [5,31], anodic aluminium oxide [31], and composites of thermal oxide and CVD silicon dioxide [5,6]. When used as a component of a metal-insulator-semiconductor (MIS) capacitor on SiC, deposited silicon dioxide, silicon nitride and aluminium oxide insulators have been shown to be of sufficient quality to deplete the SiC surface. However, SiC MIS capacitors with deposited insulators have yet to equal the low interface charge density of capacitors with thermal oxides. The quality of deposited insulators seen in these initial studies is encouraging and suggests that improvements in the electrical properties could be achieved with further research.

TABLE 2 Insulating layers (other than thermal oxide) which have been investigated for use in silicon carbide electronics.

Insulator	Comments	Ref
CVD silicon dioxide	Low resistivity and dielectric strength	[5,6]
CVD silicon nitride	Large slow trap density	[5]
Plasma-enhanced CVD silicon nitride	No change in capacitance	[31]
Anodic aluminium oxide	Successfully depleted surface	[31]
Deposited oxide/thermal oxide composite	Oxide charges comparable to thermal oxide	[5,6]

D ELECTRICAL PROPERTIES OF THERMAL OXIDE ON SiC

Significantly for eventual electronics applications, it has been demonstrated that metal-oxide-semiconductor (MOS) capacitors on SiC can be made to invert the surface of both 3C- [5-7,14,26,31-41] and 6H-SiC [19,21,42-46], while having low interface charge density. SiC MOS capacitors have been characterised by their capacitance-voltage (C-V), conductance-voltage (G-V), and current-voltage (I-V) responses. Fixed charge density can be deduced from the flatband voltage. Interface trapped charge densities have been determined by the high-frequency C-V technique [5,6,19,21,32,47], the conductance method [38,40,48], and from deep level transient spectroscopy (DLTS) [7].

Various electrical and optical properties that have been measured for the wet thermal oxide on silicon carbide are summarised in TABLE 3. In general, no systematic variation of any of the oxide properties has been reported with oxide growth temperature in the range of 1000 to 1250 °C. However, oxide grown with a wet process has been shown to have lower oxide charges than oxide grown in dry oxygen. Variations in the measured electrical properties between studies is most likely due to the quality of the SiC starting material and differences in the oxide growth processes.

TABLE 3 Electrical and optical properties of the thermal oxide on silicon carbide at room temperature from selected studies.

Property	Material	Value	Ref
Resistivity (Ω cm)	n-type 6H	2×10^{12}*	[19]
	n-type 3C	9×10^{15} $\approx 10^{16}$	[26] [40]
Dielectric strength (MV cm^{-1})	n-type 6H	2 >4**	[19] [46]
	n-type 3C	3 6 8	[26] [32] [41]
Fixed charge density ($\times 10^{11}$ cm^{-2})	n-type 6H	-20 to -50 <0.5	[19] [46]
	p-type 6H	15	[46]
	n-type 3C	4 to 7 4 to 9 2 to 4 ≈ 4	[32] [40] [41] [7]
	p-type 3C	10	[39]
Interfacial charge density at midgap ($\times 10^{11}$ cm^{-2} eV^{-1})	n-type 6H	20 <0.5	[19] [46]
	p-type 6H	30	[46]
	n-type 3C	4 0.5 to 2 0.5 to 2 0.1	[26] [32] [40] [7]
Oxide-SiC barrier height (eV)	n-type 3C	2.9 3.1	[40] [41]
Oxide refractive index	n-type 3C	1.49 ± 0.05	[12]
	n-type 6H	1.463 ± 0.003	[19]

* While resistivity is not explicitly stated in some later studies, C-V measurements are consistent with oxide resistivity on n-type 6H-SiC of greater than 10^{15} Ω cm [46].

** Estimated from maximum accumulation voltage and oxide thickness of C-V curve in [46].

There have been a number of studies of n-type 3C-SiC MOS capacitors [5-7,21,26,31,32,37,38,40,41]. Fixed charge density in the low 10^{11} cm^{-2} and interface trapped charge at midgap in the low 10^{11} cm^{-2} eV^{-1} are typical for n-type 3C-SiC MOS capacitors, although interface traps as low as $10^{10}\,cm^{-2}\,eV^{-1}$ have been reported. C-V characteristics of p-type (Al-doped) 3C-SiC MOS capacitors have also been reported [39]. Similar MOS capacitors have been reported on 6H-SiC [19,21,42,43,46]. Recently, MOS capacitors on n-type 6H-SiC with N_d of $3.4 \times 10^{15}\,cm^{-3}$ have been demonstrated with oxide fixed charge $\leq 5 \times 10^{10}\,cm^{-2}$ and interface trapped charge of $\leq 5 \times 10^{10}\,cm^{-2}\,eV^{-1}$ [46]. The same study [46] found that similar p-type 6H-SiC MOS capacitors had oxide fixed charge of $1.5 \times 10^{12}\,cm^{-2}$ and interface trapped charge of $3 \times 10^{12}\,cm^{-2}\,eV^{-1}$. The dramatic difference has been attributed to the incorporation of Al (the p-type dopant) in the oxide at the oxide-to-SiC interface.

Due to the large bandgap of SiC, the room temperature (300 K) capacitance-voltage (C-V) characteristics of SiC MOS capacitors resemble those of Si MOS capacitors at low temperatures (100 K) [5,6,14,19,26,32]. Due to the extremely low minority carrier generation rate in SiC at room temperature, when the gate voltage of SiC MOS capacitors is swept from accumulation towards inversion no inversion layer forms, but rather a deep depletion C-V characteristic is measured. For both 3C- and 6H-SiC MOS capacitors, illumination will result in inversion layer formation [5,6,19,26]. For 3C-SiC, whose bandgap is smaller than 6H by 2.2 to 2.9 eV, an inversion layer has also been observed to form gradually over several hours [5,6,32] and more quickly at high electric fields [7,32,38,39,41]. At temperatures of about 600 K, the SiC MOS C-V becomes more like Si at room temperature, with an inversion layer forming during the voltage sweep [5,6,40]. The number of active interface traps has been observed to increase with temperature [37,40]. 3C-SiC MOS capacitors have been found to be significantly more resistant to the effects of gamma irradiation than comparable devices in Si [49].

The resistivity and breakdown field of the wet thermal oxide on SiC have been measured to be comparable to those of thermal SiO_2 on Si, about $10^{16}\,\Omega\,cm$ [26,37,40] and as high as $8 \times 10^{6}\,V\,cm^{-1}$ [41], respectively. From analysis of Fowler-Nordheim tunnelling currents in the I-V characteristics of 3C-SiC MOS capacitors, the barrier height between thermal oxide and 3C-SiC has been determined as about 3.1 eV [37,40,41].

As with Si, the electrical quality of SiC MOS capacitors is dependent on the oxidation conditions and other fabrication steps. The following factors have been noted to influence the electrical properties of the SiC MOS capacitor:

(1) It has been consistently observed that wet oxidation has produced less fixed charge and fewer interface traps than dry oxidation [5-7,31]. The presence of H_2O molecules appears to be critical to the formation of high quality SiC MOS capacitors [5,7]. Dry oxide on 3C-SiC is not quite stoichiometric SiO_2 [27].

(2) The surface roughness of as-grown 3C-SiC reduces the breakdown voltage of an MOS capacitor [37]. Polishing the as-grown surface of 3C-SiC deposited on Si is essential for the highest quality MOS capacitors. 3C-SiC grown on 2°-off-axis (001) Si and containing no inversion-domain-boundaries (IDBs) has been found to produce MOS capacitors with lower numbers of interface traps than 3C-SiC grown on on-axis Si, which contains IDBs [7,38].

(3) To reduce interface traps, the oxidation of SiC should be terminated in such a manner that no oxidation can take place during the time that the SiC is cooling to room

temperature. The effects of switching to an Ar anneal immediately after oxidation have been reported in the fabrication of MOS capacitors after the dry oxidation of 3C-SiC [26] and wet oxidation of 6H-SiC [19]. The effects of switching to N_2 gas after oxidation, followed by a rapid withdrawal from the furnace, have also been reported [5].

(4) Oxidation temperature may influence the oxide charges. Wet oxidation of 3C-SiC at 1000°C has been reported to produce fewer interface traps than at higher temperatures because of reduced preferential oxidation at antiphase grain boundaries [7]. Little or no effect on the oxide charges was seen in 3C-SiC oxidised between 1050 and 1200°C [37,40] and 1050 and 1150°C [41].

(5) Microalloying in 10% hydrogen/90% nitrogen forming gas at 450°C, after aluminium gate deposition, improves capacitor electrical properties [5].

(6) For both the 3C and 6H polytypes, nitrogen-doped or undoped n-type SiC has, so far, produced MOS capacitors with much lower oxide charges than Al-doped p-type SiC [39,46].

E CONCLUSION

Procedures for thermal oxidation of SiC have been developed and shown to produce oxide layers useful in the fabrication of planar SiC microelectronic devices. The SiC oxidation rate has been studied under conditions commonly used in integrated circuit fabrication. The oxidation rate constants derived in these studies are useful for predicting the oxide thickness formed on SiC under similar conditions. The metal-oxide-semiconductor capacitors formed by thermal oxide layers on both 3C- and 6H-SiC have been shown to have low interface charge densities, suitable for transistor applications.

F ACKNOWLEDGEMENT

Contribution of the National Institute of Standards and Technology, not subject to copyright in the USA.

REFERENCES

[1] J.B. Petit, J.A. Powell, L.G. Matus [*Trans. 1st Int. High Temperature Electron. Conf.* (Albuquerque, NM, USA, 1991) p.198-204]

[2] W.-J. Lu, A.J. Steckl, T.P. Chow [*J. Electrochem. Soc. (USA)* vol.133 no.6 (1986) p.1180-5]

[3] Y. Laukhe, Y.M. Tairov, V.F. Tsvetkov, F. Shchepanski [*Sov. Inorg. Mater. (USA)* vol.17 no.2 (1981) p.177-9]

[4] L. Muehloff, M.J. Bozack, W.J. Choyke, J.T. Yates [*J. Appl. Phys. (USA)* vol.60 no.7 (1986) p.2558-63]

[5] J.J. Kopanski, D.B. Novotny [National Institute of Standards and Technology Interagency Report (USA) NISTIR 4157-89 (1989)]

[6] J.J. Kopanski, D.B. Novotny [*Extended Abstracts of the 176th Meeting of the Electrochem. Soc.* (Hollywood, Fla., USA) vol.89-2 (1989) p.722-3]

[7] S. Zaima, K. Onoda, Y. Koide, Y. Yasuda [*J. Appl. Phys. (USA)* vol.68 no.12 (1990) p.6304-8]

[8] C.H. Carter, R.F. Davis, S.R. Nutt [*J. Mater. Res. (USA)* vol.1 no.6 (1986) p.811-9]

[9] J.W. Palmour, H.J. Kim, R.F. Davis [*Mater. Res. Soc. Symp. Proc. (USA)* vol.54 (1986) p.553-9]

[10] S.R. Nutt, D.J. Smith, H.J. Kim, R.F. Davis [*Appl. Phys. Lett. (USA)* vol.50 no.4 (1987) p.203-5]

[11] R.W. Bartlett [*J. Electrochem. Soc. (USA)* vol.118 no.2 (1971) p.397-9]

[12] C.D. Fung, J.J. Kopanski [*Appl. Phys. Lett. (USA)* vol.45 no.7 (1984) p.757-9]

[13] W.-J. Lu, A.J. Steckl, T.P. Chow, W. Katz [*J. Electrochem. Soc. (USA)* vol.131 no.8 (1984) p.1907-14]

[14] J.J. Kopanski [Thesis, Case Western Reserve University, Cleveland, Ohio, USA (August 1985)]

[15] T. Narushima, T. Goto, T. Hirai [*J. Am. Ceram. Soc. (USA)* vol.72 no.8 (1989) p.1386-90]

[16] R.C.A. Harris [*J. Am. Ceram. Soc. (USA)* vol.58 no.1-2 (1975) p.7-9]

[17] W. von Münch, I. Pfaffeneder [*J. Electrochem. Soc. (USA)* vol.122 no.5 (1975) p.642-3]

[18] J.A. Costello, R.E. Tressler [*J. Am. Ceram. Soc. (USA)* vol.64 no.6 (1981) p.327-31]

[19] A. Suzuki, H. Ashida, N. Furui, K. Mameno, H. Matsunami [*Jpn. J. Appl. Phys. (Japan)* vol.21 no.4 (1982) p.579-85]

[20] J.A. Costello, R.E. Tressler [*J. Am. Ceram. Soc. (USA)* vol.69 no.9 (1986) p.674-81]

[21] H. Matsunami [*Jpn. Annu. Rev. Electron. Comput. Telecommun. (Netherlands)* vol.19 (1986) p.241-51]

[22] Z. Zheng, R.E. Tressler, K.E. Spear [*J. Electrochem. Soc. (USA)* vol.137 no.3 (1990) p.854-8]

[23] Z. Zheng, R.E. Tressler, K.E. Spear [*J. Electrochem. Soc. (USA)* vol.137 no.9 (1990) p.2812-6]

[24] R.W. Kee, K.M. Geib, C.W. Wilmsen, D.K. Ferry [*J. Vac. Sci. Technol. (USA)* vol.15 no.4 (1978) p.1520-3]

[25] A. Suzuki, K. Mameno, T. Tanaka [*J. Electrochem. Soc. (USA)* vol.125 no.11 (1978) p.1896-7]

[26] K. Shibahara, S. Nishino, H. Matsunami [*Jpn. J. Appl. Phys. (Japan)* vol.23 no.11 (1984) p.L862-4]

[27] M.I. Chaudhry [*J. Mater. Res. (USA)* vol.4 no.2 (1989) p.404-7]

[28] K.L. Luthra [*J. Am. Ceram. Soc. (USA)* vol.74 no.5 (1991) p.1095-103]

[29] B.E. Deal, A.S. Grove [*J. Appl. Phys. (USA)* vol.36 no.12 (1965) p.3770-8]

[30] J.W. Palmour, R.F. Davis, H.S. Kong, D.P. Griffis [*J. Electrochem. Soc. (USA)* vol.136 no.2 (1989) p.502-7]

[31] M.I. Chaudhry, W.B. Berry [*J. Mater. Res. (USA)* vol.4 no.6 (1989) p.1491-4]

[32] R.E. Avila, J.J. Kopanski, C.D. Fung [*Appl. Phys. Lett. (USA)* vol.49 no.6 (1986) p.334-6]

[33] Y. Kondo et al [*IEEE Electron Device Lett. (USA)* vol.7 no.7 (1986) p.404-6]

[34] K. Shibahara, T. Saito, S. Nishino, H. Matsunami [*IEEE Electron Device Lett. (USA)* vol.7 no.12 (1986) p.692-3]
[35] Y. Kondo et al [*Jpn. J. Appl. Phys. (Japan)* vol.26 no.2 (1987) p.310-11]
[36] K. Furukawa et al [*IEEE Electron Device Lett. (USA)* vol.8 no.2 (1987) p.48-9]
[37] J.J. Kopanski [National Institute of Standards and Technology Interagency Report (USA) NISTIR 4352-90 (1990)]
[38] S.M. Tang, W.B. Berry, R. Kwor, M.V. Zeller, L.G. Matus [*J. Electrochem. Soc. (USA)* vol.137 no.1 (1990) p.221-5]
[39] M. Shinohara, M. Yamanaka, S. Misawa, H. Okumura, S. Yoshida [*Jpn. J. Appl. Phys. (Japan)* vol.30 no.2 (1991) p.240-3]
[40] J.J. Kopanski, R.E. Avila [*Springer Proc. Phys. (Germany)* vol.56 (1992) p.119-24]
[41] H. Fuma, M. Kodama, H. Tadano, S. Sugiyama, M. Takigawa [*Springer Proc. Phys. (Germany)* vol.56 (1992) p.237-42]
[42] R.C.A. Harris [*Solid-State Electron. (UK)* vol.19 (1976) p.103-5]
[43] A. Suzuki, K. Mameno, N. Furui, H. Matsunami [*Appl. Phys. Lett. (USA)* vol.39 no.1 (1981) p.89]
[44] J.W. Palmour, H.S. Kong, R.F. Davis [*Appl. Phys. Lett. (USA)* vol.51 no.24 (1987) p.2028-30]
[45] J.W. Palmour, H.S. Kong, R.F. Davis [*J. Appl. Phys. (USA)* vol.64 no.4 (1988) p.2168-77]
[46] D.M. Brown et al [*Digest of Papers, Government Microcircuit Applications Conf.* (Orlando, Fla., USA) vol.XVII (1991) p.89-92]
[47] L.M. Terman [*Solid-State Electron. (UK)* vol.5 (1962) p.285-99]
[48] E.H. Nicollian, A. Goetzberger [*Bell Syst. Tech. J. (USA)* vol.46 no.6 (1967) p.1055-133]
[49] M. Yoshikawa et al [*J. Appl. Phys. (USA)* vol.70 no.3 (1991) p.1309-12]

CHAPTER 6

ETCHING

6.1　Introduction to etching of SiC
6.2　Chemical etching of SiC
6.3　Dry etching of SiC
6.4　Electrochemical etching of SiC

6.1 Introduction to etching of SiC

G.L. Harris

June 1995

The successful fabrication of SiC based devices cannot be realized without reliable etching and cleaning processes for this material. The purpose of this Chapter is to review the developments and methods of etching silicon carbide for device applications. The term etching is used to describe the processes by which SiC material can be removed. It also includes chemical machining of a semiconductor as part of the fabrication process, as well as defect delineation.

The mechanical removal of SiC can be carried out using boron carbide and/or diamond paste using a diamond coated lapping cloth. Mesh sizes of the diamond and boron carbide range from submicron to over 10 microns depending on the stage of lapping. Under normal conditions a large mesh size is used first followed by a much smaller mesh size. The lapping rates vary depending on the applied pressure. The mechanical polishing does introduce some surface damage and that damage can be removed by etching or oxidation and removal of the oxide layer.

In this Chapter, we will review the chemical, ion sputtering, plasma-assisted and reactive ion etching techniques for SiC. These processes can be used in fabricating SiC devices.

6.2 Chemical etching of SiC

G.L. Harris

September 1994

Silicon carbide is a highly refractory material and chemical etching can only be achieved successfully with the aid of hot gases and hot molten salt etches. At present, no known aqueous (acid, neutral or basic) solution attacks SiC at room temperature. (There is a brief discussion of orthophosphoric acid at 215 °C as an etchant for SiC, but the reaction is erratic [1].) The chemical etching behaviour has been studied mainly for surface treatment to reveal and investigate structural crystal damage [1-10]. This surface damage has been orientation-sensitive and has revealed defects such as dislocations, stacking faults, anti-phase boundaries (APB), point defects, double position boundaries (DPB) and other surface features.

The procedure involves placing a solid salt in a Pt (platinum) beaker and bringing it up to the required temperature. The salt crystals liquefy and the SiC sample to be etched is placed in a Pt mesh basket and is then lowered into the molten salt for the required time. All of the molten salts to be used for this procedure need oxygen for etching [1]. The oxygen can be supplied by the molten salt, as in the case of KOH, or by the surrounding atmosphere. The following molten salts have been used to etch SiC and reveal various surface features:

$KClO_3$	K_2CO_3	KOH	K_2SO_4	KNO_3
$Na_2B_4O_7$	Na_2CO_3	$NaNO_3$	Na_2O_2	$NaNO_3$
Na_2SO_4	$NaOH$	PbO	PbF_2	

A detailed list of the characteristics of these etches and the time and temperature requirements are provided in [7]. Jepps and Page [6] reported that the KOH:KNO$_3$ (50:50) molten salt at 480 °C for 5 minutes revealed grain boundaries, stacking faults, etc. Recently, Powell and co-workers [3,4] etched in fused Na_2O_2-$NaNO_2$ and observed antiphase boundaries (APD) on β-SiC grown on silicon. Nordquist et al [11] verified the results of Powell and co-workers by using a eutectic etch of NaOH-KOH at 350 °C. This etch also revealed dislocations and stacking faults. The temperature of these molten salt etches ranges from 350 to 950 °C. Various mixtures of these salts can be used to reduce the melting temperature.

In the 6H polytype, with the stacking sequence ABCACB, the $(000\bar{1})$ carbon face and the (0001) silicon face etch differently because the SiC bond is slightly ionic. Many researchers use this fact to determine the difference between the two faces. Koga et al [9] used KOH to study three kinds of etch pits in 6H-SiC single crystal grown by the sublimation method. They observed that by increasing the pressure they could reduce one type of etch pit.

Chlorine and fluorine-based gases at high temperature attack SiC [1]. If one uses pure chlorine and/or fluorine a layer of carbon is left on the surface. This layer of carbon can be removed by adding oxygen to the mixture. Fluorine and chlorine based gases are highly reactive to many materials and in many applications are not practical.

Hydrogen [1] and HCl [10] at high temperatures, above 1350 °C, etch SiC. Hydrogen etching is useful in cleaning the surface prior to epitaxial growth and appears to be non-preferential.

HCl etching at 1350°C has been used to study defects [10] and decorated defects in a similar way to that of molten salts.

REFERENCES

[1] V.J. Jennings [*Mater. Res. Bull. (USA)* vol.4 (1969) p.5199-210]
[2] G. Petzow [*Metallographic Etching* (American Society for Metals, Ohio, 1978)]
[3] P. Pirouz, C.M. Chorey, J.A. Powell [*Appl. Phys. Lett. (USA)* vol.50 (1987) p.221]
[4] J.A. Powell, L.G. Matus, M.A. Kuczmarski, C.M.C. Horey, T.T. Cheng, P. Piroz [*Appl. Phys. Lett. (USA)* vol.51 (1987) p.823]
[5] K. Shibaharn, S. Nishino, H. Matsunami [*Appl. Phys. Lett. (USA)* vol.50 (1987) p.1988]
[6] M.W. Jepps, T.F. Page [*J. Microsc. (UK)* vol.124 pt.3 (1981) p.227-37]
[7] R.C. Marshall, J.W. Faust Jr., C.E. Ryan (Eds) [*1973 Table of etchants, Silicon Carbide*, Proc. III Int. Conf. on SiC, Florida (University of South Carolina Press, 1973) Appendix I, p.657]
[8] H.S. Kong et al [*Mater. Res. Soc. Symp. Proc. (USA)* vol.97 (1987) p.233]
[9] K. Koga, Y. Fujikawa, V. Ueda, T. Yamaguchi [*Springer Proc. Phys. (Germany)* vol.71 (1992) p.96-100]
[10] J.A. Powell, P.G. Neudeck, D.J. Larkin, J.W. Yang, P. Pirouz [*Inst. Phys. Conf. Ser. (UK)* no.137 (1994) p.161-4]
[11] P.E.R. Nordquist Jr., H. Lessoff, R.J Gorman, M.L. Gripe [*Springer Proc. Phys. (Germany)* vol.43 (1989) p.119-24]

6.3 Dry etching of SiC

K. Wongchutigul

September 1994

A INTRODUCTION

SiC has several unique properties which make it a promising semiconductor for high temperature, high power, and high speed devices. Several SiC devices such as bipolar transistors [1,2], field-effect transistors [3-6], IMPATT diodes [7], Tunnel diodes [8], LEDs [9-12], and ICs [13] have been developed in recent years. If continued improvements are to be made on these devices, it is necessary to have a suitable means of etching SiC surfaces. Due to its chemical inertness, there are virtually no chemical etches that attack SiC at room temperature. Chemical etching can proceed only at high temperatures: in molten salts (Na_2O_2, NaOH, KOH, etc. at 500 °C) [14], and in a flow of chlorine, chlorine-containing gases, and hydrogen (at > 1000 °C) [14-17]. At such high temperatures, conventional photoresist, oxide, and metal masks are not usable. The main requirements of the etching process for device fabrication are as follows: (i) low temperatures, (ii) high quality of the surface treatment, (iii) high selectivity of etching with respect to the mask materials, and (iv) high resolution and anisotropy. Although etching with Ar ions [18], by means of direct physical removal of SiC, provides high anisotropy and resolution, it yields low selectivity which makes it unattractive for device fabrication. To date, the two most promising methods for etching SiC are vacuum plasma and electrochemical etching techniques. These two techniques are now briefly reviewed.

B VACUUM PLASMA ETCHING TECHNIQUES

Vacuum plasma etching techniques have been used widely for SiC device fabrication in recent years [3,5,19,20]. Plasmas of fluorine-containing gases, such as CF_4 [27,29], SF_6 [21,28], CHF_3 [25,26], and $CBrF_3$ [25], have been employed in both a plasma etching mode and a reactive ion etching (RIE) mode. While the investigation detail, using these two modes, can be found in the references, the author would like to present some interesting results obtained from these investigations.

B1 Plasma etching mode

The plasma etching mode involves (a) the use of RF energy to generate plasmas by dissociating the active gases and (b) the chemical reaction of the plasmas with C and Si and with the generation of volatile products. In this mode, the use of CF_4-containing gases was reported to have poor performance as compared to the use of SF_6-containing gases. Palmour et al [27] reported that the CF_4 plasma produced a dark surface layer with a thickness in the range of 200 - 1500 Å. This layer reportedly 'blocked' the etching process, and it limited the etching rates. The composition of this layer was suspected to be the residual SiC with a C-rich surface. Kelner [28] used SF_6 plasma and reported an improved performance with no re-deposition of C on the etched surface.

B2 Reactive Ion Etching mode

The RIE mode is similar to the plasma mode, but it includes physical removal of C and Si atoms by sputtering. It was found to be superior to the plasma etching mode because it yielded faster etch rates and could be used with a range of plasmas. Although the C-rich SiC layer was found to form during the etching process, it was believed to be removed by ion bombardment. As a result, the etch rate of the RIE mode was not limited by a formation of the C-rich SiC layer. Many authors agreed that the SF_6 plasma gave a better performance than the CF_4 plasma. Popov et al [23] reported different etch rates on different faces of 6H-SiC. They discovered that the etch rate on the [0001] face was slower than on the [000$\bar{1}$] face.

Many authors proposed the chemical reaction of SiC and the plasma, but the most interesting one was that proposed by Sugiura et al [21]. They suggested the following reactions:

$$CF_4 + O_2 = CF_x + F + O + (CO, CO_2) \quad (1)$$
$$SF_6 = S + 6F \quad (2)$$
$$SF_6 + O_2 = S + 6F + 2O \quad (3)$$
$$Si + 4F = SiF_4 \quad (4)$$
$$C + xF = CF_x \quad (5)$$
$$C + xO = CO, CO_2 \quad (6)$$

The first four steps are well known processes, but the last two steps have been proposed. They [21] explained that the CF_4 plasma yielded slow etch rates because the reaction in the CF_4-containing gases (Eqns (1), (5) and (6)) did not go to completion. On the other hand, in SF_6-based gases there exist no such limitations; therefore, the reaction is faster than that in CF_4-based gases. They also suggested that the etching of SiC could be treated as the etching of Si and C and that the formation of the C-rich SiC layer was due to a preferential removal, both physically and chemically, of Si and C atoms. This physical and chemical preferential removal was detected in their experiments. The data on etching rates and selectivity obtained in the plasma etching of SiC with different gaseous mixtures are given in TABLE 1.

C CONCLUSION

In conclusion, the etching technology for SiC has been improved in recent years. Although it is now possible to etch a mesa for device fabrication, there are still some problems that need to be solved. These problems include roughness of the etch surfaces, deposition of a C-rich SiC layer on the etched surface and slow etch rates.

6.3 Dry etching of SiC

TABLE 1 Vacuum plasma etching of SiC.

Gases	Etch rate (nm min^{-1})	Mask material	Polytype	Selectivity ratio	Ref
\multicolumn{6}{c}{Plasma etching mode}					
CF_4 + 50% O_2	15 - 55	AZ1350	3C	-	[22]
SF_6	24	Al	3C	SiC:Al = 144:1	[28]
\multicolumn{6}{c}{Reactive ion etching mode}					
CF_4	33	AZ1350J	amorphous	SiC:Si = 1:2	[21]
	500	-	6H [000$\bar{1}$]	-	[23]
	420	-	6H [0001]	-	
	6.5	AZ1350	-	SiC:AZ1350 = 1:1.8	[29]
			-	SiC:Cr = 1.3:1	
			-	SiC:Ti = 1:2	
CF_4 + 4% O_2	37.5	AZ1350J	amorphous	SiC:Si = 1:3.25	[21]
			-	SiC:SiO$_2$ = 1:1.6	
CF_4 + 40% O_2	13	AZ1350		SiC:AZ1350 = 1:1.7	[29]
			-	SiC:Cr = 4.3:1	
			-	SiC:Ti = 2.2:1	
CF_4 + 50% O_2	23	Cr/Au	3C	-	[22]
SF_6	1700	-	6H	-	[23]
SF_6 + 4% O_2	47	AZ1350J	amorphous	SiC:Si = 1:362	[21]
SF_6 + 35% O_2	52	-	amorphous	SiC:Si = 1:28	[25]
			-	SiC:SiO$_2$ = 1:1.15	
SF_6 + 90% O_2	50	-	amorphous	SiC:Si = 1:2	[25]
			-	SiC:SiO$_2$ = 1.06:1	
SF_6 + 50% He	56	AZ1350J	amorphous	SiC:Si = 1:16	[21]
			-	SiC:SiO$_2$ = 1.12:1	

REFERENCES

[1] W.V. Muench, P. Hoeck, E. Pettenpual [*Proc. Int. Electron Devices Meeting*, Washington, DC, 5 - 7 Dec. 1977 (IEEE, New York, NY, 1977) p.337-9]

[2] T. Sugii, T. Ito, Y. Furumura, H. Doki, F. Mieno, M. Maeda [*IEEE Electron Device Lett. (USA)* vol.9 (1988) p.87-9]

[3] J.W. Palmour, H.S. Kong, D.G. Waltz, J.A. Edmond, C.H. Carter [*1st HiTEC Transactions* Eds D.B. King, F.V. Thome (Albuquerque, NM, 1991) p.229]

[4] Y. Kondo et al [*IEEE Electron Device Lett.(USA)* vol.EDL-7 (1986) p.404]

[5] G. Kelner, S. Binari, K. Seger, H. Kong [*IEEE Electron Device Lett. (USA)* vol.EDL-8 (1987) p.429]

[6] S. Yoshida, H. Daimon, M. Yamanaka, E. Sakuma, S. Misawa, K. Endo [*J. Appl. Phys. (USA)* vol.60 (1986) p.2989-92]

[7] J.D. Parsons, R.F. Bunshah, O.M. Stafsudd [*Solid State Technol. (USA)* vol.28 no.11 (1985) p.133-9]

[8] V.A. Dmitriev, P.A. Ivanov, A.M. Strelchuk, A.L. Syrkin, I.V. Popov, V.E. Chelnokov [*Sov. Tech. Phys. Lett. (USA)* vol.11 (1985) p.403-4]

[9] A. Suzuki, M. Ikeda, N. Nagao, H. Matsunami, T. Tanaka [*J. Appl. Phys. (USA)* vol.47 (1976) p.4546]

[10] W.V. Muench [*J. Electron. Mater. (USA)* vol.6 (1977) p.449]

[11] V.A. Dmitriev, P.A. Ivanov, Ya.V. Morozenko, V.S. Radkin, V.E. Chelnokov [*Sov. Tech. Phys. Lett. (USA)* vol.11 (1985) p.101-2]

[12] V.A. Dmitriev, L.M. Kegan, Ya.V. Morozenko, B.V. Tsorenkov, V.E. Chelnokov, A.E. Cherenkov [*Sov. Phys.-Semicond. (USA)* vol.23 (1988) p.23-6]

[13] G.L. Harris, K. Wongchotigul, H. Henry, K. Diogu, C. Taylor [*3rd Int. Conf. Solid State and Integrated Circuit Technology (ICSICT)*, Beijing, China (1992)]

[14] R.C. Marshall, J.W. Faust, C.E. Ryan [*SiC 1973* (University of S. Carolina Press, 1974) p.659-67]

[15] W.V. Muench [*J. Electrochem. Mat. (USA)* vol.6 (1977) p.449]

[16] W.V. Muench, P. Hoeck [*Solid-State Electron. (UK)* vol.21 (1978) p.479]

[17] J.W. Faust Jr. [*Silicon Carbide, A High Temperature Semiconductor* Eds J.R. O'Connor, J. Smittens (1959) p.403]

[18] K. Wongchotigul, G.L. Harris, M.G. Spencer, K.H. Jackson, A. Gomez, A. Jones [*Mater. Lett. (Netherlands)* vol.8 no.5 (1989) p.153-5]

[19] G. Kelner, M. Shur, S. Binari, K. Sleger, H.S. Kong [*IEEE Trans. Electron Devices (USA)* vol.EDL-6 (1989) p.1045-9]

[20] M.M. Anikin, P.A. Ivanov, A.L. Syrkin, B.V. Tsarenkov, V.E. Chelnokov [*Sov. Tech. Phys. Lett. (USA)* vol.15 (1989) p.636-8]

[21] J. Sugiura, W.-Y. Lu, K.C. Cadien, A.J. Steckl [*J. Vac. Sci. Technol. B (USA)* vol.4 (1986) p.349]

[22] J.W. Palmour, R.F. David, P. Astell-Burt, P. Blackborow [*Mater. Res. Soc. Symp. Proc. (USA)* vol.76 (1987) p.195]

[23] A.L. Syrkin, I.V. Popov, V.E. Chelnokov [*Sov. Tech. Phys. Lett. (USA)* vol.12 (1986) p.99]

[24] S. Dohmae, K. Shibahara, S. Nishino, H. Mutsunami [*Jpn. J. Appl. Phys. (Japan)* vol.24 no.11 (1985) p.L873]

[25] W-S. Pan, A.J. Steckl [*Springer Proc. Phys. (Germany)* vol.43 (1989) p.217-23]

[26] A.J. Steckl, P.H. Yih [*Springer Proc. Phys. (Germany)* vol.71 (1992) p.423-9]

[27] J.W. Palmour, R.F. Davis, T.M. Wallett, K.B. Bashin [*J. Vac. Sci. Technol. A (USA)* vol.4 (1986) p.590]

[28] G. Kelner, S.C. Binari, P.H. Klein [*J. Electrochem. Soc. (USA)* vol.134 no.1 (1987) p.253-4]

[29] S. Matsui, S. Mizaki, T. Tamato, H. Aritome, S. Namba [*Jpn. J. Appl. Phys. (Japan)* vol.20 (1981) p.L38]

[30] M. Glerria, R. Memming [*J. Electroanal. Chem. (Switzerland)* vol.65 (1975) p.163]

[31] M.M. Carrabba, J. Li, J.P. Hachey, R.D. Rauh, Y. Wang [*Ext. Abstr. Electrochem. Soc. (USA)* vol.89-2 (1989) p.727]

[32] H. Morisaki, H. Ono, K. Yazawa [*J. Electrochem. Soc. (USA)* vol.131 (1984) p.2081-]

[33] I. Lauermann, D. Meissner, R. Memming, R. Reineke, B. Kastening [*Ext. Abstr. Electrochem. Soc. (USA)* vol.90-2 (1990)]

[34] G.L. Harris, K. Fekade, K. Wongchotigul [*J. Mater. Sci., Mater. Electron. (UK)* vol.3 (1992) p.1-2]

[35] J.S. Shor, X.G. Zhang, R.M. Osgood Jr. [*J. Electrochem. Soc. (USA)* vol.139 no.4 (1992) p.1213-6]

[36] J.S. Shor, R.M. Osgood Jr. [*J. Electrochem. Soc. (USA)* vol.140 no.8 (1993) p.L123-5]
[37] T.H. Nguyen, M.M. Carrabba, T.D. Plante, R.D. Rauh [*Ext. Abstr. Electrochem. Soc. (USA)* vol.92-1 (1992) p.474]
[38] G.L. Harris, K. Fekade [*J. Electrochem Soc. (USA)* vol.135 no.2 (1988) p.405-6]
[39] J.S. Shor, R.M. Osgood, A.D. Kurtz [*Appl. Phys. Lett. (USA)* vol.60 no.8 (1992) p.1001-3]

6.4 Electrochemical etching of SiC

J.S. Shor

June 1995

A INTRODUCTION

SiC is a useful material for high temperature and high frequency device applications due to its unique properties [1], such as its wide bandgap (2.2 - 3.2 eV), high melting point (2830 °C) and high breakdown electric field (4×10^6 V cm^{-1}) [1]. However, SiC is chemically inert, making patterning devices a difficult problem. There are a number of molten metals and molten salts [1] that attack SiC, but these are impractical for device fabrication due to the high temperatures involved. Device structures have been etched successfully in SiC using Reactive Ion Etching (RIE) [2]. This Datareview reviews the progress in photoelectrochemical (PEC) etching of SiC, a relatively new etching technique. Photoelectrochemical etching offers significant advantages over RIE, namely much faster etch rates and dopant selective etch-stops. These etching properties are very useful in the fabrication of electronic devices with thin multilayers, such as CMOS, bipolar and SCR devices as well as microelectromechanical sensors. Furthermore, there are some indications that electropolishing methods may be feasible for SiC.

B BASIC ELECTROCHEMISTRY OF SiC

Electrochemical etching involves the transfer of charge from a semiconductor surface to an electrolyte. Therefore, an understanding of the SiC/electrolyte interface is necessary for the development of effective etching processes. FIGURE 1 is a diagram of the n-type β-SiC/HF junction, which is similar in its charge transfer and energy band characteristics to a Schottky contact. For etching to occur, the bands of the semiconductor must be bent upwards to allow holes to reach the interface. These holes can either be supplied from the bulk, as in p-type material, or can be photogenerated in n-type or p-type samples. Most likely, the holes cause photo-oxidation of the SiC, and the oxides, SiO_x and CO_x, are removed from the surface by HF dissolution and gas phase nucleation respectively. At this time there is no direct evidence that SiO_2 or SiO is formed during the anodisation of SiC in HF, since HF dissolves the oxide before it can be detected. However, several authors [3-5] have shown that SiO_x formation is possible in alkaline electrolytes, such as NaOH [3], and H_2SO_4-based electrolytes [4,5]. Lauermann used gas chromatography to verify that CO and CO_2 form during anodisation in HF [6] and suggested the following chemical reactions:

$$SiC + 4H_2O + 8h^+ \rightarrow SiO_2 + CO_2 + 8H^+$$

$$SiC + 2H_2O + 4h^+ \rightarrow SiO + CO + 4H^+$$

It is possible, however, that under certain conditions etching occurs by direct dissolution without the intermediate oxidation step.

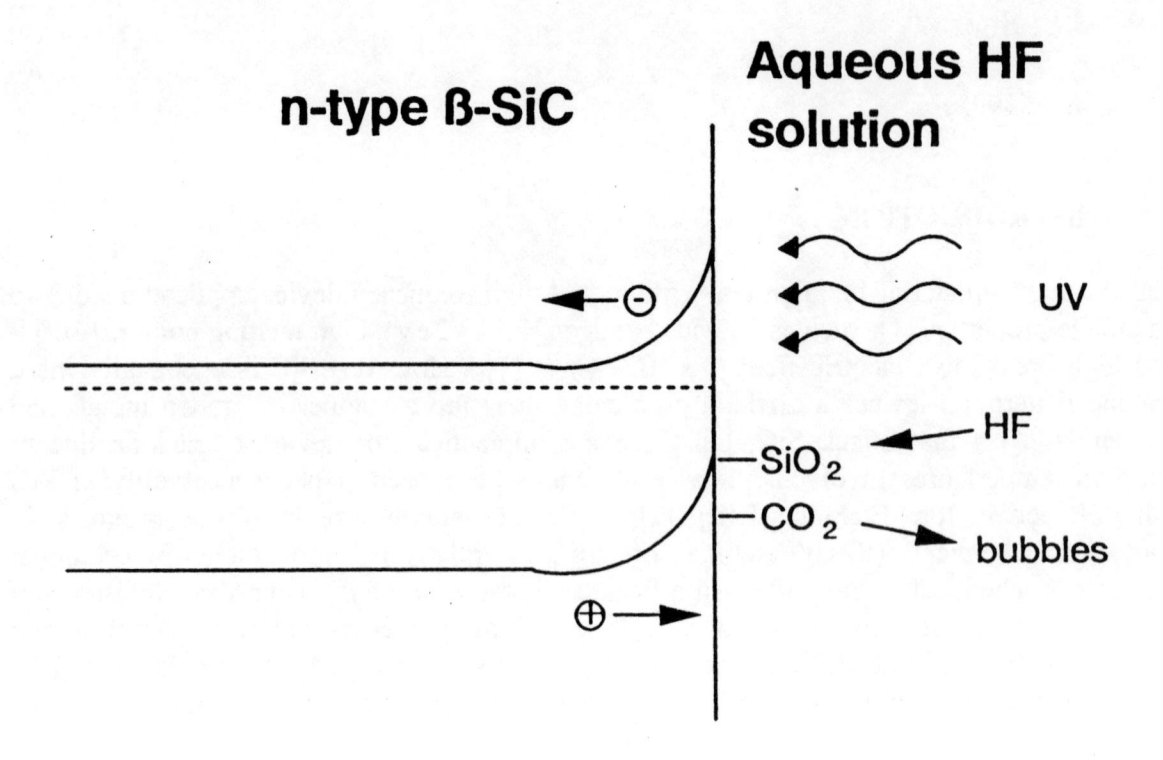

FIGURE 1 Energy band diagram of the n-type β-SiC/HF interface under the etching conditions.

FIGURE 2 is a photocurrent I-V curve of the n-type β-SiC/HF interface. Large photocurrents occur above $\approx 0.5\,V_{sce}$ (sce refers to the saturated calomel reference electrode) and increase steadily up to $1.6\,V_{sce}$ where they level off. The etch rates, being proportional to the current, will be maximised at $1.6\,V_{sce}$. In n-type β-SiC, the flat-band voltage of the semiconductor/HF interface has been measured to be $-0.5\,V_{sce}$ [5]. This is consistent with the etching potential of p-type β-SiC [7] but deviates somewhat from the etch-potential of n-SiC as described in [5]. Reported flat-band voltage measurements of 6H-SiC taken in non-HF electrolytes have been inconsistent [3,4,8]. Morisaki attributes this to an SiO_x layer forming on the surface and changing the capacitance [3].

C ETCHING PROCESSES

Photoelectrochemical etching is commonly used to etch holographic images, such as diffraction gratings, into semiconductors. Silicon carbide is a particularly attractive material for this application because of its high reflectance in the deep UV and X-ray regions. Carrabba et al [9] used photoelectrochemical etching to form diffraction gratings in single crystal n-type β-SiC and polycrystalline α-SiC. The electrolytes used were 1 - 5 M HF and 0.1 - 1.0 M ammonium bifluoride. Photocurrents of 2 - 4 mA cm^{-2} were reported using a focused Xe lamp at $I = 3\,W\,cm^{-2}$ and an anodic potential of $1.2\,V_{sce}$. These currents correspond to etch rates of approximately 200 - 400 Å min^{-1} assuming 8 equivalents mol^{-1} for the dissolution of SiC.

6.4 Electrochemical etching of SiC

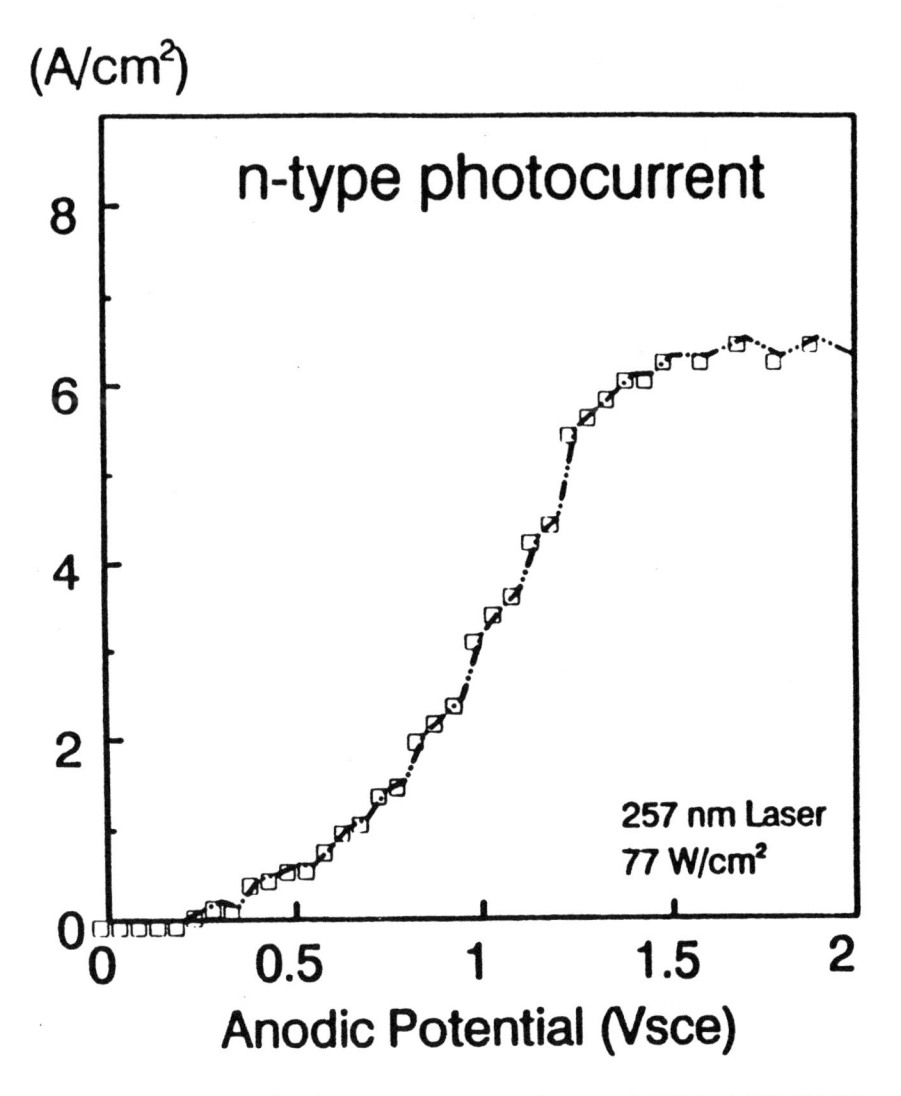

FIGURE 2 Anodic photocurrent curves of n-type β-SiC in 2.5% HF [5].

Shor et al [5] reported high etch rates in β-SiC (100 µm min^{-1}) using a frequency doubled Ar* laser at 257 nm. FIGURE 3 shows the variation of etch rate with laser intensity. It is apparent that much higher etch rates can be obtained when UV rather than visible light is used. The etch rates tended to increase with laser intensity up to the point where they saturated. The strong wavelength dependence for the photo-etching of β-SiC seen in FIGURE 3 is in accord with earlier reports on the optical absorption of SiC [10], which show that although the bandgap of SiC allows visible light to be absorbed, the absorption of such light is very weak (e.g. $1/\alpha = 1000$ µm for green light). From the reported flat-band potential of β-SiC in HF [5,8], the thickness of the space charge layer is calculated to be $d_{sc} \approx 1300$ Å at 1 V$_{sce}$ (assuming no contributions from defects). Thus, most photocarriers will be generated deep in the field free bulk when visible light is used, resulting in a large percentage of holes recombining before reaching the surface. In contrast, at 257 nm, the absorption depth in β-SiC is ≈ 1000 Å, allowing most carriers to be photogenerated in the space charge layer and collection is efficient. Therefore, for large etch rates to be obtained in SiC, UV radiation must be used. Large area etching of β-SiC has been reported using a UV lamp, with reasonably large etch-rates of 1000 - 6000 Å min^{-1} [11].

6.4 Electrochemical etching of SiC

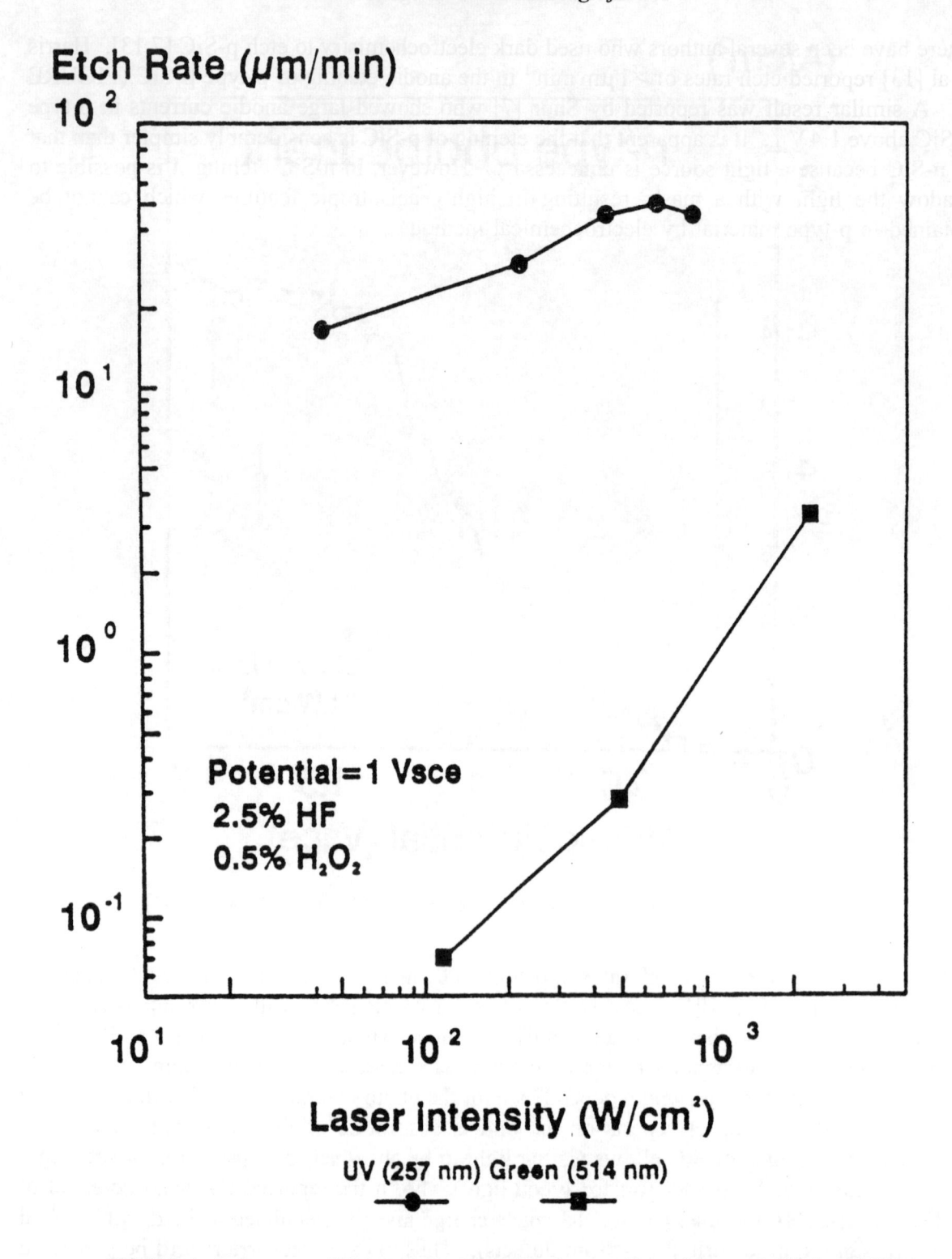

FIGURE 3 Etch rate vs. UV laser intensity for the laser etching of n-type β-SiC in HF [5].

The etched surface morphologies reported in [5,9] all exhibited a surface roughness of ≈ 3000 Å. This has been attributed to defects contained in the material causing local non-uniformities in the etching [5]. However, it has been shown that increasing the F⁻ concentration while decreasing the pH level can result in much smoother surfaces as will be discussed further below [12].

6.4 Electrochemical etching of SiC

There have been several authors who used dark electrochemistry to etch p-SiC [7,13]. Harris et al [13] reported etch rates of > 1 μm min^{-1} in the anodic etching of p-type β-SiC (FIGURE 4). A similar result was reported by Shor [7] who showed large anodic currents in p-type β-SiC above 1.4 V$_{sce}$. It is apparent that the etching of p-SiC is considerably simpler than that of n-SiC because a light source is unnecessary. However, in n-SiC etching it is possible to shadow the light with a mask, resulting in highly anisotropic features which cannot be obtained in p-type material by electrochemical methods.

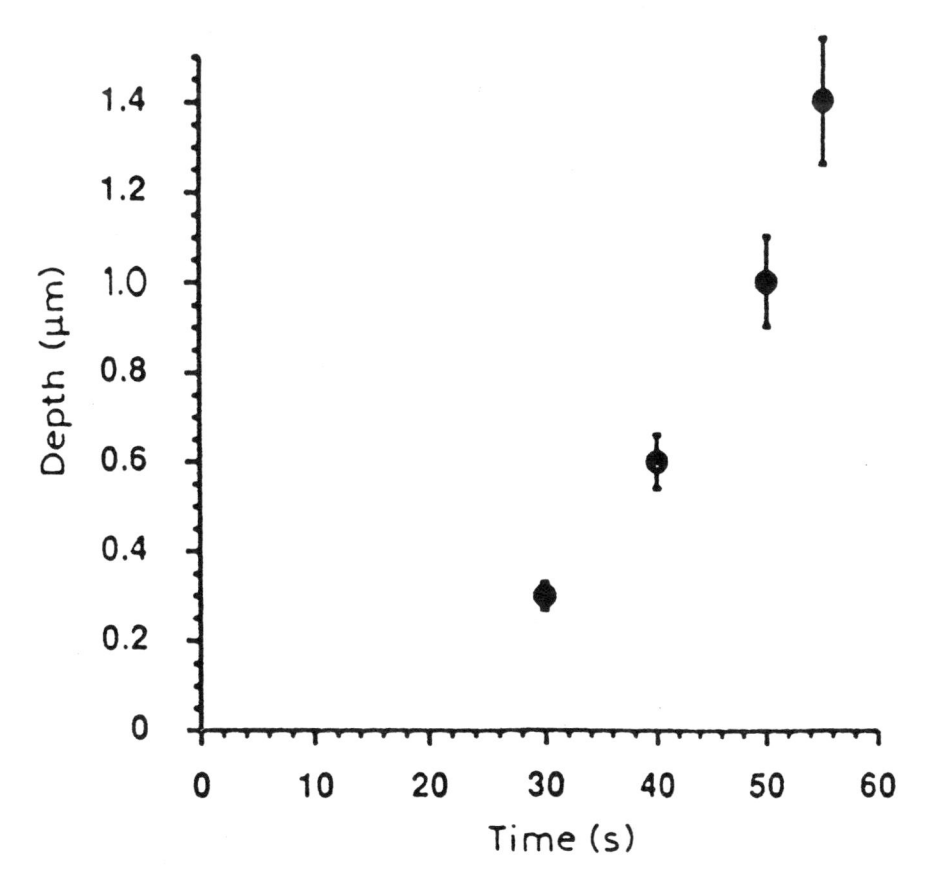

FIGURE 4 Etched depth vs. anodisation time for the etching of p-type β-SiC in HF at a constant current density of 96.4 mA cm^{-2} [13].

D   ELECTROPOLISHING OF SiC

Electropolishing of SiC has been investigated by several authors [12,14]. Brander et al [14] reported electropolishing of p-type 6H-SiC Lely crystals, but were not able to repeat the results on n-SiC. In a recent study, Nguyen et al [12] investigated the effect of F$^-$ concentration and pH on the morphology of etched features for epitaxial β-SiC grown on Si and CVD grown polycrystalline 6H-SiC. It was found that solutions which had a high F$^-$ concentration and a low pH, such as solutions of HF:NH$_4$F, achieved much smoother morphologies than observed in other studies [5,9]. FIGURE 5 is a contour plot of the surface morphology vs. pH and F$^-$ concentration. TABLE 1 shows the best values for RMS roughness for photoelectrochemical and electrochemical polishing of SiC compared to what

6.4 Electrochemical etching of SiC

has been obtained by mechanical polishing. It is clear that electropolishing methods achieve similar surface roughnesses to mechanical polishing, probably without introducing as much surface damage.

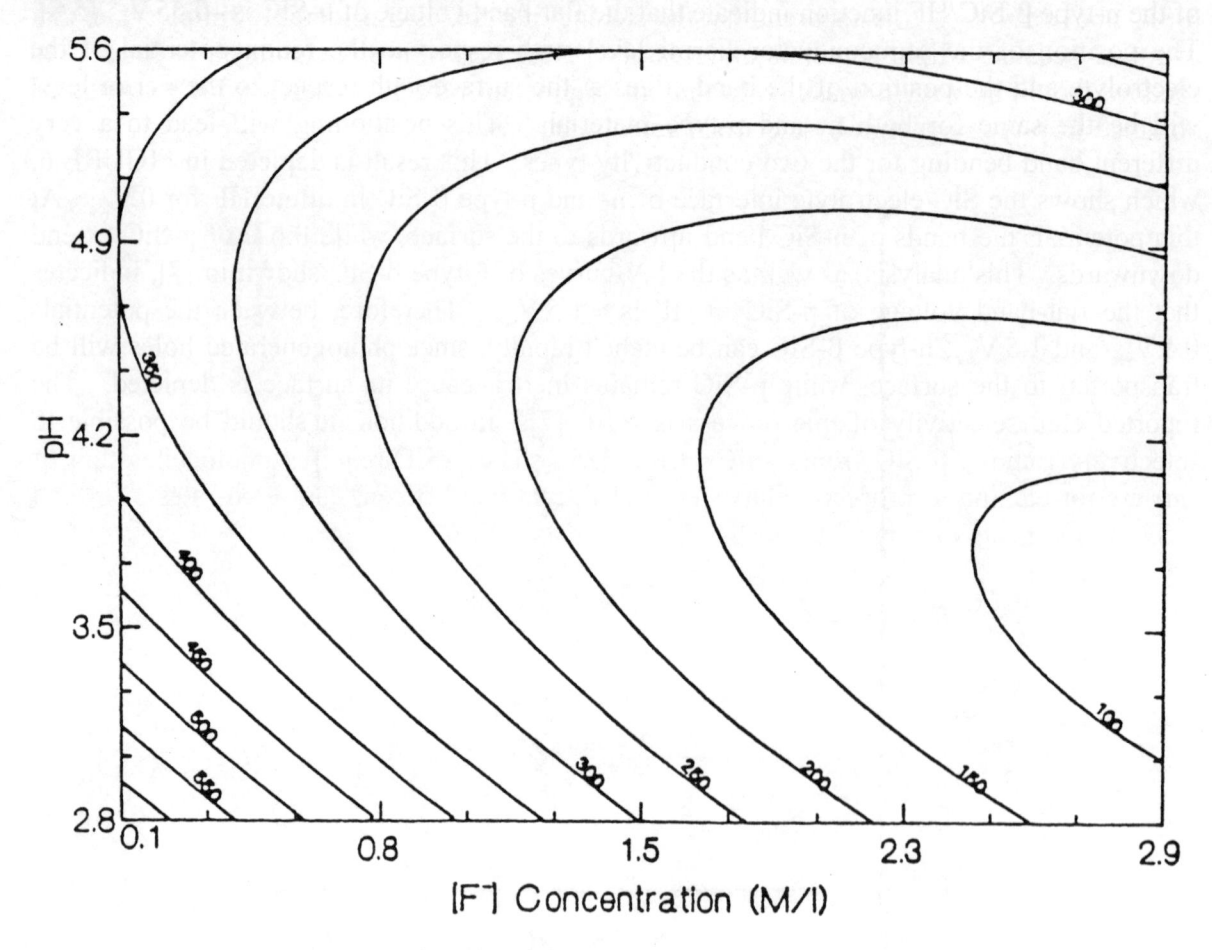

FIGURE 5 Contour plot of electropolishing surface roughness RMS (Å) for n-type β-SiC at 2.0 V_{sce} [12].

TABLE 1 Roughness measurements of SiC surfaces for epitaxial β-SiC on Si and CVD grown polycrystalline α-SiC [12].

Material	RMS (Å)		
	Photoelectrochemical polishing	Electrochemical polishing	Mechanical polishing
EPI SiC	172	201	197
CVD SiC	396	64	92

E   DOPANT SELECTIVE ETCH-STOPS

One of the most useful features of this etching method for SiC is the ability to etch one conductivity type rapidly, while stopping at a layer of opposite conductivity [7]. Etch stops

6.4 Electrochemical etching of SiC

are very useful for the fabrication of electronic devices with thin multilayers and electromechanical microsensors [15]. SiC etch-stop capability can be explained in terms of the band bending of n- and p-type materials in solution. Capacitance-voltage measurements of the n-type β-SiC/HF junction indicate that the flat-band voltage of n-SiC is - 0.55 V_{sce} [5,8]. The position of the semiconductor Fermi level with respect to the redox potential of the electrolyte and the position of the band edges at the surface with respect to the Fermi level will be the same for both n- and p-type materials. This positioning will lead to a very different band bending for the two conductivity types. This result is depicted in FIGURE 6, which shows the SiC/electrolyte interface of n- and p-type β-SiC in dilute HF for $0\,V_{sce}$. At this potential, the bands of n-SiC bend upwards to the surface, while those of p-SiC extend downwards. This analysis, as well as the I-V curves of p-type β-SiC shown in [7], indicates that the flat-band voltage of p-SiC in HF is $\approx 1.5\,V_{sce}$. Therefore, between the potentials $0.5\,V_{sce}$ and $1.5\,V_{sce}$, n-type β-SiC can be etched rapidly, since photogenerated holes will be transported to the surface, while p-SiC remains inert because its surface is depleted. The reported etch selectivity of this process is $> 10^5$ [7]. In addition, it should be possible to selectively remove p-SiC from n-SiC in the dark, since n-SiC requires photogeneration of carriers for etching to proceed. Harris et al [16] used the different dark etch-rates of n- and p-SiC to delineate p-n junctions in β-SiC.

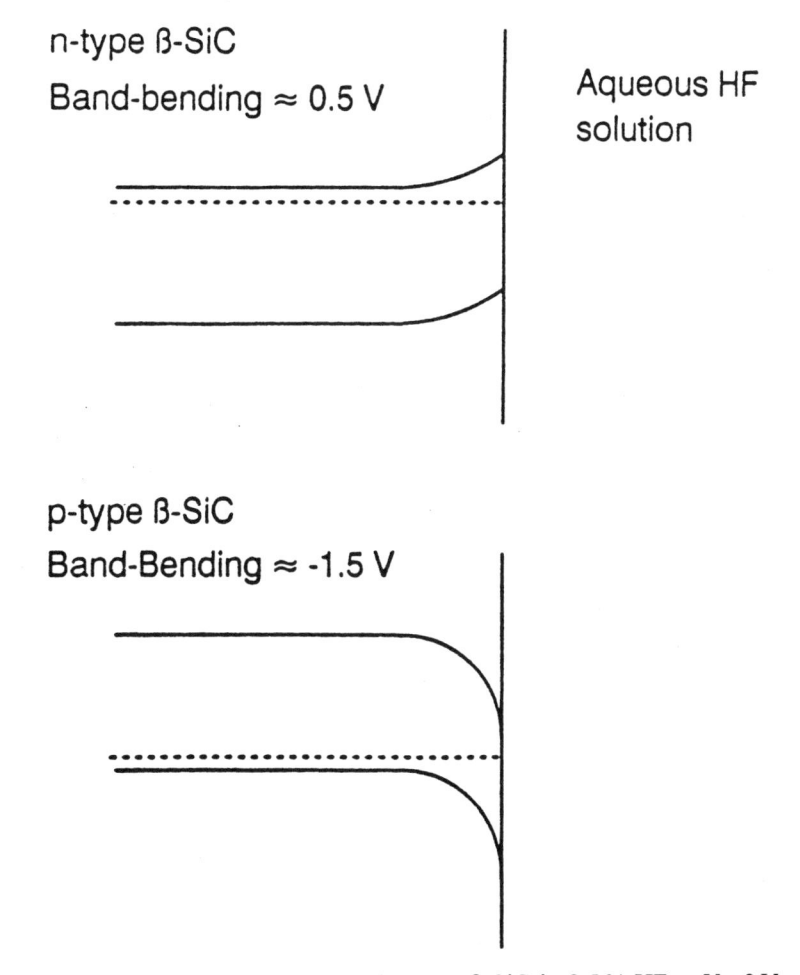

FIGURE 6 SiC/HF interface for n- and p-type β-SiC in 2.5 % HF at $V = 0\,V_{sce}$ [7].

Similar etch-stop principles can be applied to all polytypes of SiC which can be etched photoelectrochemically. Etching and etch stops have been demonstrated in 6H-SiC recently [17].

F POROUS SILICON CARBIDE

Recently, investigators have observed the formation of porous SiC from single-crystal wafers during the anodizing of 6H-, 4H- and 3C-SiC in HF and other electrolytes [18-20]. This effect was first reported by Shor et al [18] while investigating n-type 6H-SiC in HF under UV illumination. TEM analysis indicates that the pore size varies from 10 to 50 nm. The porous material is also discoloured when compared to the SiC and appears to be highly resistive. The porous material can be removed by placing the SiC sample in an oxidizing environment. Cathodoluminescence spectra of the porous SiC illustrate a UV peak above the 3.0 eV bandgap of SiC [21]. Clearly this result needs to be investigated further, but this formation is interesting and may lead to new device technology.

G CONCLUSION

Photoelectrochemical etching has been demonstrated in several polytypes of SiC. TABLE 2 compares the properties of photoelectrochemical etching to those of the more conventional reactive ion etching. The main advantages of photoelectrochemical etching are higher etch rates, high dopant selectivity, less surface damage and the ability to electropolish. One large disadvantage is that an ohmic contact is required on the SiC for electrochemical etching to proceed adding an additional processing step. Further research and development is needed to measure the etch rates and surface morphologies over different conductivities, polytypes, and anodisation conditions. It is also necessary to improve the masking methods so that fine linewidths can be achieved.

TABLE 2 Comparison of the properties of photoelectrochemical etching and reactive ion etching of SiC.

SiC	Reactive ion etching	PEC etching
Etch rates:	100 - 2000 Å min^{-1}	50 Å min^{-1} - 100 µm min^{-1}
Anisotropy:	Very high	Very high
Masks:	Aluminium	Photoresist, Cr
Morphology:	Good	6H-good, 3C-fair
Dopant selectivity:	None	Excellent ($> 10^5$)
Etch rate controllability:	Good; controlled by operating parameters	Excellent; measured and adjusted, in-situ
Polytypes:	3C, 6H, etc.	3C, 6H so far

REFERENCES

[1] R.C. Marshall, J.W. Faust, C.E. Ryan (Eds) [*SiC 1973* (University of South Carolina Press, USA, 1974) p.666]

[2] J.W. Palmour, R.F. Davis, P. Astell-Burt, P. Blackborow [*Science and Technology of Microfabrication* Eds R.W. Howard, E.L. Hu, S. Namba, S.W. Pang (Materials Research Society, Pittsburgh, 1987) p.185]

[3] H. Morisaki, H. Ono, K. Yazawa [*J. Electrochem. Soc. (USA)* vol.131 (1984) p.2081]

[4] M. Glerria, R. Memming [*J. Electroanal. Chem. (Switzerland)* vol.65 (1975) p.163]

[5] J.S. Shor, X.G. Zhang, R.M. Osgood [*J. Electrochem. Soc. (USA)* vol.139 (1992) p.1213]

[6] I. Lauermann, D. Meissner, R. Memming, R. Reineke, B. Kastening [*Dechema-Monographien Band 124* (VCH Verlagsgesellschaft, 1991) p.617]

[7] J.S. Shor, R.M. Osgood, A.D. Kurtz [*Appl. Phys. Lett. (USA)* vol.60 (1992) p.1001]

[8] I. Lauermann, D. Meissner, R. Memming, B. Kastening [*Ext. Abstr. Electrochem. Soc. (USA)* vol.90-2 (1990)]

[9] M.M. Carrabba, J. Li, J.P. Hachey, R.D. Rauh, Y. Wang [*Ext. Abstr. Electrochem. Soc. (USA)* vol.89-2 (1989) p.727]

[10] H.R. Philipp, R.A. Taft [in *Silicon Carbide* Eds J.R. O'Connor, J. Smiltens (Pergamon Press, 1960) p.366]

[11] J.S. Shor, X.G. Zhang, R.M. Osgood, A.D. Kurtz [*Springer Proc. Phys. (Germany)* vol.71 (1992) p.356-61]

[12] T.H. Nguyen, M.M. Carrabba, T.D. Plante, R.D. Rauh [*Ext. Abstr. Electrochem. Soc. (USA)* vol.92-1 (1992)]

[13] G.L. Harris, K. Fekade, K. Wongchotigul [*J. Mater. Sci., Mater. Electron. (UK)* vol.3 (1992) p.162]

[14] R.W. Brander, A.L. Boughey [*Br. J. Appl. Phys. (UK)* vol.18 (1967) p.905]

[15] K.C. Lee [*J. Electrochem. Soc. (USA)* vol.137 (1991) p.2556]

[16] G.L. Harris, K. Fekade [*J. Electrochem. Soc. (USA)* vol.135 (1988) p.405]

[17] J.S. Shor, I. Grimberg, L. Morcos, A.D. Kurtz [presented at the ONR Workshop on SiC Materials and Devices, Charlottesville, VA, 9-11 Sept. 1992]

[18] J.S. Shor, I. Grimberg, B. Weiss, A.D. Kurtz [*Appl. Phys. Lett. (USA)* vol.62 no.22 (1993) p.2836-8]

[19] A.O. Konstantinov, C.I. Harris, E. Janzen [*Appl. Phys. Lett. (USA)* vol.65 no.21 (1993) p.2699-701]

[20] D.M. Collins [M. Eng. Thesis, Howard University, USA, August 1995]

[21] J.S. Shor et al [*Inst. Phys. Conf. Ser. (UK)* no.137, Chapter 3 (1993) p.193-6]

CHAPTER 7

DIFFUSION OF IMPURITIES AND ION IMPLANTATION

7.1 Diffusion and solubility of impurities in SiC
7.2 Ion implantation and anneal characteristics of SiC

7.1 Diffusion and solubility of impurities in SiC

G.L. Harris

August 1994

A INTRODUCTION

In the commercial development of Si devices, diffusion is an important semiconductor fabrication process. This process does not play a major role (except in the case of the sublimation growth of SiC discussed in Chapter 8) in the development of SiC, because the diffusion coefficients for the most part are negligible at temperatures below approximately 1800 °C. As a result of this commercial insignificance, the diffusion process in SiC and its various polytypes has not received a great deal of scientific attention and diffusion data are incomplete. It does, however, appear that the solubility of impurities and their diffusive mobilities in different SiC polytypes are very similar.

B SOLUBILITY LIMITS OF IMPURITIES IN SiC

The solubility limits of impurities in SiC and its various polytypes have not been studied in great detail. The most systematic study has been performed by Vodakov et al [1] in conjunction with the growth of 6H-SiC by using the sublimation 'sandwich' [2,3] technique. No stronger scientific research has been done on the variation of the solubility with polytype. The data presented in this section can probably be used for other polytypes, but no clear study has shown this to be the case.

Al, N, B, Ga and P solubilities in 6H have shown a crystal face dependence [1]. For Al, Ga and B the solubility limit is smaller on the [0001]C face. For P and N the solubility limit is smaller on the [0001]Si face. Vodakov et al [1] suggest that this dependence is caused by the very strong orientation doping anisotropy of 6H-SiC. The data for Al, B and Ga showed very good agreement when measured by the neutron-activation method and electrophysical measurements employing the Hall effect. The dependence on crystal direction of the solubility for N, B, Al, Ga and P in 6H-SiC is listed in TABLE 1. The maximum solubility limits for many of the impurities in 6H-SiC are listed in TABLE 2.

C DIFFUSION COEFFICIENTS IN SiC

In SiC, the high band energy and small interatomic distances affect the migration of impurities. Boron, nitrogen and possibly beryllium substitute for carbon [4]. However, for recent isotopically enriched samples of 3C-, 4H- and 6H-SiC atoms migrate via the carbon or silicon vacancy with the exception of small atoms like beryllium, lithium and hydrogen that can migrate via an interstitial mechanism. A series of theoretical calculations by Bernhok et al [5] has provided some insight into the native defects, doping and diffusion in SiC.

7.1 Diffusion and solubility of impurities in SiC

The results of diffusion studies are summarized in TABLE 3. Some of the impurities can best be characterized by a fast and slow branch type diffusion and the information about each branch is provided where appropriate. Additional more recent studies [6] have been completed but provide no major new significant information about actual diffusion coefficients in SiC.

TABLE 1 Solubility of impurities in 6H-SiC.

N_3 (cm^{-3})	Temperature (°C)	Impurity and crystal face
1.2×10^{20}	2225	N (0001) silicon face
3.0×10^{20}	2000	
4.0×10^{20}	1800	
3.0×10^{20}	2225	N (0001) carbon face
4.8×10^{20}	2000	
6.5×10^{20}	1800	
8×10^{20}	2225	Al (0001) silicon face
7×10^{20}	2000	
5×10^{20}	1800	
5×10^{20}	2225	Al (0001) carbon face
9×10^{19}	2000	
3×10^{19}	1800	
7×10^{19}	2225	B (0001) silicon face
6×10^{19}	2000	
4.8×10^{19}	1800	
6×10^{19}	2225	B (0001) carbon face
4.8×10^{19}	2000	
2×10^{19}	1800	
1.9×10^{19}	2225	Ga (0001) silicon face
1.7×10^{19}	2000	
1.5×10^{19}	1800	
7×10^{18}	2225	Ga (0001) carbon face
5×10^{18}	2000	
3×10^{18}	1800	
7×10^{17}	2225	P (0001) silicon face
8×10^{17}	2000	
9.2×10^{17}	1800	
1.07×10^{18}	2225	P (0001) carbon face
2.3×10^{18}	2000	
4.8×10^{18}	1800	

7.1 Diffusion and solubility of impurities in SiC

TABLE 2 Maximum solubility limits for impurities in 6H-SiC (at temperatures above 2500 °C).

Element	Concentration (cm^{-3})
Li	1.2×10^{18}
Be	8.0×10^{20}
B	2.5×10^{20}
N	6.0×10^{20}
Al	2.0×10^{21}
P	2.8×10^{18}
Sc	3.2×10^{17}
Ti	3.3×10^{17}
Cr	3.0×10^{17}
Mn	3.0×10^{17}
Cu	1.2×10^{17}
Ga	2.8×10^{19}
Ge	3.0×10^{20}
As	5×10^{16}
Y	9.5×10^{15}
In	9.5×10^{16}
Sn	1.0×10^{16}
Sb	8.0×10^{16}
Ho	6.0×10^{16}
Ta	2.0×10^{17}
W	2.5×10^{17}
Au	1.0×10^{17}

TABLE 3 Diffusion coefficients for various impurities in SiC.

Impurity	N	O	Be	B
Temperature range (°C)	1800 - 2450	1800 - 2450	1800 - 2300	1800 - 2300
Effective diffusion coefficient ($cm^2 s^{-1}$) Fast branch Slow branch	- 5×10^{-12}	- $1.5 \times 10^{-16} - 5 \times 10^{-13}$	$2 \times 10^{-9} - 1 \times 10^{-7}$ $3 \times 10^{-12} - 1 \times 10^{-9}$	$2 \times 10^{-9} - 1 \times 10^{-7}$ $2.5 \times 10^{-13} - 3 \times 10^{-11}$

Impurity	Al	Ga	Sc
Temperature range (°C)	1800 - 2300	1800 - 2300	1800 - 2300
Effective diffusion coefficient ($cm^2 s^{-1}$) Fast branch Slow branch	- $3 \times 10^{-14} - 6 \times 10^{-12}$	- $2.5 \times 10^{-14} - 3 \times 10^{-12}$	- $< 10^{-13}$

D   CONCLUSION

Solubility with temperature data for nitrogen, aluminium, boron, gallium and phosphorus, on both Si and C faces, and maximum solubilities for a wider range of impurities at temperatures >2500°C in 6H-SiC have been detailed. No data for other polytypes are available. Diffusion rates of impurities in SiC are very slow (for temperatures between 1800 and 2300°C) whether for those species, such as boron and nitrogen, which migrate via Si/C vacancies or for those, such as beryllium, lithium and hydrogen, which diffuse interstitially. Some impurities show 2-component diffusion profiles.

REFERENCES

[1]   Y.A. Vodakov, E.N. Mokhov, M.G. Ramm, A.D. Roenkov [*Springer Proc. Phys. (Germany)* vol.56 (1992) p.329-34]

[2]   Y.A. Vodakov, E.N. Mokhov, M.G. Ramm, A.D. Roenkov [*Krist. Tech. (Germany)* vol.I4 (1979) p.729]

[3]   E.N. Mokhov, I.L. Shulgina, A.S. Tregubova, Y.A. Vodakov [*Cryst. Res. Technol. (Germany)* vol.I6 (1981) p.879]

[4]   Yu. A. Vodakov, E.N. Mokhov [*Silicon Carbide - 1973* Eds R.C. Marshall, J.W. Faust, E.E. Ryan (University of South Carolina Press, Columbia, SC, 1974) p.266-83]

[5]   J. Bernhok, S.A. Kajihara, C. Wang, A. Antonelli, R.F. Davis [*Mater. Sci. Eng. B (Switzerland)* vol.11 (1992) p.265-72]

[6]   A.G. Zubatov, V.G. Stepanov, Yu. A. Vodakov, E.N. Mokhov [*Sov. Tech. Phys. Lett. (USA)* vol.8 (1982) p.120]

7.2 Ion implantation and anneal characteristics of SiC

K. Wongchotigul

September 1994

A INTRODUCTION

In the commercial semiconductor world of silicon (Si) devices, great importance is attached to diffusion and ion implantation. These processes play a key role in the local doping performance through windows in a thermally grown oxide film. Because the diffusion coefficients for most dopants in silicon carbide (SiC) are negligible, at temperatures below 1800 °C, the development of ion implantation into SiC for microelectronic technology is of major importance.

B IMPLANTATION AND ANNEALING CHARACTERISTICS

Many ion species implanted in SiC have been experimented with in the past two decades. However, there are a few species that have been studied in detail. The behaviour of these ions is now briefly reviewed.

B1 Boron (B)

This is a light ion, with the experimental projection range of 0.1 - 0.45 µm for the implanted energy of 100 - 200 keV [1,9]. The p-type conduction in B-implanted SiC has been very weak, although it was realized in samples which were doped during growth at 1630 K with a concentration of B near the solubility limit ($\sim 2 \times 10^{20}$ cm^{-3}) [1]. The annealing characteristic of B-implanted SiC reported in [1] indicated that the annealing temperature between 1273 and 1573 K caused solid phase epitaxial regrowth of the amorphous layer, and there was no detectable formation of coarse precipitation near the surface and dislocation loops near the centre of the original amorphous layer. Weak p-type conduction was observed in the temperature range of 1473 - 1773 K; however, n-type conduction was observed after annealing above 1973 K. This phenomenon could be a result of the out-diffusion of B.

B2 Aluminium (Al)

Implantation of SiC with Al has been carried out with more success than implantation with B in converting the undoped SiC to p-type SiC. In fact, almost all of the ion-implanted p-n junction diodes were processed using Al$^+$. Spitznagel et al [5] reported a relatively low annealing temperature (573 K at a pressure of 2×10^{-6} torr) to obtain p-type conductivity in Al-implanted β-SiC. The experimental projection range was reported to be approximately 0.08 - 0.38 µm for the implanted energy of 90 - 200 keV [1,5,8,9,18]. Gudkov et al [3] also reported a critical dose at which the implanted layer could be recrystallized. This critical dose is reported to be approximately 3×10^{15} cm^{-2}.

B3 Gallium (Ga)

This ion has not been used extensively. There are only a few reports of Ga-implanted SiC. The experimental projection range was reported to be 0.08 - 0.2 µm for the implanted energy of 90 - 360 keV [4,14]. Burdel et al [4] reported a critical dose of 1.8×10^{14} cm^{-2}, at which the implanted layer could be recrystallized. For a dose higher than the critical value, the structural perfection of the layer increases with the amount of Ga evaporated from the layer during annealing.

B4 Nitrogen (N)

This ion has been used quite extensively for forming local n$^+$-layers by implantation, but very little annealing behaviour of N-implanted SiC has been studied in detail. Spitznagel et al [5] reported the critical value of 10^{16} cm^{-2} at which the implanted layer could be recrystallized. Suzuki et al [6] reported that up to 50% of implanted ions could be activated by either annealing at temperatures as high as 1300°C for 0.5 hr or annealing at 1100°C for as long as 5 hr.

B5 Phosphorus (P)

This is an alternative ion for forming an n$^+$-layer by implantation. This ion is, however, heavier than N; hence, the diffusion coefficient is extremely small. Suzuki et al [6] reported that no change in the depth profile occurred after annealing, but the activation is much lower than that of N (approximately half that of N). Davis [1,18] reported the critical dose of 10^{15} cm^{-2} at which the implanted layer could be recrystallized.

C FURTHER READING

There are many articles regarding the use of ion implantation in SiC. However, most of these articles describe the use of ion implantation for device fabrication only. The author has summarized the data from some of these articles in TABLE 1.

D CONCLUSION

There are still many problems associated with the use of ion implantation to dope SiC. In particular, a low annealing temperature is required to obtain a high percentage of activated ions in order to produce material suitable for device application.

7.2 Ion implantation and anneal characteristics of SiC

TABLE 1 Ion implantation in SiC (E_{imp} and T_{imp} are the energy and temperature of implantation respectively, cc is the carrier concentration and R_p is the range parameter).

Ion	Dose (cm^{-2})	E_{imp} (keV)	T_{imp} (K)	Annealing method	cc (cm^{-3})	R_p (Å keV^{-1})	Ref
Al	9×10^{14}	190	300	RTA in Ar at 1000 - 1900 °C		10.5	[1]
	4.8×10^{14}	100	823	30 min. in O$_2$ followed by 30 min. in Ar at 1473 K			[2]
	$0.625 - 5 \times 10^{16}$	40 - 70	300	Static annealing in Ar at 2050 °C followed by pulse e-beam in vacuum (5 ms pulse width, e-beam power $\sim 2 \times 10^4$ W cm^{-2})			[3]
	$1.8 - 13 \times 10^{14}$	130 - 300	623 - 1053	30 min. in O$_2$ followed by 30 min. in Ar at 1473 K	2×10^{16}	10.8	[4]
	$0.1 - 5 \times 10^{15}$	150	300	2×10^{-6} torr at 573 K		8.3	[5]
	3×10^{15}	90	300	15 - 180 s at 1500 - 1950 °C			[7]
	5×10^{16}	90	300	15 s at 1800 °C		8.8	[8]
	10^{15}	100	300	30 min. in Ar at 1365 °C		12.5	[9]
	5×10^{16}	200	300	Diffusion furnace at 1750 °C			[10]
	6×10^{14}	100	300	RTA in Ar at 1000 - 1900 °C	0.5 %	9.1	[18]
B	2×10^{15}	200	300	RTA in Ar at 1000 - 1900 °C		17.5	[1,18]
	1.5×10^{15}	100	300	RTA in Ar at 1000 - 1900 °C		18	[1,18]
	$0.4 - 10 \times 10^{14}$	30, 50, 100	300	30 min. in Ar at 1365 °C			[9]
Ga	$1.4 - 6 \times 10^{16}$	40 - 90	300	5 s in Ar at 1900 - 2200 °C		11.1	[4,12]
N	0.9×10^{14}	90	823	30 min. in dry O$_2$ followed by 30 min. in Ar at 1473 K			[2]
	1.3×10^{14}	180	823	30 min. in dry O$_2$ followed by 30 min. in Ar at 1473 K			[2]
	$5 \times 10^{14} - 10^{16}$	75	300	30 min. in vacuum 2×10^{-6} torr		17.3	[5]
	10^{14}	25	300	Diffusion furnace for 7 min. at 1470 °C followed by RTA for 4 min. at 1050 °C	90 %	20	[16]
	$0.11 - 3.6 \times 10^{15}$	50 - 200	195	RTA in Ar for 300 s at 1000 - 1800 °C	40 %	20	[1,18]
	$10^{14} - 10^{15}$	30	300	500 - 1650 °C			[19]
P	7.7×10^{13}	110	300	RTA in Ar for 300 s at 1000 - 1900 °C	20 %	9.1	[1,18]
	$0.3 - 1 \times 10^{15}$	30 - 100	300	30 min. 1300 °C	50 %	9	[6]
	$0.3 - 1 \times 10^{15}$	25 - 100	300	IR radiative heating furnace for 60 min. at 1080 °C			[17]
Sb	2.3×10^{13}	40		500 - 1650 °C	50 %		[19]
	10^{14}			500 - 1650 °C	50 %		
Kr	$2 \times 10^{12} - 4 \times 10^{15}$	500	300	1300 - 1900 °C in vacuum 10^{-5} torr			[21]

RTA = Rapid Thermal Anneal

REFERENCES

[1] R.F. Davis [*Semiannual Progress Report on Fundamental Studies and Device Development in β-SiC* (North Carolina State University, Department of Materials Engineering, NC, March-August 1985)]

[2] J.A. Edmond, K. Dass, R.F Davis [*J. Appl. Phys. (USA)* vol.63 no.3 (1988) p.922-9]

[3] V.A. Gudkov, G.A. Krysov, V.V. Makarov [*Sov. Phys.-Semicond. (USA)* vol.18 no.6 (1984) p.684-5]

[4] K.K. Burdel, A.V. Suvorov, N.G. Chechenin [*Sov. Phys.-Solid State (USA)* vol.32 no.6 (1990) p.975-7]

[5] J.A. Spitznagel, S. Wood, W.J. Choyke, N.J. Doyle [*Nucl. Instrum. Methods (Netherlands)* vol.B16 (1982) p.237-43]

[6] A. Suzuki, K. Furukawa, Y. Fuji, M. Shigeta, S. Nakajama [*Springer Proc. Phys. (Germany)* vol.56 (1990) p.101-9]

[7] E.V. Kalinina, A.V. Suvorov, G.F. Kholuyanov [*Sov. Phys.-Semicond. (USA)* vol.14 no.6 (1980) p.652-4]

[8] E.V. Kalinina, N.K. Prokof'eva, A.V. Suvorov, G.F. Kholuyanov, V.E. Chelnokov [*Sov. Phys.-Semicond. (USA)* vol.12 no.12 (1978) p.3469-71]

[9] R.E. Avila, J.J. Kopanski, C.D. Fung [*J. Appl. Phys. (USA)* vol.62 no.8 (1987) p.3469-71]

[10] V.E. Oding et al [*Sov. Phys.-Semicond. (USA)* vol.18 no.4 (1984) p.434-6]

[11] O.J. Marsh, H.L. Dunlap [*Radiat. Eff. (UK)* vol.6 (1970) p.301-10]

[12] K.K. Burdel, P.V. Varakin, V.N. Makarov, A.V. Suvorov, N.G. Chechenin [*Sov. Phys.-Solid State (USA)* vol.30 no.2 (1988) p.364-5]

[13] E.E. Violin, E.A. Gorin, E.N. Potapov, Yu.M. Tairov [*Sov. Tech. Phys. Lett. (USA)* vol.10 (1984) p.645-6]

[14] J.A. Edmond, S.P. Withrow, W. Wadlin, R.F. Davis [*Proc. Symp. Interfaces, Superlattices and Thin Films* Boston, MA, USA, 1 - 6 Dec. 1986, Eds J.D. Dow, I.K. Schuller (Mater. Res. Soc., Pittsburgh, USA, 1987) p.193-8]

[15] B.I. Vishnevskaya et al [*Sov. Phys.-Semicond. (USA)* vol.22 (1988) p.414-7]

[16] P. Glasow, G. Ziegler, W. Suttrop, G. Pensl, R. Helbig [*Proc. SPIE (USA)* vol.868 (1987) p.41]

[17] K. Shibahara, T. Saito, S. Nishino, H. Matsunami [*IEEE Electron Device Lett. (USA)* vol.EDL-7 (1986) p.692-3]

[18] R.F. Davis [*Annual Letter Report on Fundamental Studies and Devices Development in Beta SiC* (North Carolina State University, Department of Materials Engineering, 1986) Feb. 1985 - Jan. 1986]

[19] R.R. Hart, H.L. Dunlap, O.J. Marsh [*Radiat. Eff. (UK)* vol.9 (1971) p.261-6]

[20] N.V. Kodrau, V.V. Makarov [*Sov. Phys.-Semicond. (USA)* vol.11 no.5 (1977) p.572-3]

[21] H. Matzke, M. Koniger [*Phys. Status Solidi A (Germany)* vol.1 (1970) p.469]

[22] V.M. Gusev, K.D. Demakov, M.G. Kosaganova, M.B. Reifman, V.G. Stolyarova [*Sov. Phys.-Semicond. (USA)* vol.9 no.7 (1976) p.820-2]

CHAPTER 8

GROWTH

8.1 Bulk growth of SiC
8.2 Sublimation growth of SiC
8.3 Chemical vapour deposition of SiC
8.4 LPE of SiC and SiC-AlN

8.1 Bulk growth of SiC

S. Nishino

June 1995

A INTRODUCTION

Production of large sized bulk SiC has been long expected to realise electronic devices using this material. Lely's work was a guide to produce reasonable sized crystals for academic research [1,2]; however, large sized bulk material was not realised using this method. For a long time, sublimation growth was carried out without using seed crystals [3-5]. Surface morphology and impurity effects of the substrate were studied using the sublimation furnace [6-10]. Recently, wafers of SiC about 2" in diameter were made from SiC bulk material produced by sublimation growth [11]. Many data are brought together in this Datareview.

B SINGLE CRYSTAL BULK GROWTH

In sublimation growth, sublimed SiC is transported in the vapour phase to a SiC seed crystal at a lower temperature. Typical parameters used to control sublimation growth are substrate temperature, source temperature, temperature gradient between substrate and source and ambient pressure. These parameters affect growth rate and the quality of the crystal. Dissociation of SiC at high temperature in high vacuum was studied using a mass spectrometer. These data are useful in considering the partial pressures of chemical species [12-14].

Two types of crucible are used for sublimation growth. One is composed of a single wall crucible made of graphite as shown in FIGURE 1(a) [15-23] and the other is a double wall or coaxially arranged cylinder type crucible as shown in FIGURE 1(b) [24-26]. In the former case, source material at high temperature is directly diffused to the seed substrate at a lower temperature. However, in the latter case, the sublimed SiC clusters must be diffused through a porous thin-graphite wall. This porous graphite should filter out the extraneous sublimation.

In 1978, Tairov and Tsvetkov [24] reported a sublimation technique to produce SiC boules for device application. They produced an 8 mm diameter by 8 mm long boule of SiC on a seed crystal placed within a graphite crucible. In a further study, growth of SiC boules up to 14 mm diameter and 18 mm in length were obtained [15].

Many researchers then joined the sublimation growth field and much progress has been reported. In 1983, Ziegler reported a 20 mm diameter, 24 mm long 6H-SiC crystal grown by a sublimation method [25,26]. In 1987, Carter reported the seeded growth of 15 mm diameter, 8 mm thick polytype SiC single crystals [30,31]. In 1989, Nakata reported 6H-SiC polytype crystals 33 mm in diameter and 14 mm long [20].

Transport mechanisms in sublimation growth are complicated. Growth rate increases with increasing source temperature, increasing source to seed distance, decreasing pressure, and decreasing crystal-source distance [19,20,27-33].

8.1 Bulk growth of SiC

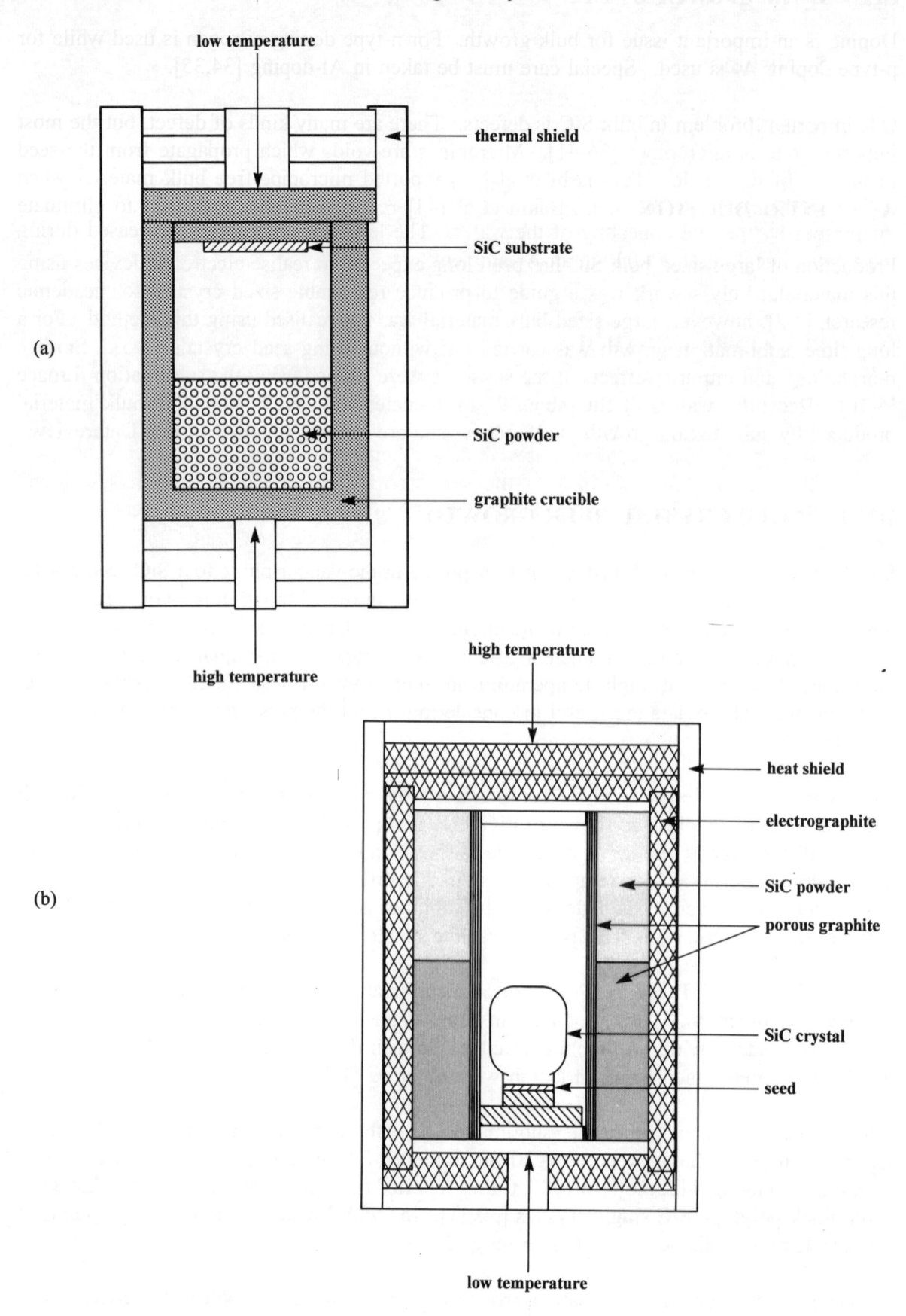

FIGURE 1 Sublimation growth crucibles (a) single-wall graphite type and (b) double-wall type.

Doping is an important issue for bulk growth. For n-type doping nitrogen is used while for p-type doping Al is used. Special care must be taken in Al-doping [34,35].

One important problem in bulk SiC is defects. There are many kinds of defect, but the most important one is micropipes [36-41]. Micropipes are voids which propagate from the seed to the top of the boule. Takahashi et al [43] reported micropipe-free bulk material when α-face seed crystals were used. Bakin et al [44] reported another approach to eliminate micropipes by thermal annealing of the wafer. The length of micropipes decreased during long time annealing at 2290 K in an Ar ambient.

C   POLYTYPE CONTROL

C1   6H-SiC

Most of the reports described above involve 6H-SiC bulk growth on 6H-SiC seed substrates. Yoo et al [46] used 3C-SiC(100) as the substrate for 6H-SiC bulk growth and found 6H-SiC(01$\bar{1}$4) bulk material was produced. They reported crystals with diameters of 15 mm and 6 mm long. By using different planes of 6H-SiC, polytype control will be easily achieved.

C2   4H-SiC

Polytype control of SiC is a difficult problem because so many parameters, such as pumping speed of the reaction chamber, distance between source and substrate and the temperature gradient between source and substrate are involved. Tairov [18] pointed out that the initial stage of crystal growth is key to controlling the polytypes, as shown in FIGURE 2. Koga and co-workers [47,48] reported intentional polytype control by selecting specific temperature ranges of substrates (T_s) and source temperatures (T_o). They obtained 6H-SiC at $T_o = 2200\,°C$ and 4H-SiC material at $T_o = 2300\,°C$. However, Kanaya et al [22] reported different combinations of T_o and T_s. These are 4H-SiC at $T_o = 2280\,°C$ and $T_s = 2200\,°C$, and 6H-SiC at $T_o = 2400\,°C$ and $T_s = 2350\,°C$. In these two reports the C-face of 6H-SiC(0001) was used as the substrate for 4H-SiC growth. The initial nucleation stage and consequent growth process are key to controlling these polytypes [49].

C3   3C-SiC

3C-SiC bulk material is required for device applications. However, usable sizes of 3C-SiC bulk are not produced. In 1987, Yoo et al [50] obtained 3C-SiC films 100 μm thick on 3C-SiC CVD grown substrates (20 μm thick) by the sublimation method. In 1993, Furukawa [51] reported 10 mm diameter, 4 mm long 3C-SiC crystals grown by a sublimation method. Several attempts followed, but reproducible bulk crystals like 6H-SiC polytype have not been reported as yet [52-55].

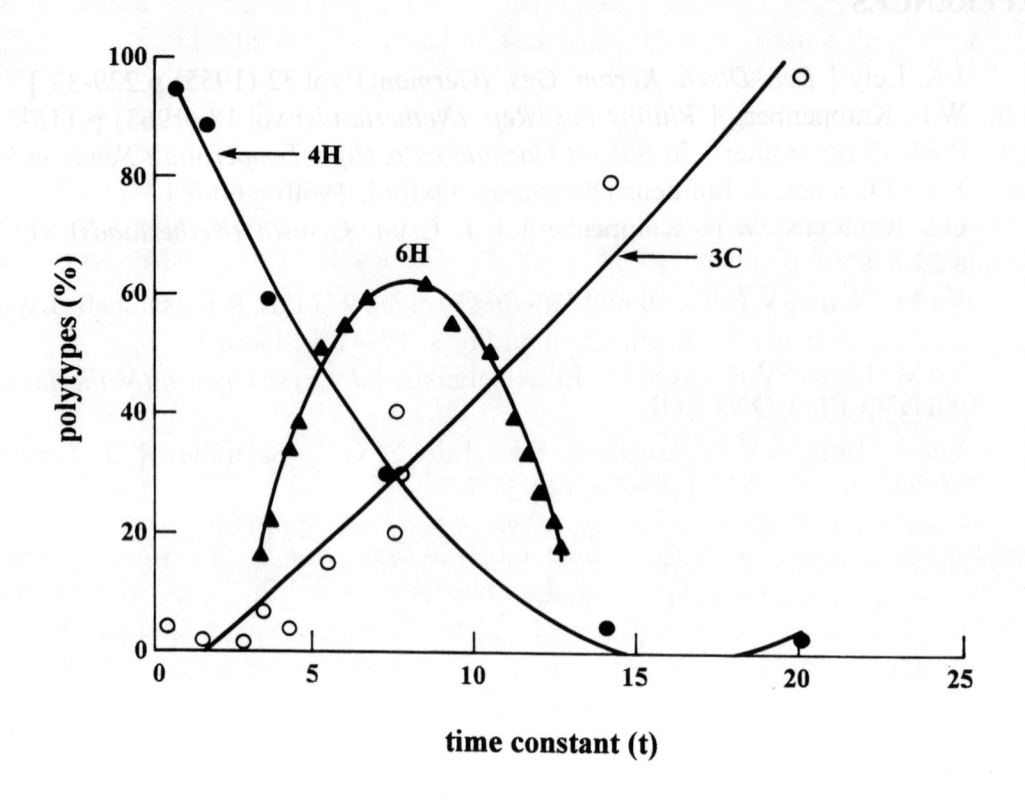

FIGURE 2 Dependence of SiC crystal yield of 3C, 6H and 4H polytypes on the deposition kinetics in the initial stage of growth. After [64].

D EPITAXIAL GROWTH

Epitaxial growth was carried out by modifying the sublimation method, the so-called sandwich method [8,56-60]. In this method, source and substrate are separated by a small gap of about 1 mm. The growth is carried out close to equilibrium in this arrangement. In this method, the substrate is thermally etched before sublimation begins. This makes it possible to grow, at low rates, thin homo-epitaxial layers on 6H and 4H α-SiC with residual impurity concentration. This method is used for device fabrication.

E CONCLUSION

Bulk growth has progressed from a research stage to a production stage, although the serious and fundamental problem of defects still remains. By gathering many research results this problem will be solved in the near future. Another important issue is the wafer polishing technique. As far as mechanical polishing is concerned, surface damage remains and this produces poor interfaces for subsequent CVD growth and device application. Some mechano-chemical polishing techniques, which are well established in Si and GaAs wafer processing, should be established for SiC. There are many good reviews of sublimation growth [18,61,64,65].

REFERENCES

[1] J.A. Lely [*Ber. Dtsch. Keram. Ges. (Germany)* vol.32 (1955) p.229-33]
[2] W.F. Knippenberg [*Philips Res. Rep. (Netherlands)* vol.18 (1963) p.116-274]
[3] R.M. Hergenrother [in *Silicon Carbide as a High Temperature Semiconductor* Eds J.R. O'Conner, J. Smiltens (Pergamon, Oxford, 1960) p.60-6]
[4] C.J. Kapteyns, W.F. Knippenberg [*J. Cryst. Growth (Netherlands)* vol.7 (1970) p.20-8]
[5] Yu.M. Tairov, V.F. Tsvetkov [*Silicon Carbide 1973* Eds. R.C.Marshall, J.W.Faust Jr., C.E. Ryan (Univ. of South Carolina Press, 1974) p.146-60]
[6] Yu.M. Tairov, V.F. Tsvetkov, I.I. Khlebnikov [*J. Cryst. Growth (Netherlands)* vol.20 (1973) p.155-7]
[7] Yu.M. Tairov, V.F. Tsvetkov, S.K. Lilovg, G.K. Safaraliev [*J. Cryst. Growth (Netherlands)* vol.36 (1976) p.147-51]
[8] I. Swiderski [*J. Cryst. Growth (Netherlands)* vol.32 (1976) p.350-6]
[9] S.K. Lilov, Yu.M. Tairov, V.F. Tsvetkov, M.A. Chernov [*Phys. Status Solidi A (Germany)* vol.37 (1976) p.143-50]
[10] J. Auleytner, I. Swiderski, W. Zahorowski [*J. Cryst. Growth (Netherlands)* vol.38 (1977) p.192-6]
[11] J.W. Palmour, V.T. Tsvetkov, L.A. Lipkin, C.H. Carter Jr. [*Proc. ISCS'94 (USA)* (1994) San Diego]
[12] J. Drowart, G. de Maria, M.G. Inghram [*J. Chem. Phys. (USA)* vol.29 (1958) p.1015-21]
[13] J. Drowart, G. de Maria [SiC conference]
[14] R.E. Honig [*J. Chem. Phys. (USA)* vol.22 (1954) p.1610-1]
[15] Yu.M. Tairov, V.F. Tsvetkov [*J. Cryst. Growth (Netherlands)* vol.52 (1981) p.146-50]
[16] V.I. Levin, Yu.M. Tairov, M.G. Travadzhyan, V.F. Tsvetkov [*Inorg. Mater. (USA)* vol.14 (1978) p.830-3]
[17] S. Lilov, Yu.M. Tairov, V.F. Tsvetkov [*J. Cryst. Growth (Netherlands)* vol.46 (1979) p.269]
[18] Yu.M. Tairov, V.F. Tsvetkov [*Prog. Cryst. Growth Charact. (UK)* vol.7 (1983) p.111-62]
[19] K. Koga, T. Nakata, T. Niina [*Extend. Abst. 17th Conf. Solid State Devices and Materials* Tokyo (Business Center for Academic Societies, Tokyo, Japan, 1985) p.249-52]
[20] T. Nakata, K. Koga, Y. Matsushita, Y. Ueda, T. Niina [*Springer Proc. Phys. (Germany)* vol.43 (1989) p.26-34]
[21] W.S. Yoo, S. Nishino, H. Matsunami [*Springer Proc. Phys. (Germany)* vol.43 (1989) p.35-9]
[22] M. Kanaya, J. Takahashi, Y. Fujiwara, A. Moritani [*Appl. Phys. Lett. (USA)* vol.58 (1991) p.56-8]
[23] N.L. Buchan, D.N. Henshall, W.S. Yoo, P.A. Mailloux, M.A. Tischler [*Inst. Phys. Conf. Ser. (UK)* no.137 (1994) p.113-6]
[24] Yu.M. Tairov, V.F. Tsvetkov [*J. Cryst. Growth (Netherlands)* vol.43 (1978) p.209-12]
[25] G. Ziegler, P. Lanig, D. Theis, C. Weyrich [*IEEE Trans. Electron Devices (USA)* vol.ED-30 (1983) p.277-81]

[26] G. Ziegler [German Patent: Offenlegungsshrifft, DE3220727 A1 (February 23 1984)]
[27] D.L. Barrett, R.G. Seidensticker, W. Gaida, R.H. Hopkins, W.J. Choyke [*J. Cryst. Growth (Netherlands)* vol.109 (1991) p.17-23]
[28] D.L. Barrett et al [*J. Cryst. Growth (Netherlands)* vol.128 (1993) p.358-62]
[29] D.L. Barrett, R.G. Seidensticker, W. Gaida, R.H. Hopkins, W.J. Choyke [*Springer Proc. Phys. (Germany)* vol.56 (1992) p.33]
[30] C.H. Carter Jr., L. Tang, R.F. Davis [presented at 4th Natl. Review Meeting on the Growth and Characterisation of SiC, Raleigh, NC, USA (1987)]
[31] R.F. Davis, C.H. Carter Jr., C.E. Hunter [US Patent No.4,866,0005 (September 12 1989)]
[32] H.J. Kim, D.W. Shin [*Springer Proc. Phys. (Germany)* vol.56 (1992) p.2328]
[33] I. Garcon, A. Rouault, M. Anikin, C. Jaussaud, R. Madar [*Mater. Sci. Eng. B (Switzerland)* vol.29 (1995) p.90-3]
[34] K. Koga, T. Nakata [*Kokai Tokkyo Koho* (Publication of Unexamined Patent Application) (A) No.1-48495 (April 25 1989)]
[35] Yu.M. Tairov, V.P. Rastegaev [*Trans. 2nd Int. High Temperature Electronics Conf. (HiTEC)* Charlotte, NC, USA, 5-10 June 1994, p.159-63]
[36] E.N. Mokhov, E.I. Radovanova, A.A. Sitnikova [*Springer Proc. Phys. (Germany)* vol.56 (1992) p.231]
[37] D. Black, L. Robin [*Inst. Phys. Conf. Ser. (UK)* no.137 (1994) p.337-40]
[38] R.C. Glass, C.I. Harris, V.F. Tsvetkov, P.F. Fewster, J.E. Sundgren, E. Janzen [*Inst. Phys. Conf. Ser. (UK)* no.137 (1994) p.165-8]
[39] J.A. Powell, P.G. Neudeck, D.L. Larkin, J.W. Yang, P. Pirouz [*Inst. Phys. Conf. Ser. (UK)* no.137 (1994) p.161-4]
[40] Yu.M. Tairov, V.F. Tsvetkov [*Springer Proc. Phys. (Germany)* vol.56 (1992) p.41]
[41] S. Nishino, Y. Kojima, J. Saraie [*Springer Proc. Phys. (Germany)* vol.56 (1992) p.15-21]
[42] S. Nishino, T. Higashino, T. Tanaka, J. Saraie [*J. Cryst. Growth (Netherlands)* vol.147 (1995) p.339]
[43] J. Takahashi, M. Kanaya, Y. Fujiwara [*J. Cryst. Growth (Netherlands)* vol.135 (1994) p.61-70]
[44] A.S. Bakin, S.I. Dorozhkin [*Trans. 2nd Int. High Temperature Electronics Conf. (HiTEC)* Charlotte, NC, USA, 5-10 June 1994 p.169-73]
[45] R.A. Stein [*Physica B (Netherlands)* vol.185 (1993) p.211-6]
[46] W.S. Yoo, S. Nishino, H. Matsunami [*J. Cryst. Growth (Netherlands)* vol.99 (1990) p.278-83]
[47] K. Koga et al [*Ext. Abstracts Electrochemical Society Fall Meeting* Hollywood, Florida, USA, no.427 (Electrochemical Society, Pennington, NJ, USA, 1989) p.698]
[48] K. Koga, Y. Ueda, T. Nakata [*Kokai Jitsuyoushinan Koho* (Publication of Unexamined Utility Model Application) no.2-48495 (February 19 1990)]
[49] R.A. Stein, P. Lanig [*J. Cryst. Growth (Netherlands)* vol.131 (1993) p.71-4]
[50] W.S. Yoo, S. Nishino, H. Matsunami [*Mem. Fac. Eng. Kyoto Univ. (Japan)* vol.49 (1987) p.21-31]
[51] K. Furukawa, Y. Tajima, H. Saito, Y. Fujii, A. Suzuki, S. Nakajima [*Jpn. J. Appl. Phys. (Japan)* vol.32 (1993) p.L645-7]. Diameter of bulk crystal is 10 mm (private communication).
[52] M. Omori, H. Takei, T. Fukuda [*Jpn. J. Appl. Phys. (Japan)* vol.28 (1989) p.1217-20]

[53] V. Shields, K. Fakade, M. Spencer [*Appl. Phys. Lett. (USA)* vol.62 (1993) p.1919-21]

[54] V. Shields, K. Fakade, M. Spencer [*Inst. Phys. Conf. Ser. (UK)* no.137 (1994) p.21-4]

[55] J.W. Yang, S. Nishino, J.A. Powell, P. Pirouz [*Inst. Phys. Conf. Ser. (UK)* no.137 (1994) p.25-8]

[56] R.T. Yakimova, A.A. Kalnin [*Phys. Status Solidi A (Germany)* vol.32 (1975) p.297-300]

[57] M.M. Anikin, A.A. Lebedev, S.N. Pyatko, A.M. Strel'chuk, A.L. Syrkin [*Mater. Sci. Eng. B (Switzerland)* vol.11 (1992) p.113-5]

[58] Yu.M. Vodakov, E.N. Mokhov, M.G. Ramm, A.D. Renkov [*Springer Proc. Phys. (Germany)* vol.56 (1992) p.323]

[59] Yu.A. Vodakov, E.N. Mokhov, M.G. Ramm, A.D. Roenkov [*Springer Proc. Phys. (Germany)* vol.56 (1992) p.329]

[60] M.M. Anikin, A.A. Lebedev, M.G. Rastegaeva, A.M. Strel'chuk, A.L. Syrkin, V.E. Chelnokov [*Inst. Phys. Conf. Ser. (UK)* no.137 (1994) p.99-100]

[61] J.A. Powell, L.G. Matus [*Springer Proc. Phys. (Germany)* vol.34 (1989) p.2-12]

[62] P.A. Gassow [*Springer Proc. Phys. (Germany)* vol.34 (1989) p.13-33]

[63] H. Matsunami [*Springer Proc. Phys. (Germany)* vol.43 (1989) p.2-7]

[64] Yu.M. Tairov [*Mater. Sci. Eng. (Switzerland)* vol.29 (1995) p.83]

[65] Yu.M. Tairov, Y.A. Vodakov [*Top. Appl. Phys. (Germany)* vol.17 (1977) p.31-61]

8.2 Sublimation growth of SiC

A.O. Konstantinov

January 1995

A   INTRODUCTION

Sublimation is one of the main methods of growing silicon carbide. This method is employed for growth of the material for abrasive applications as well as for the growth of single crystals and epitaxial layers for use in semiconductor electronics. The idea of the method is fairly simple, and is based on material transport from a hot source of material to a substrate which rests at a somewhat lower temperature. The transport is performed by the intrinsic vapour of the material at a high temperature, usually in the range 1600 - 2700 °C.

The high efficiency of the method is evidenced by the fact that the first sublimation method of growing the crystalline material, the Acherson method, was proposed at the beginning of the 20th century and it is used today with only small variations. The Acherson process yields material for abrasive use and the rate of production is really very high, more than half a million tons per year [1]. No other technique of growing silicon carbide can be compared with sublimation in its productivity and efficiency.

The crystalline material, obtained in the Acherson process, has two major drawbacks which hindered its electronic applications. The crystals are of a small size and they are contaminated. This is related to the spontaneous nucleation of the crystals and to the poorly controlled growth conditions. The problem of doping control has been partially solved [2,3]. The latter, [3], proposed a laboratory method of growing crystals similar to those obtained in the Acherson process. The Lely method has led to a degree of progress both in crystal growth and in device fabrication [4]. However, it is virtually impossible to obtain large crystals by the Lely method, as in the Acherson furnaces. Attempts to employ directed crystallisation in the sublimation growth of SiC have been undertaken by several groups. However, they were rather unsuccessful due to uncontrolled nucleation and twinning [5].

A new approach to the directed crystallisation of silicon carbide has been proposed by Vodakov and Mokhov [6]. Their idea was to exclude the conditions which could permit any uncontrolled nucleation. They employ a nearly flat source positioned close to the substrate and perform the growth under near-equilibrium conditions. This has ensured a high quality of the grown material. The method was named the sublimation 'sandwich' method and it appeared to be very effective.

The studies of growth by the 'sandwich' method have provided a better understanding of the sublimation growth peculiarities and they have formed the basis of the new approach to the bulk crystal growth of silicon carbide. The first successful results in this direction were reported by Tairov and Tsvetkov [7,8]. Currently, similar studies are being performed by a number of research groups and rather impressive progress has been achieved thus far: see Datareview 8.1.

The possibilities of the sublimation growth methods are not limited, however, to ingot growth. They are also very effective in obtaining doped epitaxial layers and device structures. Since the growth proceeds under near-equilibrium conditions, the growth temperature range is extremely broad, from 1600 to 2500 °C. Thus the optimum conditions can be chosen for obtaining high quality grown layers or device structures. The sublimation growth technique appeared to be the first method to provide fabrication of high quality p-n junctions in SiC which could be free from the defects which led to early electric breakdown [9]. The defect-free p-n junctions in SiC have a unique electric strength, greatly exceeding that of a p-n junction in any other semiconductor material. This opens up opportunities of employing silicon carbide in semiconductor electronics.

B EQUIPMENT AND METHODS OF GROWING SILICON CARBIDE BY SUBLIMATION

There exist no materials which are fully inert to the products of silicon carbide evaporation at the high temperatures required for the sublimation growth. Thus the furnace, the thermal insulation, the crucible and the internal parts of the crucible have to be made either of graphite or of graphite-based materials: pyrolytic graphite or carbon, glassy carbon, graphite felt and graphite foam. Refractory metals are, in principle, capable of withstanding the temperatures employed for the growth of SiC: however, their vapour pressures at high temperatures are not negligible. Traces of the metals, reaching the surface of the growing crystal, are sufficient for the formation of metal carbide inclusions. These inclusions deteriorate crystalline perfection. The only exception known thus far is tantalum which can still be used for those parts which experience high temperatures [10,11].

Furnaces of two types are employed for the sublimation growth: resistively heated furnaces and induction heated ones. Schematic diagrams of typical furnaces are shown in FIGURES 1(a) and 1(b). The peculiarities of their construction can be found in [12-17] as well as in the reviews [1,4]. We shall outline only the basic features of their design.

The resistively heated furnaces do not require a high frequency generator. However, they have a more complicated design due to the necessity to provide the input of a high current with low power losses and the additional non-uniformity of thermal fields arising from the heat sink to the cooled current leads. Another drawback of the resistance furnaces is related to the graphite used for their fabrication which should have a very high uniformity of specific resistivity. This parameter of graphite is generally not checked. The practice of industrial growth of silicon carbide crystals shows that the main cause of heater element failure is regions of increased specific resistivity which are rather abundant in graphite [17]. These regions are overheated and they have an increased erosion rate. Their presence also produces inhomogeneities of the thermal field.

The design of induction type furnaces depends primarily on the frequency of the current employed for heating. Usually either a low frequency of about 10 kHz or a high frequency (about 0.5 or 1 MHz) is employed. The value of the operation frequency determines the depth of the electromagnetic field penetration into the graphite. For the frequency of 10 kHz the penetration depth is about 2 cm, so the susceptor tube should have sufficiently thick walls. At high frequencies the skin layer depth is only a few millimetres. Low frequency furnaces

8.2 Sublimation growth of SiC

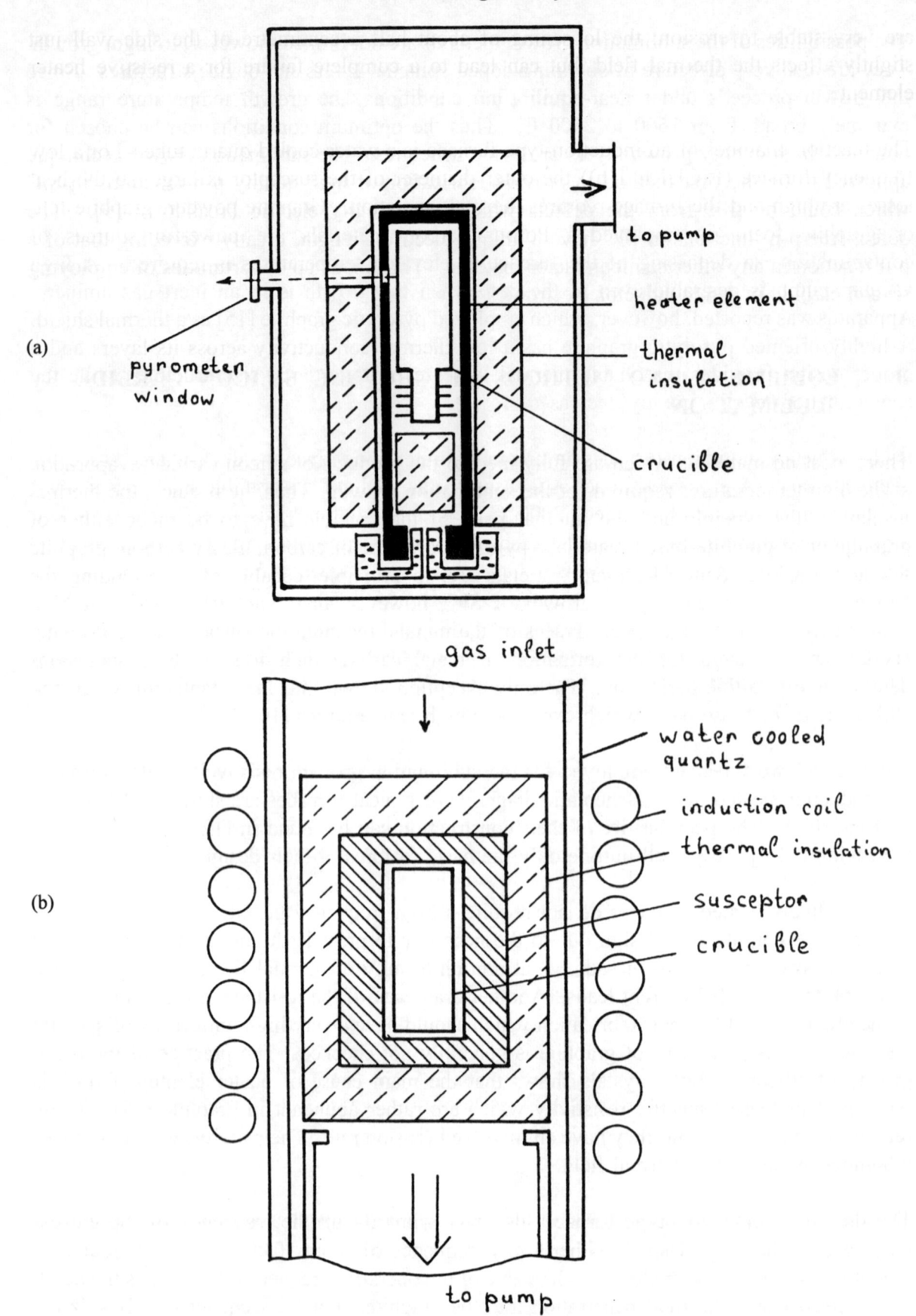

FIGURE 1 Schematic diagrams of furnaces for sublimation growth of SiC: (a) resistive type furnace; (b) induction type furnace.

are very stable to erosion: the loosening of about half a centimetre of the side wall just slightly affects the thermal field, but can lead to a complete failure for a resistive heater element.

The reaction chamber of an induction-type furnace is a water-cooled quartz tube. For a low frequency furnace (FIGURE 1(b)) the outer diameter of the susceptor is large, so the heat losses are high and the furnace requires thermal insulation. Graphite powder, graphite felt, or graphite foam is employed. Porous carbon materials are powerful sources of contamination, so degassing of the furnace at elevated temperatures is required. A high vacuum pump is desirable even in the case when the growth is in an inert gas ambient. Apparatus was reported, however, which employed pyrolytic graphite [15] as a thermal shield. A highly oriented pyrolytic graphite has a low thermal conductivity across its layers and a proper design provides reasonable thermal losses even for a large size crucible, while the contamination level is greatly decreased [15].

High frequency induction furnaces have smaller dimensions, since the thick-walled susceptor is unnecessary. A crucible can act as a susceptor as well. If the crucible diameter is small and the temperatures do not exceed 2000 °C the graphite thermal shields are quite sufficient for reducing radiation power losses to reasonable values. High frequency furnaces, however, have a lower temperature stability than the massive low frequency furnaces. The important advantage of induction type furnaces is the convenience of temperature field regulation. Both value and sign of the temperature gradient in the crucible can be varied by displacement of the inductor coil along the axis. In addition, all the high temperature graphite block can be easily replaced, and thus the heavily doped and the undoped layers can be grown in the same furnace.

The intrinsic vapour of silicon carbide is strongly enriched by silicon [18,19], so the heating of SiC crystals in vacuum produces a dense graphite film on the crystal surface. If this happens with a seed crystal or with a substrate crystal the further growth of single crystalline material is usually impossible. Certain measures should be taken both to prevent high losses of silicon and to compensate for the evaporated silicon. The crucibles for the sublimation growth of SiC are fabricated either from a dense graphite or from pyrolytic graphite or from glassy carbon. This helps to prevent the loss of silicon through the crucible side walls. The growth cells (i.e. the SiC substrates with their sources of SiC) are positioned not in the region of maximal temperature in the crucible, but at a somewhat lower temperature (see FIGURE 2). A subsidiary graphite crucible with polycrystalline silicon carbide powder is placed in the region of the highest temperature. A large effective surface area of the powder ensures that the silicon vapour pressure is close to equilibrium, in spite of partial decomposition of the SiC powder. Since the main sources of SiC and the substrates rest at a lower temperature, the silicon vapour pressure in the cells should not fall below the equilibrium value. Additional containers with SiC powder are placed also in positions where high losses of silicon can be expected, i.e. at the top and on the bottom of the crucible. Simultaneous growth on a number of wafers can be performed with use of the sublimation 'sandwich' method. The 'sandwich' cells are displaced along the crucible one upon another.

The design of 'sandwich' cells (FIGURE 3) is defined, firstly, by the direction of mass transfer. In the upper part of the crucible the material transport occurs upwards and the substrate is mounted on a graphite or tantalum holder. Polycrystalline SiC powder is the most

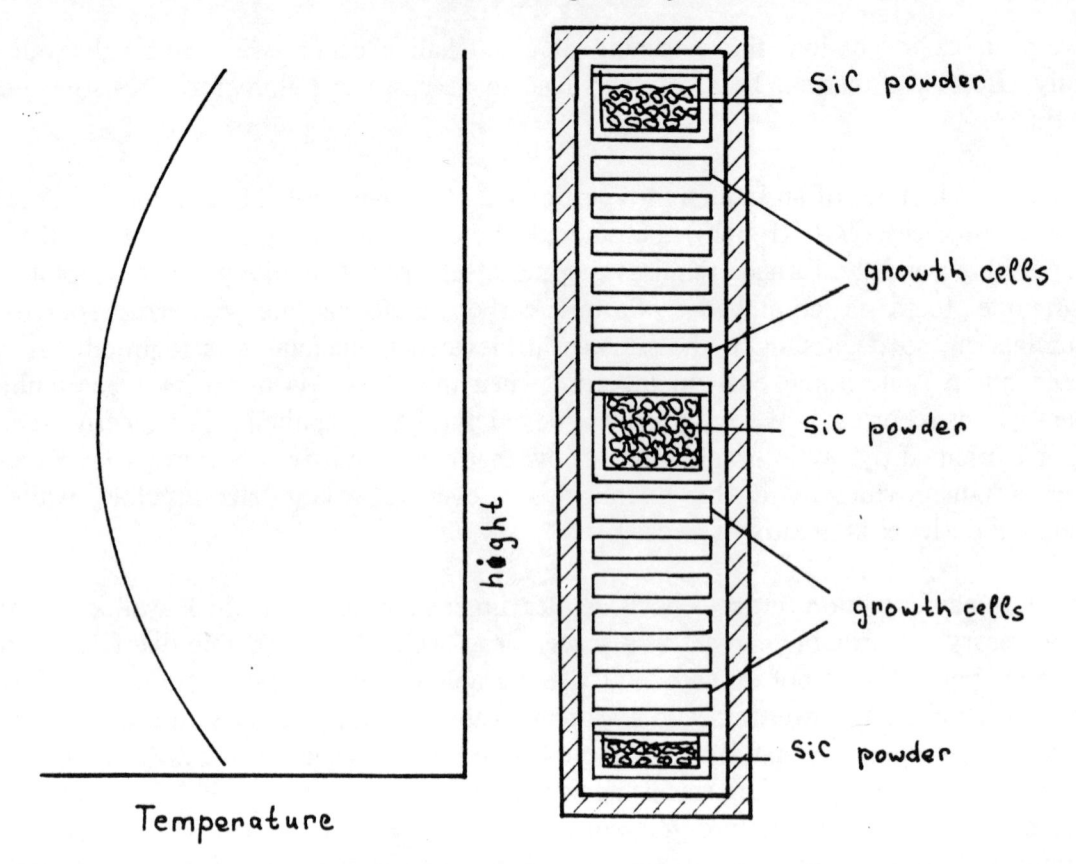

FIGURE 2 The positioning of silicon sources and of 'sandwich' cells with respect to the temperature field in the crucible.

reliable source for maintaining the equilibrium silicon vapour pressure. Additionally, if the silicon pressure is ensured by an external silicon source, the best crystalline quality is obtained with the use of dense SiC sources. The number of stacking faults and pores decreases noticeably with sources of this type. The dense SiC materials suitable for use as sources are SiC polycrystalline boules grown by sublimation, CVD-grown polycrystalline material and single crystal wafers of SiC. The use of single crystal wafers as a source is reasonable for growing thin layers, since a source can be used several times. Dense sources are necessary for growing the layers in the lower region of the crucible, where the source should be placed above the substrate. Of all the source materials the purest is the CVD-deposited SiC, obtained by methyltriclorosilane decomposition (see Datareview 8.3).

The growth of SiC crystals is carried out either at low temperatures, below 2000°C [7,8,20], in vacuum or at a higher temperature in a low pressure argon ambient [21-25]. Two basic modifications of the growth chamber are used. The first is a modification of the Lely process [1,3] (see also Datareview 8.1). In the modified Lely process the growing crystal is separated from the vapour source by a thin-walled cylinder of porous graphite, FIGURE 4(a). This growth chamber design was employed in [7,21,22,24]. The alternative bulk crystal growth method is essentially a modification of the sublimation 'sandwich' method. The seed crystal is placed directly above the polycrystalline SiC source, FIGURE 4(b). The latter method was employed in [8,20,23]. Much higher growth rates at a given temperature and temperature gradient are observed for the 'sandwich'-type system, since diffusion through graphite does not retard the vapour flow and the source to substrate distance is small. The growth of

8.2 Sublimation growth of SiC

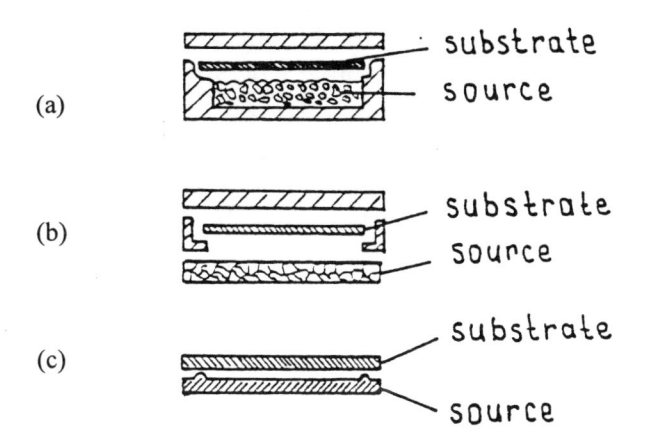

FIGURE 3 The modifications of 'sandwich' cells employed for the growth of SiC (a) with powder SiC source, (b) with dense polycrystalline SiC source, (c) with single crystal source.

polycrystalline material on the crucible walls is also not as significant for the 'sandwich' system. However, the separation of source and crystal by the graphite cylinder diminishes the probability of inclusion formation in the grown crystal. Considerable temperature gradients are employed for ingot growth. A high temperature gradient can result in liquid silicon condensation on the surface of the growing crystal. The silicon is evaporated from the hotter part of the source. This results in poor crystal formation. Another source of inclusions is the contamination of the source by metal traces. Refractory metals tend to form highly stable carbides and the precipitates are incorporated in the crystal. Metal contamination is especially high in the abrasive SiC powders, often used as sources for the SiC ingot growth. In both cases the presence of a graphite separator suppresses the second phase condensation.

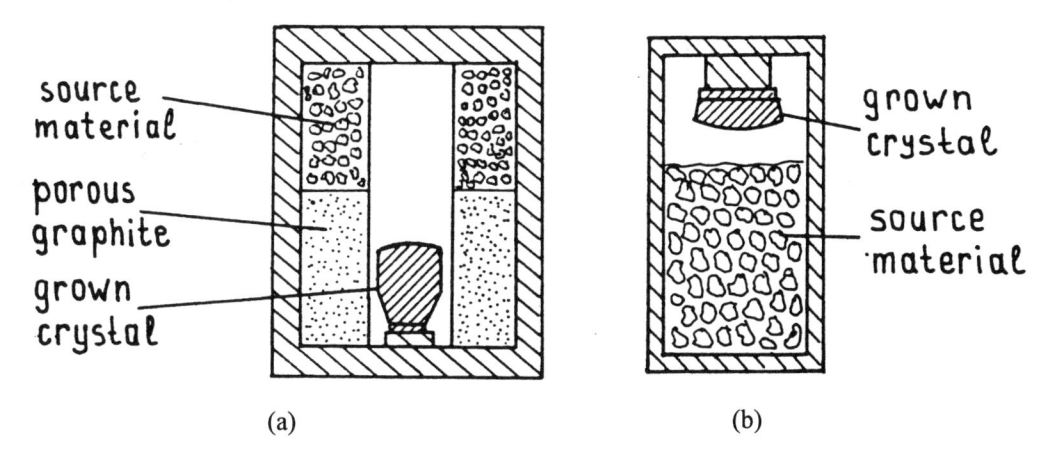

FIGURE 4 The types of growth cavities employed for bulk crystal growth: (a) for the modified Lely process, (b) for the modified 'sandwich'-type growth system.

C   PHASE EQUILIBRIA IN THE SYSTEM SiC - INTRINSIC VAPOUR

The experimental studies of the composition and component pressures of the silicon carbide intrinsic vapour have been performed by Drowart and De Maria [18] and by Behrens and Rinehart [19]. Both groups employed the mass-spectrometer analysis of the vapour, flowing out from an effusion cell. The cell had been filled by SiC powder. Drowart and De Maria have also studied the component pressures in the equilibrium of SiC with liquid silicon.

8.2 Sublimation growth of SiC

Elemental silicon was introduced for this in addition to SiC. It was established that the main product of silicon carbide evaporation is monoatomic silicon vapour. The most important carbon-containing components are the molecules SiC_2 and Si_2C. The qualitative results of the two groups are in good agreement, although some discrepancies in vapour pressures are still observed. The dependence of a component vapour pressure p_j in the SiC-C equilibrium on temperature can be represented by an activation type dependence:

$$p_j = A_j \exp(-Q_j/RT) \tag{1}$$

where R is the gas constant, T is the absolute temperature, A_j is the pre-exponential factor, Q_j is the activation energy. The values of A_j and Q_j are presented in TABLE 1. The results of Drowart and De Maria [18] for the equilibrium of SiC with graphite are plotted in FIGURE 5. The conditions of SiC growth are improved at higher temperatures, since the ratio of silicon to carbon content in the vapour decreases with temperature. Thus, the problems related to the vapour phase non-stoichiometry are easier to solve if the growth is performed at a higher temperature.

TABLE 1 Activation energies and pre-exponential factors for the SiC evaporation products.

Component	Drowart [18] A (atm)	Q (kcal mol^{-1})	Behrens [19] A (atm)	Q (kcal mol^{-1})
Si	7.8 x 10^8	133	7.4 x 10^8	111
SiC$_2$	1.2 x 10^{11}	165	3.4 x 10^{10}	146
Si$_2$C	5.1 x 10^{10}	163	8.5 x 10^9	137

We shall first consider the range of conditions which permit quality growth of SiC from the vapour. At a given temperature, a system composed of two components (silicon and carbon) and of two phases (crystal and vapour) has one degree of freedom. In other words, in the SiC-vapour equilibrium the pressure of one component determines the pressures of all the others. The highest vapour pressure in the system is that of silicon, so it is convenient to take this pressure (or the ratio of the pressure to the equilibrium vapour pressure in the system SiC-C) as the parameter of the system state. The pressure of silicon p_1 can be varied in the limits from the decomposition point of SiC (SiC-C equilibrium), $p_1 = p_1^*$, to the point of liquid silicon condensation, the SiC-Si equilibrium, $p_1 = p_1^s$. The dependences of p_1^* and of p_1^s on temperature are plotted in FIGURE 6. The pressure interval between the two dependences is the region where SiC can be deposited homogeneously: no other condensed phase can coexist with the SiC. The interval is actually rather large, the pressure ratio p_1^s/p_1^* changing from about 50 at 1600 °C to 11 at 1900 °C.

Within the interval $p_1^* < p_1 < p_1^s$ the pressures of SiC_2 and of Si_2C are given by the mass-action law for the surface reactions:

$$2\ SiC_{sol} = SiC_{2,vap} + Si_{vap}; \qquad p_2 = p_2^*\ p_1^*/p_1 \tag{2}$$

$$SiC_{sol} + Si_{vap} = Si_2C_{vap}; \qquad p_3 = p_3^*\ p_1/p_1^* \tag{3}$$

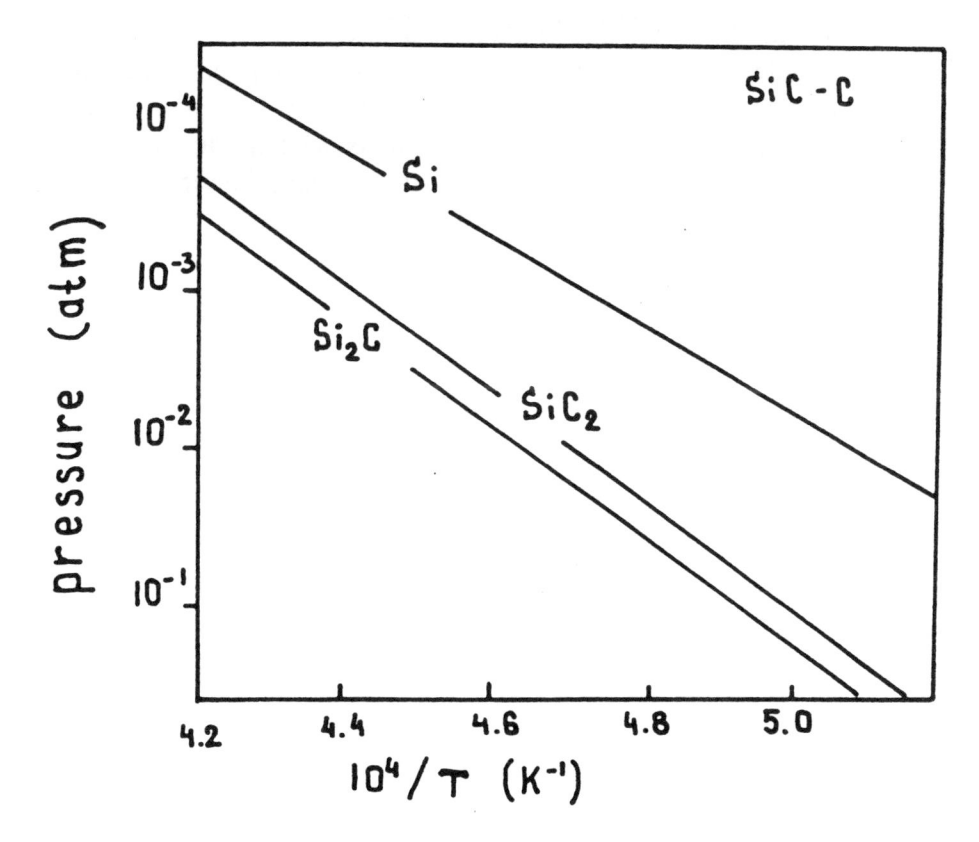

FIGURE 5 Temperature dependence of the vapour component pressures in the equilibrium of SiC with carbon.

where p_2 and p_3 are the vapour pressures of SiC_2 and of Si_2C respectively, and p_2^* and p_3^* are the values corresponding to the SiC-C equilibrium. The vapour pressure of SiC_2 decreases with increasing silicon pressure, whereas that of Si_2C increases. In the SiC-C equilibrium the pressures of SiC_2 and of Si_2C are of the same order of magnitude, and thus the total carbon content in the vapour is much higher for the SiC-Si system than for the SiC-C system.

Now we shall consider the methods of silicon vapour pressure control during the sublimation. One of them was discussed in the previous section. The method is to provide the required silicon vapour by decomposing the SiC powder, which rests at a higher temperature. It is the most common method of silicon vapour control in the sublimation growth process. In some cases, however, particularly during low temperature vacuum sublimation, the rate of silicon evaporation from the crucible is very high and a source of silicon of this sort can appear to be insufficient. Obviously, the alternative is to introduce excessive elemental silicon into the growth cell. Elemental silicon, however, has a high vapour pressure and it reacts rapidly with the graphite parts of the crucible. To preserve liquid silicon for lengthy periods is possible only if the internal parts of the crucible are made of special dense sorts of graphite, of pyrolytic graphite, or of glassy carbon, i.e. of the materials which lack a developed network of pores. A thin film of SiC is formed on the surface of these materials which prevents the further reaction of silicon and carbon. However, in the presence of liquid silicon the probability of welding the parts of the crucible to each other is rather high even when the dense carbon materials are used. A common way to maintain a moderate excessive pressure of silicon in the growth cell is to employ powder sources, enriched by silicon. Such sources

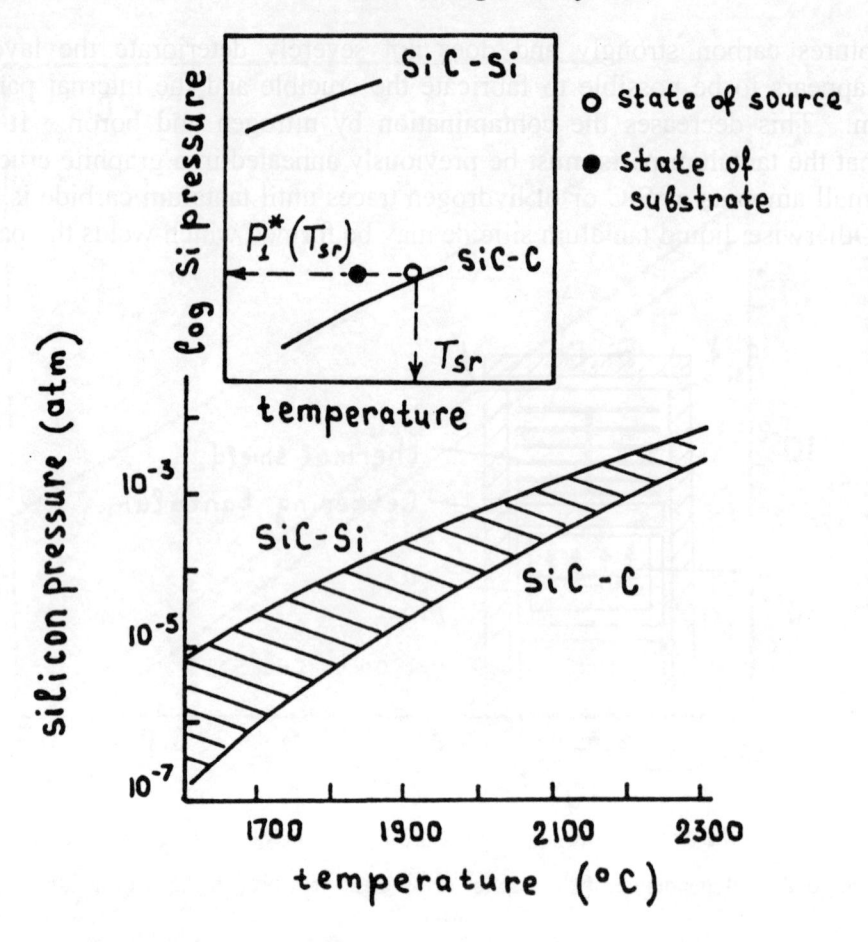

FIGURE 6 The domain of SiC equilibrium with vapour. In the insert: the states of source and of substrate during the sublimation growth. The substrate is enriched by silicon.

are made by synthesising SiC from a Si+C mixture in vacuum at a temperature below 1900 °C. A full conversion of the mixture to SiC does not occur at low temperatures, so silicon remains partially as a separate phase. The surface tension and the porous structure of the powder prevent rapid evaporation of the silicon. The increased values of silicon pressure can be obtained by increasing the silicon to carbon ratio in the initial powder over the stoichiometric ratio 8:3. For a high purity growth system silicon powder should be added to the powder of pyrolytic SiC. The sources with excessive silicon, however, have a tendency to deposit liquid silicon onto the substrate surface when using high temperature gradients in the crucible.

Besides the method of maintaining the equilibrium based on the direct introduction of excessive silicon, another method exists based on carbon gettering. This method was first reported in [10,11]. Refractory metals form very stable carbides at high temperatures. If the rate of carbide formation is sufficiently high and the SiC is placed in the vicinity of the metal, then the carbon-containing components will be strongly adsorbed by the metal. The excess silicon will be released to the vapour phase. Maintaining the silicon excess in the vapour requires the rate of carbon adsorption q_C to be equal to the rate of silicon evaporation q_{Si} for a pressure, exceeding p_1^*, $q_C = q_{Si}$ ($p_1 > p_1^*$) (see FIGURE 7). Since the getter surface can be much larger than the area of the slots left for residual gas evacuation, this condition can be easily fulfilled. Fairly good results have been obtained using the tantalum getter [10,11].

8.2 Sublimation growth of SiC

Tantalum captures carbon strongly and does not severely deteriorate the layer quality. Moreover, it appears to be possible to fabricate the crucible and the internal parts entirely from tantalum. This decreases the contamination by nitrogen and boron. It should be emphasised that the tantalum parts must be previously annealed in a graphite crucible in the presence of small amounts of SiC or of hydrogen traces until tantalum carbide is formed on the surface. Otherwise, liquid tantalum silicide may be formed which welds the parts to each other.

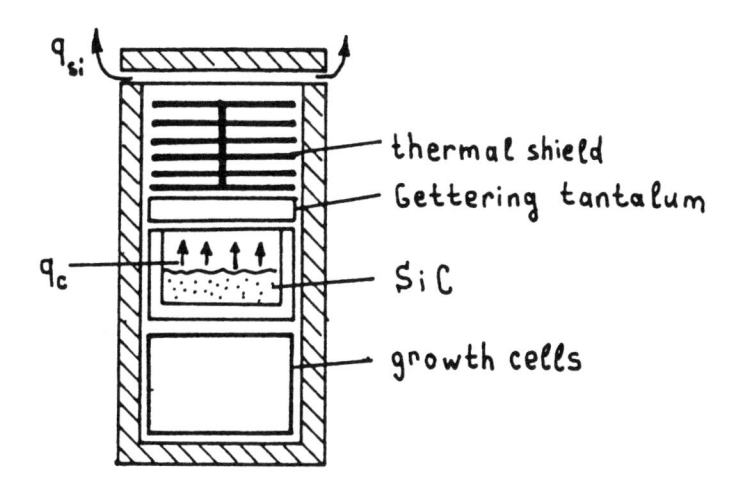

FIGURE 7 Maintaining the equilibrium silicon pressure in the crucible with use of gettering carbon by tantalum.

The vapour phase composition can be significantly changed by the introduction of impurities. In the presence of an impurity an additional degree of freedom is attained by the system, the impurity pressure. At a high temperature the carbides are generally more stable than the silicides. Thus, in the presence of impurities the content of carbon in the vapour can significantly exceed that of silicon. Varying the impurity pressure, one can obtain the desired Si to C content ratio.

Thermodynamic analyses of the composition and vapour pressures in the systems SiC-H and SiC-N were performed by Lilov et al [26,27]. The most important carbon transporters are C_2H_2 for hydrogen and CN for nitrogen. In both cases the atmospheric pressure of the gas yields a higher content of carbon in the vapour than of silicon. The volatile products, formed in the presence of Al, Ga, Be, Sn and Ge, were analysed by Chupka et al [28] and Drowart et al [29].

D   MASS TRANSFER UNDER A TEMPERATURE GRADIENT

If a temperature drop exists between the source and the substrate, a pressure difference occurs for all three major components of the vapour (for Si, SiC_2 and Si_2C) and the vapour flows from the source to the substrate. The source evaporates, whereas an epitaxial layer is deposited on the substrate. Most methods of growing SiC by sublimation are based on material transfer in a temperature gradient. It has been noted by Vodakov and Mokhov [6], however, that the most effective material transport occurs when the source and substrate are separated by only a small spacing. To provide effective transport the clearance should be smaller than the diameter of the source by 5 or more times. The method of growing SiC this way was named the sublimation 'sandwich' method. A small clearance is required to provide the near-equilibrium conditions in the growth cell.

For properly chosen conditions the epitaxial layer growth rate is limited only by the rate of vapour transport from source to substrate and it is hardly affected by surface phenomena kinetics [30-33]. Since the silicon vapour pressure sufficiently exceeds the pressures of carbon-containing species, the vapour flow from source to substrate is limited by the transport of carbon, and the flux q_{SiC} is given by:

$$q_{SiC} = 2\Re_2 \Delta p_2 + \Re_3 \Delta p_3 \qquad (4)$$

where Δp_2 and Δp_3 are the SiC_2 and Si_2C pressure drops between the source and the substrate, and $\Re_2$ and $\Re_3$ are the corresponding transport coefficients. For growth in an argon ambient the transport is limited by diffusion of carbon containing species through argon,

$$\Re_j = D_j/RT \, \Delta z \qquad (5)$$

where D_j is the diffusivity of the species in argon and Δz is the source to substrate clearance. For transport in vacuum the flux is given by the well-known Knudsen-Langmuir relation, and thus

$$\Re_j = (2\pi M_j RT)^{-1/2} \qquad (6)$$

where M_j is the molar mass of the species.

The dependence of the epitaxial growth rates on the temperature difference between source and substrate has been studied experimentally by Vodakov et al [31]. They have established that in a pure, uncontaminated growth system the growth rate depends linearly on the temperature difference. Under equilibrium conditions the deviations from a linear dependence appeared only when impurities were introduced into the growth cell. The introduction of impurities suppressed the growth at low supersaturations, most probably due to poisoning of the growth centres by impurities.

The effect of inert gas pressure on the growth rate was studied in [30,31]. At low pressures the growth rate remained unchanged up to a certain pressure, as expected from Eqn (6) for $\lambda < \Delta z$ (λ is a mean free path in argon). At higher vapour pressures the growth rate decreased with the argon pressure, $V_g \propto 1/p_{Ar}$, in accordance with Eqn (5). The transition point from molecular flow to diffusion limited flow was dependent upon the spacing between source and substrate and it was shifted to higher pressure values with decreased spacing (see FIGURE 8) [31].

The dependence of growth rate on temperature has been studied in [31,33,34]. Vodakov et al [31] found the apparent activation energy of mass transfer to be close to that of SiC_2 and Si_2C evaporation. A calculation was performed of the SiC epitaxial growth rate assuming the

8.2 Sublimation growth of SiC

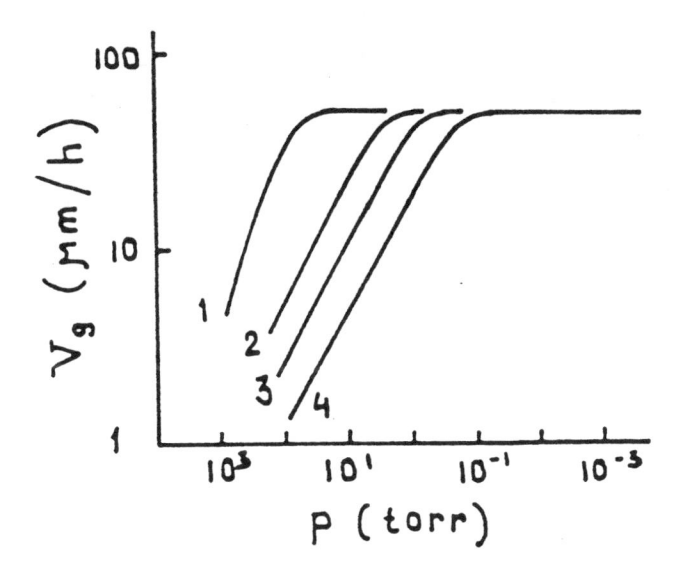

FIGURE 8 Dependence of growth rate on argon pressure in the growth chamber for the source to substrate clearances: curve 1: 10 μm, curve 2: 150 μm, curve 3: 0.5 mm, curve 4: 3 mm.

growth is limited only by transport in the vapour phase. The source and the substrate were assumed to be in the SiC-C equilibrium; then the pressure drop between source and substrate Δp_j is:

$$\Delta p_j = p_j^* \, Q_j \, \Delta T / RT^2 \tag{7}$$

where ΔT is the temperature drop. The experimentally observed values of growth rate appeared to be even somewhat higher than those calculated with the use of the data of Drowart and De Maria [18]. A possible reason for the discrepancy was attributed to the underestimation of equilibrium pressures by [18] due to graphitisation of the SiC powder in the effusion cell. Rather similar results were obtained by Tairov et al [33,34]. It was established in [34] that the activation energy of SiC growth is close to Q_2 and Q_3, about 150 kcal mol^{-1}, for growth with low temperature drops between the source and substrate. At high temperature differences, corresponding to the conditions of bulk crystal growth, the activation energy fell to about 70 kcal mol^{-1}.

Phenomenological relations of the form of Eqns (4) and (7) predict a sufficient increase of growth rate for growing SiC in excess silicon ambient, since the total pressure of carbon-containing species is much higher for the SiC-Si system than for SiC-C. This prediction was discussed in detail in [31,33], but no experimental confirmation was reported. The analysis of silicon vapour pressure effect on the growth rate was performed in [35]. It was noted by the authors of [35] that during the normal sublimation growth the substrate rests in a state of relative silicon enrichment with respect to the source. The relative silicon pressure drop is negligible, since its pressure is high, see FIGURES 5 and 6. This produces a decrease of equilibrium vapour pressure at the substrate for the SiC_2 and an increase for the Si_2C. The actual pressure drop for SiC will be higher for the SiC_2 complex than that predicted by Eqns (4) and (7), whereas for Si_2C it will be substantially lower. Thus Si_2C appears to be an ineffective carbon transporter. According to [35] the growth rate V_g is given by the relation:

8.2 Sublimation growth of SiC

$$V_g = \frac{M_{SiC}}{\rho_{SiC}} \frac{\Delta T}{RT^2} \frac{1}{\sqrt{2\pi M_2 RT}} [2p_2(Q_2+Q_1) + p_3(Q_3-Q_1)\sqrt{M_2/M_3}] -$$

(8)

$$(2p_2 + p_3 \frac{M_2}{M_3}) \frac{p_2(Q_1+Q_2)\sqrt{M_1/M_2} - p_3(Q_3-Q_1)\sqrt{M_1/M_3}}{p_1 + p_2\sqrt{M_1/M_2} + p_3\sqrt{M_1/M_3}}$$

where M_{SiC} and ρ_{SiC} are molar mass and density of SiC, and M_1, M_2 and M_3 are the molar masses of Si, SiC_2 and Si_2C respectively. For moderate growth temperatures, when the content of silicon in the vapour is much higher than that of carbon, the growth rate is given by:

$$V_g = 0.17 \, (2000/T)^{5/2} \, (1/\beta + a\beta) \, \Delta T \, p_2^* \; [cm/s] \tag{9}$$

where

$$a = \frac{1}{2}\sqrt{M_2/M_3} \frac{(Q_3-Q_1)}{(Q_2+Q_1)} \tag{10}$$

and $\beta = p_1/p_1^*$ is the relative silicon excess in the system. It follows from Eqns (9) and (10) that for the same partial pressures the carbon transport by Si_2C is 20 times less effective than that by SiC. Thus the introduction of excessive silicon into the growth cell first results in a decrease of growth rate, due to the decrease of the more effective carbon transporting content, of the SiC. Only at a very high pressure of excessive silicon does a slow increase of V_g with the silicon pressure occur, FIGURE 9.

The experimental study of the growth rate dependence on silicon vapour pressure was performed by comparison of growth rates in the systems SiC-C and SiC-Si [35]. The temperature dependence of the ratio of growth rates in the two systems is plotted in FIGURE 10. The calculated dependence is given in the same figure. One can see from the figure that a great enrichment of vapour phase both by silicon and carbon in the SiC-Si system can result in somewhat lower growth rates than in the SiC-C system. A strong silicon excess is not, therefore, favourable for SiC growth. A high material content in the vapour combined with a low rate of mass transfer makes the growth system vulnerable to temperature fluctuations and other perturbations. The conditions are preferable which provide substrate homogeneity reliably, $p_1 > p_1^*$; however the p_1 to p_1^* ratio is not high.

A substantial (twofold) decrease of growth rate in silicon excess has been observed by Anikin et al [36]. Excessive silicon was introduced by a powder SiC source enriched by silicon. It has been noted above that such a source produces a silicon vapour pressure noticeably lower than the saturated value. Most probably the state of the growth system in the experiments of [36] was close to the minimum of the growth rate dependence on p_1/p_1^*, see FIGURE 9. The

8.2 Sublimation growth of SiC

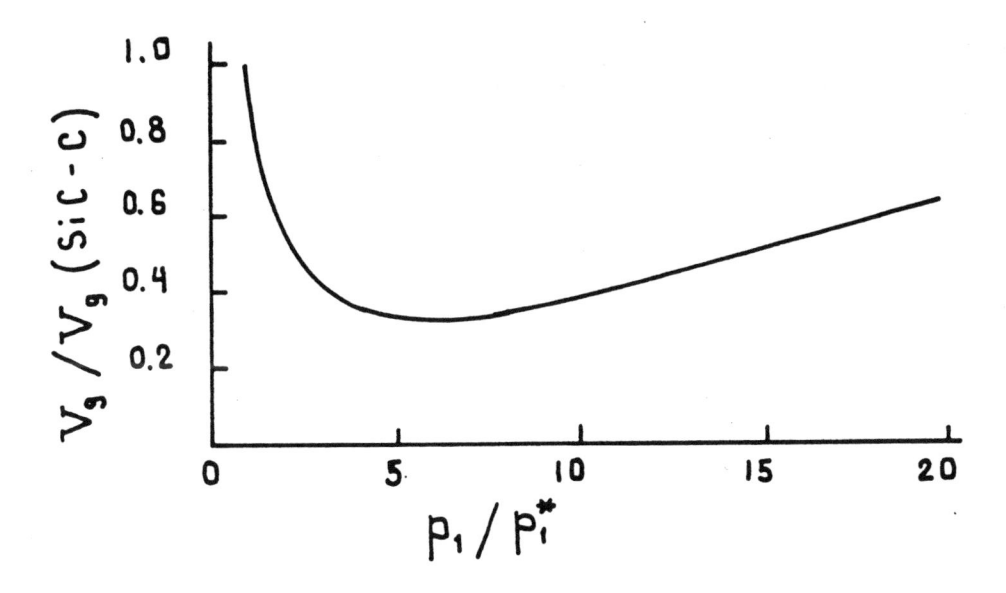

FIGURE 9 Calculated dependence of growth rate on the relative silicon excess in the growth cell.

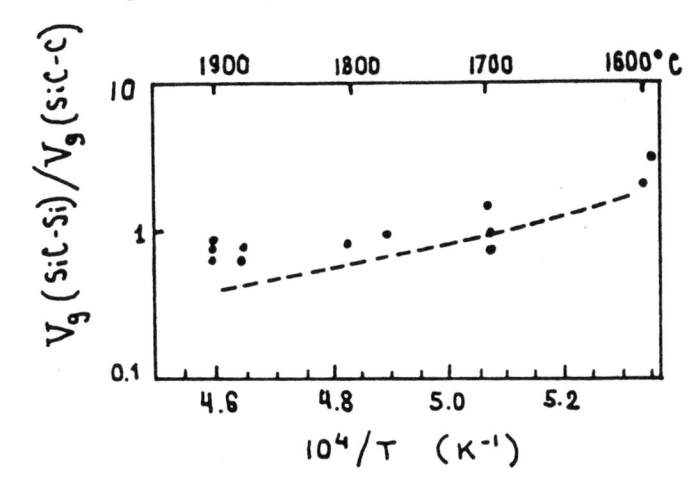

FIGURE 10 Dependence of the ratio of growth velocities in the systems SiC-Si and SiC-C on temperature. Dashed line indicates the theoretical dependence.

authors noted a decrease of layer quality for a high silicon excess, and polycrystalline β-type material was often formed.

A similar decrease of growth rate under excessive silicon pressures was observed for bulk crystal growth in [37]. The authors introduced excessive silicon by crushing the source SiC powder to a fine grain size. Fine grained material has a higher vapour pressure [37] and has a higher effective surface per mass unit, so graphitisation has a smaller effect on the evaporation rate. Since for the bulk crystal growth the substrate is at a substantially lower temperature than the hot zone of the source, a fine grained powder could provide a substantially higher silicon supersaturation. The value of the silicon excess was monitored in [37] by the mass of decomposed carbon in the source. At a high silicon supersaturation the growth rate reduced by a factor of 2-3.

E   THE MAIN BACKGROUND IMPURITIES

The most common background impurities in the SiC crystals grown by sublimation are nitrogen, boron and aluminium. These impurities are also the principal dopants employed in SiC device fabrication. Their behaviour and properties have been studied in a number of papers.

If the SiC source, the substrate, the crucible and the parts of the furnace are materials of sufficient purity, the grown epitaxial layers and crystals will have n-type conductivity. The donor which provides n-type conductivity is nitrogen. There exists a difference between the levels of nitrogen doping for the crystals grown at a low temperature in vacuum and those grown in argon at a high temperature. The lowest doping level is usually obtained in vacuum. The removal of nitrogen, degassing from the crucible and from the furnace, occurs more effectively in vacuum than in argon. The growth of epitaxial layers has been reported with a nitrogen concentration below 10^{16} cm^{-3} and with a low compensation level [33,39]. Growth from the same sources and in the same furnace in purified argon usually results in donor concentrations of about 10^{18} cm^{-3}.

The level of nitrogen doping is somewhat lower for layers grown on the Si-face (the plane (0001) Si) than on the C-face (plane (000$\bar{1}$) C). The doping level generally decreases with increasing growth temperature. In addition, the authors of [38,39] established a very strong kinetic dependence of nitrogen solubility, see FIGURE 11. An increase in growth rate radically decreases the nitrogen contamination. The kinetic dependence of solubility occurs for impurities with very low diffusivities in the crystal, when the time of impurity diffusion through a monolayer exceeds the period of growth step motion. In this case the impurity incorporation is governed by the kinetics of capture at growth steps or at kinks rather than by the interaction thermodynamics of surface adsorbed atoms with the crystal bulk. Nitrogen is actually a very slow diffuser in SiC (see Chapter 7), so the model of nitrogen dynamic solubility [38,39] appears to be quite realistic. The reduction of donor concentration by employing high growth rates has certain limits. High growth rates result in the generation of donor-type intrinsic point defects [39] which make a major contribution to the layers grown in [39] at nitrogen concentration below 1×10^{16} cm^{-3}.

A very effective way of decreasing nitrogen contamination is the replacement of the graphite crucible by a tantalum one. A substantial decrease in nitrogen concentration was observed in [10] for the tantalum crucible. The dependences of nitrogen concentration on its concentration in the source material for the graphite and tantalum crucibles observed in [10] are shown in FIGURE 12. A lower compensation level is also observed for the case of tantalum components.

Aluminium contamination is seldom observed for low temperature vacuum sublimation. Aluminium has a low capture coefficient at low temperatures and it does not form refractory carbides with a low vapour pressure. Therefore, traces of aluminium can be easily removed by annealing the furnace in vacuum even if contamination occurs. However, if the material source is insufficiently pure, it can result in noticeable aluminium contamination, especially at elevated growth temperatures. For the bulk crystal growth, aluminium contamination is always observed when abrasive silicon carbide is used as source material [20,22]. The abrasive material usually is highly contaminated [1,22].

8.2 Sublimation growth of SiC

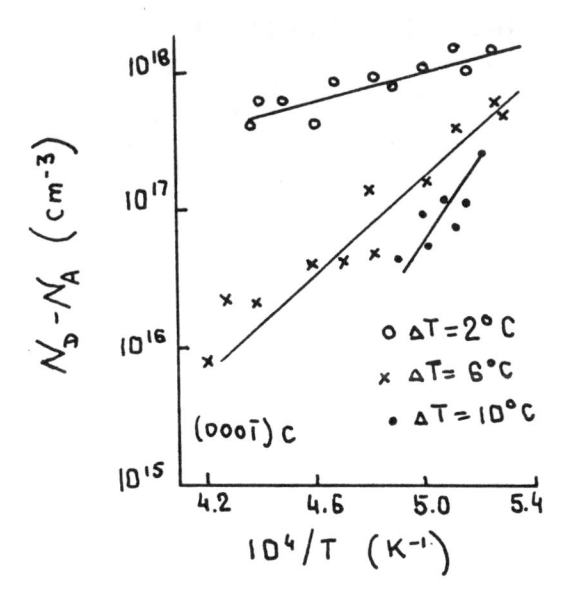

FIGURE 11 Dependence of the uncompensated donor concentration in sublimation grown layers on temperature and on temperature drop between source and substrate.

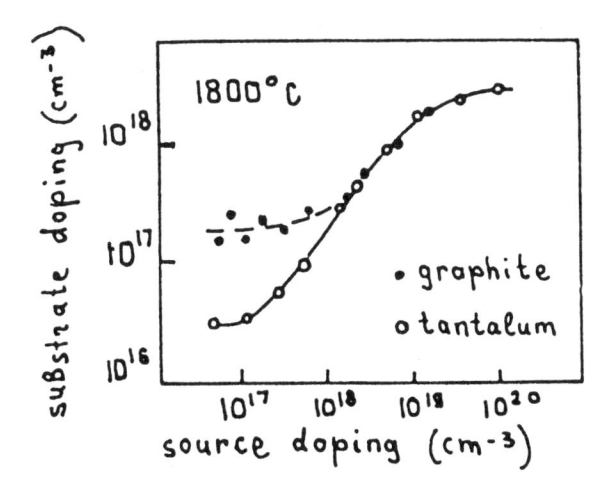

FIGURE 12 Dependence of the uncompensated donor concentration on nitrogen concentration in the source material for a graphite and a tantalum crucible.

Alternatively, boron contamination can cause serious problems. Unintentional boron doping is observed both for epitaxial layer growth and for bulk crystal growth. Boron forms a highly stable carbide with a very low vapour pressure. Once introduced into the furnace, it is difficult to remove. Boron contamination can result from the graphite parts of the furnace, from the source material and from evaporation from the reverse side of the substrate as well as from occasional sources. The boron content can be decreased by long anneals at elevated temperatures, although this method is not always effective.

The acceptor impurities have a much weaker dependence of solubility on growth rate than nitrogen [39]. This is consistent with the fact that the diffusion mobility is higher for the acceptors in silicon carbide (see Chapter 7). In contrast to the donors, acceptor impurities tend to increase their solubility at elevated temperatures [10,40,41]. The increase of growth temperature produces, therefore, a lower donor concentration and a higher degree of compensation. Nitrogen concentration is usually evaluated by the uncompensated donor concentration. For the layers grown on substrates of the opposite conductivity type, the leakage of the p-n junctions being low, the concentration of compensating acceptors can be determined by Hall mobility measurements (see Chapter 4). If the substrate is of n-type conductivity, then the uncompensated donor concentration is determined by C-V measurements on Schottky barriers. The concentration of compensating acceptors should be determined with the use of deep level transient spectroscopy or by neutron activation analysis [42]. Neutron activation analysis is very effective for the determination of boron concentration, the sensitivity of boron determination in SiC being about 10^{16} cm^{-3}. A convenient method of rapid evaluation of donor concentration has been proposed in [43]. The authors found that the breakdown voltage of a metal contact attached to the crystal surface remains essentially unchanged for a number of metals, and they have established relationships between the uncompensated donor concentration and the breakdown voltage.

Low concentrations of compensating acceptors are efficiently detected via luminescence methods. The pairs of nitrogen and aluminium yield characteristic photoluminescence spectra and the luminescence efficiency at low temperatures (80 K and lower) is fairly high [44]. Boron is nearly a single phosphor, providing efficient photoluminescence at room temperature. Hence, the detection of the compensating boron does not require any spectral measurements. The room temperature photoluminescence of boron is a broad band with a maximum in the yellow region for the 6H polytype and in the green region for the 4H polytype [45]. It should be noted, however, that no yellow luminescence can be observed in the boron compensated crystals grown at elevated temperatures (2300 - 2500 °C). A very weak red luminescence is observed for such crystals [46].

It is highly probable that the metal contamination strongly affects the electrical properties of silicon carbide crystals and reduces the carrier lifetimes, as in other semiconductors. Unfortunately, no detailed studies of this effect have been reported thus far.

F COEFFICIENTS OF DOPANT CAPTURING AND THE LIMITS OF SOLUBILITY

The sublimation 'sandwich' method provides an opportunity to study impurity capture phenomena, since the growth conditions are well-controlled. The source containing a known amount of impurity and the known growth cell configuration enables one to determine the probability of impurity capture per single collision with the surface of the growing crystal. A study of impurity capturing processes has been performed by Vodakov, Mokhov and co-authors [10,39,47]. The authors conducted a set of experiments for different clearances of source and substrate and calculated the dependence of the geometric factor on the clearance. It is essential to account for the growth cell configuration, since for a small distance between source and substrate, $\Delta z \ll d$ (d is the substrate diameter), an impurity can have multiple collisions with the substrate surface and the total capture probability is increased. The authors observed good agreement between calculation and experiment and they determined the capture coefficients (probabilities) for the principal dopants N, B, Al and Ga. The capture coefficient is constant at low dopant concentrations and decreases at higher ones. The low concentration values of the coefficients are given in TABLE 2. It can be seen from the table that nitrogen is captured better by the C-face of the crystal, whereas the acceptor dopants are more effectively captured by the Si-face. At low temperatures only nitrogen and boron have high capture coefficients. Obviously, these properties are consistent with the peculiarities of background doping discussed above.

TABLE 2 Impurity capture coefficients at 1800 °C.

Impurity	C-face	Si-face
N	0.1	0.03
B	0.025	0.07
Al	4×10^{-4}	2×10^{-3}
Ga	3×10^{-5}	10^{-3}

The doping of growing layers by nitrogen is usually performed by filling the growth chamber with gaseous N_2. According to [10], the nitrogen concentration C_N increases as the square root of the N_2 pressure $C_N \propto (p_{N_2})^{1/2}$ (see FIGURE 13). This is attributed to the inefficiency of molecular nitrogen capture. The formation of atomic nitrogen follows the equation $N_2 = 2N$, which yields the square root dependence of the active nitrogen content on its total pressure. The solubility limit is not reached at a pressure of about 1 atm, so lower temperatures are desirable to obtain the highest doping level. The maximal value of donor concentration, obtained in [10], is 4×10^{20} cm^{-3}, which corresponds to an N_2 pressure of 1 atm and a growth temperature of 2100 °C. The dependence of nitrogen solubility on its

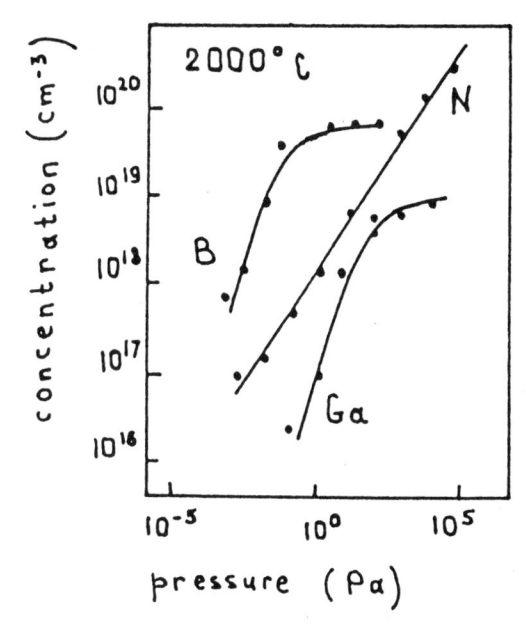

FIGURE 13 The dependences of dopant concentration on their vapour pressure.

partial pressure in the growth chamber has been also studied by Lilov et al [48]. They obtained somewhat lower nitrogen concentrations under similar growth conditions and a different exponent of the donor concentration dependence on the pressure. The exponent of the dependence was about 0.8. For chemical vapour deposition the efficiency of doping can be significantly increased by replacing nitrogen gas with ammonia [49]. This method does not work, however, for the sublimation growth, since ammonia is decomposed in the crucible.

The maximum phosphorus concentration achievable in sublimation grown crystals is about 10^{18} cm^{-3} at high temperatures and it increases slightly as the growth temperature decreases [40,50]. Phosphorus appears to be an inefficient dopant for sublimation growth.

The acceptor concentration in grown layers tends to saturate with increasing dopant vapour partial pressure [10] (FIGURE 13). The solubility limit of Al and Ga is only slightly dependent on temperature for the Si-face [40] (FIGURE 14). For the C-face it can approach

8.2 Sublimation growth of SiC

the solubility on the Si-face at high temperature, whereas it can be much lower at lower temperatures. Thus, heavy aluminium or gallium doping requires the use of the Si-face if the growth temperatures are not high. The boron solubility limit does not exhibit a dependence on crystal orientation (see FIGURE 14) [40].

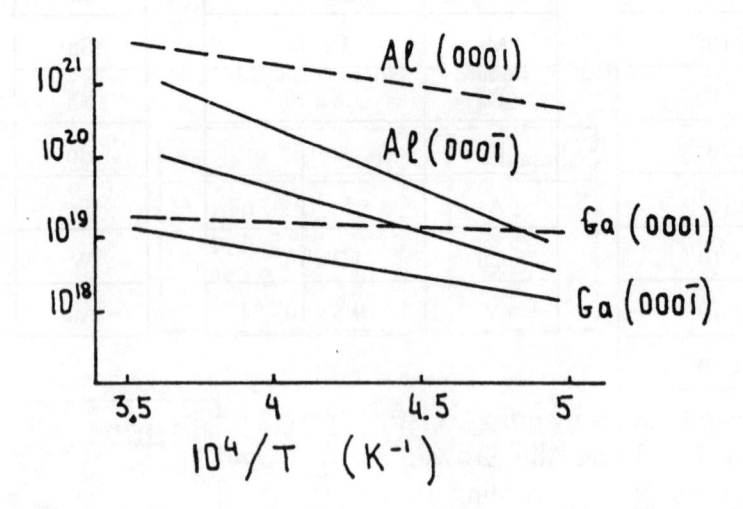

FIGURE 14 Temperature dependence of the acceptor dopant solubility limit.

Acceptor dopants are introduced in the crucible either in elemental form or in the form of carbides. If a dopant is introduced in elemental form, it is placed in a special internal crucible with carbon or silicon carbide powder. This is required to prevent the dissolution of the crucible, in the case of aluminium doping, and to reduce the boron vapour pressure to the equilibrium value for the SiC-C system, in the case of boron doping. If elemental boron is placed in the vicinity of the substrate, this results in the formation of boron carbide on the crystal faces of SiC [46]. For moderate doping of crystals, grown at high temperatures, doped SiC sources also can be employed.

Boron sources are not depleted during the crystal growth time since the boron pressure is very low. Aluminium sources may be rapidly depleted during vacuum growth if the dopant is placed at the same, or nearly the same, temperature as the crystals. Double temperature zone crucibles are desirable, therefore, when growth proceeds in vacuum. In such a crucible, the dopant is placed at a temperature substantially lower than that of the substrates. Gallium has a higher vapour pressure than aluminium, so the use of double temperature zone crucibles is necessary for gallium doping.

A high solubility limit is appropriate to beryllium, the group II acceptor. It reaches 10^{20} cm^{-3} [40]. However, beryllium is usually not employed as a dopant in the sublimation growth. Firstly, it has an exceptionally high diffusivity in SiC (see Chapter 7). For the typical growth time of about 10 - 60 min the diffusion front will penetrate as far as 10 - 100 μm. In addition, the activation energy of Be acceptors is considerable and this dopant has a tendency to self-compensation [51]. The results of a comprehensive study of impurity solubility in SiC performed by Vodakov, Mokhov and co-authors [40] are presented in TABLE 3.

TABLE 3 Solubility limits for impurities in SiC (cm^{-3}).

Impurity	Concentration
Li	1.2×10^{18}
Be	8×10^{20}
B	2.5×10^{20}
N	6×10^{20}
Al	2×10^{21}
P	2.8×10^{18}
Sc	3.2×10^{17}
Ti	3.3×10^{17}

Impurity	Concentration
Cr	3×10^{17}
Mn	3×10^{17}
Ga	1.8×10^{19}
Ge	3×10^{20}
As	5×10^{16}
Cu	1.2×10^{17}
Y	9.5×10^{15}

Impurity	Concentration
In	9.5×10^{16}
Sn	1×10^{16}
Sb	8×10^{15}
Ho	6×10^{16}
Ta	2×10^{17}
W	2.5×10^{17}
Au	1×10^{17}

G   THE FORMATION OF DEFECTS AND OF CUBIC MATERIAL INCLUSIONS AT THE LAYER-SUBSTRATE INTERFACE

Defects of crystalline structure can appear at different stages of crystal growth. The majority of defects, however, are generated at the initial growth stage, immediately after heating the furnace to the operating temperature or even during the heating process. Pores and inclusions of the cubic material appear at this stage. The inclusions of the cubic material have, generally, a polycrystalline structure. They strongly degrade the quality of the grown material. The problems related to twinning and simultaneous growth of several polytypes are the main obstacles to obtaining high quality single crystals of silicon carbide.

The problem of interphase transitions of α and β modification materials has been studied by a number of authors [1,14,52-56]. β-type material is known to be generally of a lower stability. This conclusion is based on a number of observations, mainly on the polytype structure changes, resulting from high temperature annealing. If the conditions in the furnace are close to equilibrium, then the annealing of the β-SiC powder results in its conversion to α-SiC. The process is rather rapid for annealing temperatures above 2000 °C, although a noticeable conversion occurs at temperatures as low as 1200 °C [14]. These are the main experiments underlying the conclusion that the β-phase is metastable in the entire temperature range [1,14]. Additionally, according to the Ostwald's rule of steps, it is the metastable phase which is formed first under non-equilibrium conditions (see discussion in [52]). The self-nucleated crystals will, therefore, have the structure of α-SiC provided the conditions are near-equilibrium and there is a sufficient time for conversion to a more stable modification. Alternatively, if the supersaturation is high the most probable result will be β-SiC. A high silicon excess has the same effect since it implies a high supersaturation for the reaction of graphite and silicon. The crystals grown by the Lely method are mainly of the α-SiC structure. The ratio of abundance of 6H to 15R and to all the other polytypes is about 84:15:1. This is consistent with the fact that growth by the Lely method proceeds under near-equilibrium conditions. If the growth is performed at a low temperature, the vapour pressures are low, so a high supersaturation is usually employed. The rate of β-SiC to α-SiC transition is also low at lower temperatures; thus the self-nucleated crystals have the β-SiC structure. The amorphised SiC regions recrystallise first in the β-SiC structure, as follows from the experiments of [58] on recrystallisation of the ion implanted layers. Only at temperatures above 1800 °C are they converted to α-SiC. This means that the surface layers, damaged by chemical etching or by other pre-epitaxial treatment, have a high probability of transformation to 3C-SiC.

The mechanism of the β-SiC to α-SiC transformation has been discussed by several authors [2,53-57] and there exist rather controversial points of view on the subject. One of them is that the transformation is essentially a solid phase process [53-56]. This is based primarily on the observations of such transitions in the β-SiC crystals grown at low temperatures by pyrolysis of methyltrichlorosilane [53,54]. As the crystals are heated to or above 1800 °C, some regions of the crystals are transformed to 6H-SiC. According to [53], the transformation does not proceed over the entire bulk simultaneously; it is assisted by the motion of dislocations or other extended defects. The rate of the solid phase process is, therefore, strongly dependent on the perfection of the crystal. According to [59] the pyrolytically grown β-SiC crystals actually have disordered regions. In these regions a great number of twinning planes is observed. The average distance between them is so small that the chemical etch delineates a disordered region as a whole, not the particular twinning planes. Obviously, the rate of solid phase transformation in such a region will be much higher than in a perfect crystal. This explains why certain regions of β-SiC crystals are converted to 6H-SiC rapidly, whereas the other ones appear to be quite stable, as observed in [54]. The authors of [53] also noticed that some regions of the β-SiC crystals and some crystals are extremely stable.

In addition to the solid phase mechanism, the α-SiC to β-SiC transformation can proceed by vapour phase transport or by surface diffusion. It is highly probable that some results attributed to the solid phase transformation are actually related to vapour phase transport. Particularly, Yoo et al [55] performed high temperature anneals of thin β-SiC platelets. The platelets were β-SiC epitaxial layers grown on silicon wafers which were removed by a chemical etch. The transition to α-SiC could be observed only when the platelets became thicker after the anneal, i.e. when sublimation growth occurred. Vodakov et al [56] have performed prolonged anneals of β-SiC at high temperatures under conditions which excluded the possibility of sublimation transport. They found that in the β-SiC crystals which lacked disordered regions no phase modifications occurred at all. The solid phase recrystallisation mechanism is, therefore, effective only for crystals with an extremely high defect density. In crystals with a moderate defect density a phase transformation is possible only under the condition of deposition on the surface and only in the case when the nuclei of the new phase have been formed on the surface. The present author has performed the growth of p-n junctions in the cubic material [59] with use of the preliminary sublimation etch-back in an inverted temperature gradient. The etch-back removed all the foreign nuclei which appeared as the furnace was heated to the growth temperature. Only growth of 3C-SiC has been observed on the substrates of 3C-SiC, grown on the 6H-SiC by preliminary sublimation epitaxy. For the β-SiC substrates, obtained by methyltrichlorosilane pyrolysis, some regions were actually converted to 6H-SiC. A transition over the entire surface of the substrate was never observed.

The main causes of structural defect and foreign polytype formation are related to the imperfection of chemically etched surfaces and to the surface contamination. At a high temperature the crystal surface is recrystallised, and the growth steps are formed on an initially mirror-flat surface. Such a modification of the surface has been observed after performing diffusion in SiC, even if the diffusion proceeds under essentially temperature gradient-free conditions [61] (see FIGURES 15(a) and 15(b)). This surface modification is observed also at the substrate-layer interface by etching away the layer electrochemically [9]. This can be easily accomplished if the substrate is of n-type conductivity, whereas the layer is a heavily doped p-type material. Since the recrystallisation process starts at a rather low temperature, the probability of β-SiC nuclei formation is rather high. Further growth on these nuclei severely deteriorates crystalline perfection. Another mechanism of defect formation is related to the surface contamination. The graphite furnaces employed for sublimation growth are actually a powerful source of contamination. As they are heated up to growth temperatures, they release large amounts of hydrocarbons and carboxides, which are capable of depositing carbon onto the crystal surface. The same deposits arise from diffusion pump oil decomposition. The presence of carbon on crystal faces has been observed directly in specially designed experiments, when the furnace was switched off immediately after it had reached the growth temperature [11]. The traces of carbon are removed rapidly as soon as the temperature is sufficient for the silicon sources to become effective. However, the direct reaction of silicon and carbon is a highly non-equilibrium process and new polytypes may appear. Even in the case when carbonisation is weak enough not to produce new polytypes, it results in stacking fault formation, and the perfection of the layers is degraded [11].

Several practical methods of improving the crystalline perfection are known. The first and the simplest one is based on the fact that the number of β-SiC inclusions decreases as the crystal grows, since the cubic material has a lower growth rate [61] (see also [8,52]). The inclusions of β-SiC do not disappear completely, however, even for a substantial length of the grown crystal.

A rather effective method of eliminating the defects and β-SiC inclusions has been proposed by Mokhov et al [32]. After evacuation of residual gases they filled the growth chamber with pure argon and heated the furnace up to the growth temperature. The growth rate in argon is several orders of magnitude less than in vacuum, and thus hardly any growth takes place in the argon ambient at temperatures of about 1700 - 1900 °C. During the surface reconstruction, FIGURE 15(c), the nuclei and carbon traces are dissolved by the surface diffusion process. After annealing for 5 - 20 min at a high argon pressure the gas is evacuated and growth is started. The use of this method has led to the industrial fabrication of sublimation grown layers with a good yield of high quality layers [17].

The growth of epitaxial layers in an argon ambient at elevated temperatures is usually accompanied by the formation of cubic phase inclusions at the interface. At a moderate value of temperature gradients in the furnace the surface reconstruction proceeds more rapidly than the growth of cubic material nuclei. The same applies to the pore and local graphitisation region formation. For bulk crystal growth, however, both the pore formation and the β-phase inclusions often appear [20,21,25]. This results from the high temperature gradients employed in bulk crystal growth. The means of improving bulk material quality are essentially the same as for the epitaxial growth. The surface recrystallisation at an elevated argon pressure and the decrease of temperature gradients during the initial stage of crystal growth are the main

8.2 Sublimation growth of SiC

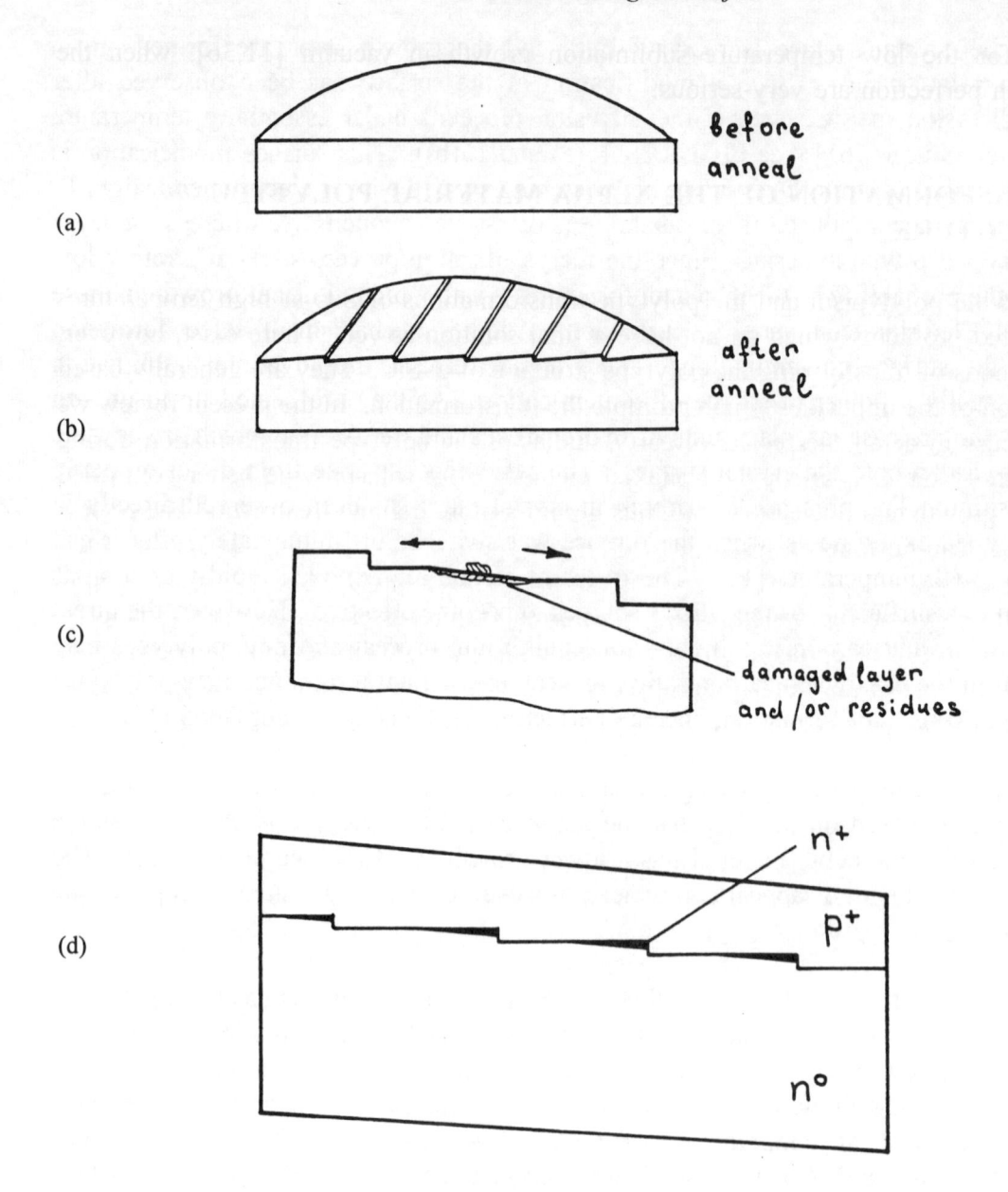

(a) before anneal

(b) after anneal

(c) damaged layer and/or residues

(d) n+ / p+ / n°

FIGURE 15 The modification of crystal surface on high temperature anneal (a) initial flat surface, (b) the terraced surface structure after the anneal, (c) dissolution of the surface damaged layer and of the deposits during step formation, (d) the heavily doped n-type interface layer formation during p-n junction growth.

means of improving the crystalline perfection. Heating of the furnace to the growth temperature at high argon pressures has actually proved its efficiency [8]. The regulation of temperature gradients during the growth process is, certainly, another important method, since the best results are obtained by those who allowed for this possibility in the design of their equipment (see [24]).

A more radical way of creating a quality interface is to evaporate the transition layers by the sublimation etch of the substrate surface. It is performed at the growth temperature via a reverse temperature gradient. This method was initially proposed for the growth of quality p-n junctions in SiC [9]. However, it has also proved its efficiency for other applications, particularly for the low temperature sublimation growth in vacuum [11,36], when the problems with perfection are very serious.

H TRANSFORMATION OF THE ALPHA MATERIAL POLYTYPE STRUCTURE

The origin of the polytypism and the polytype transformations of SiC at high temperatures are complicated problems which do not have a final solution as yet. There exist, however, practical methods of transforming the polytype structure of α-SiC. They are generally based on introduction of the impurities which promote the transformation. In the present review we shall not discuss in detail the results of early studies of the polytype transformation during crystal growth by the Lely method. Epitaxial methods of growth provide better controlled conditions of growth and nucleation, and thus permit a more definite treatment. The review of early studies can be found in [1,14].

A study of the polytype reproduction and transformation processes has been performed by Safaraliev et al [62]. The epitaxial growth proceeded in an argon ambient at temperatures of 1800-2000°C. For the substrates precisely oriented in the (0001)Si or (0001)C planes, the authors could observe only the growth of 6H and 15R polytypes. The probability ratio of the two polytypes to occur was the same as for the Lely method grown crystals, i.e. 85:15. On the off-axis substrates (the off-axis angles were 0.5-90°) the polytype of the substrate was always reproduced. The authors have attributed the effect to the fact that the plane (0001) (or (000$\bar{1}$)) carries no information on the polytype structure. The atom positioning in this plane is identical for all the polytypes. The information on the polytype structure of the substrate is transferred only in the cases when the growth occurs by a mechanism controlled by growth steps and not in the case when the growth centres are nucleated on the surface.

Somewhat different results were obtained by Vodakov et al [56]. They studied the polytype transformations for growth in the temperature range 1600-2200°C. Growth at temperatures between 1600 and 1950°C was carried out in vacuum, whereas an argon ambient was employed for higher temperatures. Polytype transformation was observed for somewhat higher off-axis angles, about 3°. The new polytype inclusions were exclusively of 3C structure for the low temperature growth and 6H or 15R structure for the high temperature growth in argon. The transformation occurred, however, only for growth on the C-face. For growth on the Si-face the polytype of the substrate was always reproduced with the exception of the cases when a high supersaturation induced the growth of 3C. The authors attribute the orientation dependence to the differences in the nature of the chemical bond for Si- and C-faces.

The transformation of the epitaxial layer polytype to a structure different from 3C, 6H or 15R was first obtained with the use of liquid phase epitaxy by the travelling solvent method [63]. The authors employed scandium-based solutions and they observed the 4H-SiC layer growth on substrates of the 6H polytypes. Later, the detailed studies of Vodakov et al [64] showed that the 6H to 4H polytype transition occurs when the growth is performed onto the carbon face and only if the melt contains excess carbon. No polytype transformation occurred if the melts were free from excess carbon or contained excess silicon. In [63] no carbon was added to the solvent intentionally, so most probably it entered the melt via reaction with the graphite

8.2 Sublimation growth of SiC

crucible. The presence of scandium or other rare-earth metals in the vapour phase was observed to promote the 6H to 4H transition during sublimation growth [56]. The transition could also be promoted by aluminium and boron impurities. The same effect was later discovered for isoelectronic impurities, germanium and tin. Transition to the 3C structure could be stimulated by Ba, N, P and H [32,65,66].

A general property of the impurities which promoted growth of the 4H polytype is that they enrich the surface of the growing crystal with carbon. Prolonged anneals of the SiC crystals in vapours of these elements resulted in enhanced carbonisation. Alternatively, the impurities Ba, N, P and H had a tendency to remove carbon if it had been previously deposited onto the crystal face [64]. The studies of diffusion phenomena and other results show an increase in carbon vacancy concentration [67] and of the degree of deviation from stoichiometric Si to C ratio [68] with decreasing hexagonality of the polytype, i.e. for the sequence 4H-6H-3C. Thus, the trend to increasing hexagonality of the grown polytype with carbon enrichment of the surface and the inverse effect for the silicon enrichment, point to the fact that the polytypism is probably governed by crystal stoichiometry. Enrichment of the growth surface by silicon promotes the formation of silicon-rich phases, whereas enrichment by carbon is favourable for those having the greatest C to Si ratio.

Further studies [65,66] have shown the possibility of growing the 4H polytype without any impurities providing a moderate excess of carbon is provided on the crystal face. Controlled introduction of barium and phosphorus has ensured the transformation of 6H to 15R or to 21R. Development of the polytype transformation methods has permitted the growth of epitaxial layers of 4H-SiC free from foreign polytype inclusions on 6H substrates [65,66]. The growth of 4H ingots on seeds of other polytypes has been reported in [22,25,52]. These authors did not introduce any impurities in the growth system. The probable reason for the polytype transformation is the vapour depletion of silicon.

The 4H polytype has a wider gap compared to 6H and a much higher electron mobility, making the growth of the 4H polytype of significant practical interest. Favourable properties of this polytype could not be employed in the early studies since the abundance of the 4H polytype in crystals grown by the Lely process is below 1%. Currently methods of governing the polytype structure permit the intentional growth of this polytype.

I GROWTH OF HIGH QUALITY INTERFACES AND P-N JUNCTIONS BY SUBLIMATION

High quality p-n junctions in silicon carbide exhibit an exceptional electrical strength, nearly an order of magnitude higher than the p-n junctions in silicon (see Chapter 9). However, the appearance of defects in p-n junctions results in early breakdown and can negate all the advantages the fundamental material properties provide. The attempts to create p-n junctions without these defects were unsuccessful until the early 1980s. The creation of high quality p-n junctions [9] was based on development of the sublimation growth methods.

The majority of SiC-based devices employ lightly doped regions of n-type material whereas the heavily doped ones are p-type. Aluminium is strongly preferred as a p-type dopant due to its lower acceptor activation energy and its high solubility. The main advantage of sublimation growth as a method of p-n junction formation is the possibility of employing very high temperatures, not achievable with CVD. High growth temperatures promote the growth of high quality p-n junctions.

It is difficult to grow both the p^+ and the n-type layers in a single process due to memory effects of the graphite furnaces for the acceptor dopants. The heavily doped layers should be grown in a separate equipment which has been specially designed for doping. If the formation of a transition layer takes place at the initial stages of growth, the quality of the p-n junction might be seriously degraded. Although the method of surface recrystallisation (see Chapter 6) allows the growth of epitaxial aluminium-doped layers of high crystalline perfection, it appeared to be unsuitable for the formation of p-n junctions. Surface recrystallisation results in the formation of an interface layer with a very high donor concentration, above 10^{19} cm^{-3}. The p-n junctions grown by this method have very high leakage currents and a breakdown voltage of about 15 V irrespective of the substrate doping [9,69]. There exist two possible reasons for high donor concentrations in the recrystallised layer. Firstly, the recrystallisation process starts at rather low temperatures, much lower than those used for the growth. Low temperatures promote donor doping and inhibit acceptor doping. The donor concentration in the layer grown during the furnace heating process can be very high due to nitrogen degassing from the graphite [9]. Thus, portions of a heavily doped layer appear just at the p-n junction interface (FIGURE 15(d)). Secondly, the intrinsic point defects in silicon carbide are, predominantly, of the donor type, and thus an incomplete recrystallisation suggests incorporation of large amounts of donor type defects [68].

The depth of the recrystallised layer is very small, of the order of 0.01 - 0.1 μm. Thus, a possible way to eliminate its effect on p-n junction properties is to perform a drive-in diffusion of aluminium from the epitaxial layer. Unfortunately, the diffusivity of aluminium from an epitaxial layer is extremely slow in SiC. The diffusivity is 3 - 5 orders of magnitude lower than that observed for diffusing aluminium from the vapour phase [70]. The authors of [69] had to employ very high diffusion temperatures, over 2500 °C. The anneal produced a shift of the p-n junction into the crystal bulk and the electrical properties were substantially improved. However, this could not provide the elimination of the 'weak points' of the junctions. The characteristics of the p-n junctions were worse than those with the recrystallised layer removed by sublimation etching. In addition, the surface evaporation and graphitisation at temperatures above 2500 °C severely reduces the reproducibility of the results.

The removal of transition layers by an etch-back in the inverted temperature gradient provides a means of solving the problems related to interface quality. The experiments performed have shown, however, that the standard methods employed for the sublimation growth result in heavily short-circuited p-n junctions after the sequential etch-back and growth. The short-circuiting is produced by graphite residues, always present on the SiC crystal face, evaporated under the near-equilibrium conditions. In order to remove the graphite from the surface, the state of the 'sandwich' cell should be shifted to that of silicon excess. This can be done easily by placing small amounts of silicon in the vicinity of the substrate. Silicon reacts readily with the graphite parts of the crucible and it cannot provide a high excess pressure for a long time. However, since the surface cleaning requires only a few minutes, this method is quite sufficient for growth in an argon ambient. After the temperature gradient is inverted and the growth is started, the necessity for a high silicon excess disappears. For

8.2 Sublimation growth of SiC

growth in vacuum, the silicon evaporation rate is very high, so either silicon-enriched sources [36] or carbon gettering [11] is required.

The p-n junctions grown with the use of the preliminary sublimation etch-back usually have neither short-circuiting nor noticeable leakage currents. The avalanche breakdown occurs uniformly over the area of the diode structure. The breakdown voltages are very high: a reverse voltage above 1 kV can be obtained for diodes with the base width below 10 µm and base doping $> 10^{16}$ cm^{-3}. The dependence of the avalanche breakdown voltage on the concentration of uncompensated donors in the base for abrupt p-n junctions is plotted in FIGURE 16 [9,71]. Also plotted in the figure are results of a previous paper (von Muench and Pfaffeneder, [72]) and the recent results obtained by Edmond et al [73].

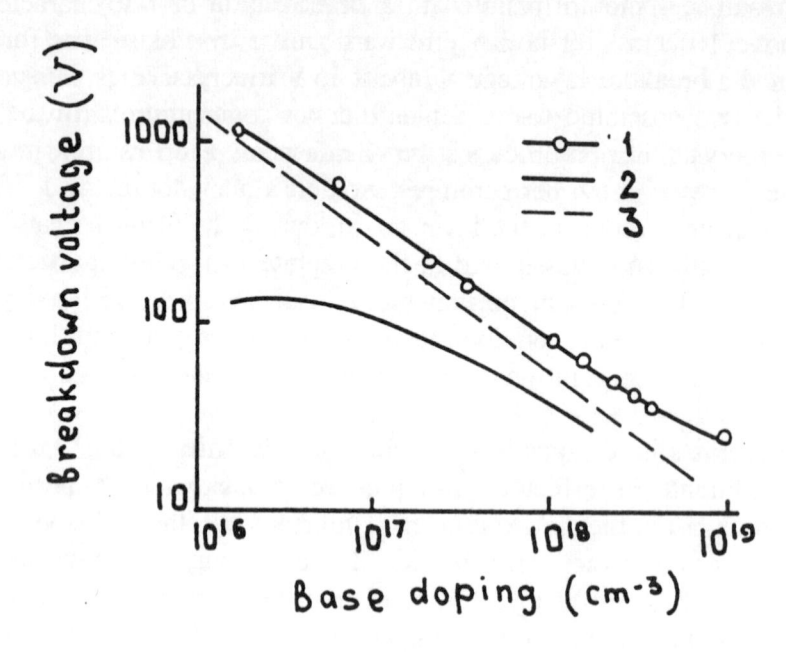

FIGURE 16 Dependence of the avalanche breakdown voltage on the dopant concentration in the diode base: (1) for the sublimation p-n junctions, uniform breakdown [9], (2) CVD-grown diodes [72], (3) in rectifier diodes [73].

A study of electric field inhomogeneities in the p-n junctions has been performed with use of avalanche multiplication of the electron beam induced current (EBIC) [74]. It has been shown that the regions of electric field concentration have a linear configuration. The microplasma appears at the end of the region at the point of the highest electric field. The most probable cause of electric field concentration is related to the effect of the residual damage introduced by mechanical processing, lapping and polishing. It is necessary to remove 40 - 80 µm of the damaged surface layer to eliminate the microplasmas for factory-processed crystals. A more thorough laboratory performed processing yields a somewhat lower thickness of the damaged layer.

An abrupt or nearly abrupt doping profile can be obtained for growth temperatures above 2500 °C. The reason for a small diffusion smearing is related to the effect of diffusion suppression for the solid state source [70]. The diffusion is actually proceeding only within a short period of time, which corresponds to the inversion of the temperature gradient. Before the gradient is inverted the diffused layers are being removed by etching, whereas after the

8.2 Sublimation growth of SiC

crystallisation of the first p-type layers the diffusion is essentially stopped. C-V studies delineate, however, a thin compensated layer in the vicinity of the interface. For the p-n junctions grown at temperatures of about 2450 °C the donor concentration in this layer may be about 1.2 times lower than in the bulk of the substrate. A decrease of growth temperature increases the extent of compensation in the interface layer. A similar effect has been previously observed by van Opdorp [70]. The p-n junctions studied had a doping profile close to a p-i-n structure, if they were grown or diffused at an intermediate temperature. He attributed the cause of the phenomenon to the effect of pulling the electric field of the p-n junction via dopant diffusion. A temperature increase raises the intrinsic carrier concentration and this suppresses the pulling electric field effect.

A decrease of growth temperature resulted in a deterioration of I-V characteristics of the epitaxially grown p-n junctions for both the forward and reverse bias. The formation of the strongly compensated interface layer often results in an increase in series resistance of a forward-biased diode. Under the forward bias the low temperature grown diodes exhibit microplasma breakdown. For the 0.4 - 0.5 mm diode size very few diodes are free from microplasmas, when grown at temperatures below 2200 °C.

Since formation of the diffused layers and appearance of microplasmas are correlated, it is reasonable to suppose the cause of electric field inhomogeneities is related to impurity aggregation due to diffusion in non-perfect crystals. This phenomenon is well-known for diffusion of highly mobile impurities in germanium and in silicon, particularly for gold and copper in germanium and silicon [74]. If the mobility of the impurity is high and/or it diffuses at a high concentration, it breaks the equilibrium distribution of the intrinsic point defects. The crystal bulk is depleted of vacancies, since nearly all of them have captured interstitial impurity atoms. The vacancy undersaturation slows down the diffusion since the interstitial to substitutional conversion of impurity atoms is limited by the supply of vacancies from the surface and/or by the sink of self-interstitials from the bulk to the surface. The crystal imperfections under non-equilibrium conditions act as additional sources of vacancies (compare FIGURES 17(a) and 17(b)). Thus, impurities crowd around the imperfection and, if a dislocation is normal to the surface, diffusion in the vicinity of the dislocation is enhanced.

It has been shown recently [75] that the diffusion of boron and aluminium in SiC occurs by a mechanism very similar to that of gold in silicon, i.e. it breaks the equilibrium of intrinsic point defects in the system. Thus, if aluminium concentration is high and the diffusion temperature is not too high, the enhanced diffusion will take place along dislocations. In the vicinity of an imperfection the p-n junction profile will be distorted, and a needle-shaped projection will appear (FIGURE 17(c)). This suggests a very strong electric field concentration and microplasma formation in a reverse-biased p-n junction.

The suppression of this deleterious effect can be attained by one of the three methods:

(i) By growing the p-n junction at a very high temperature, when the equilibrium concentrations of vacancies and self-interstitials are high. The non-equilibrium phenomena are suppressed and the effect of imperfections is negligible.

(ii) By suppressing the diffusion, thus minimising the deleterious effects.

8.2 Sublimation growth of SiC

(a)

(b)

extended defect

(c)

dislocation

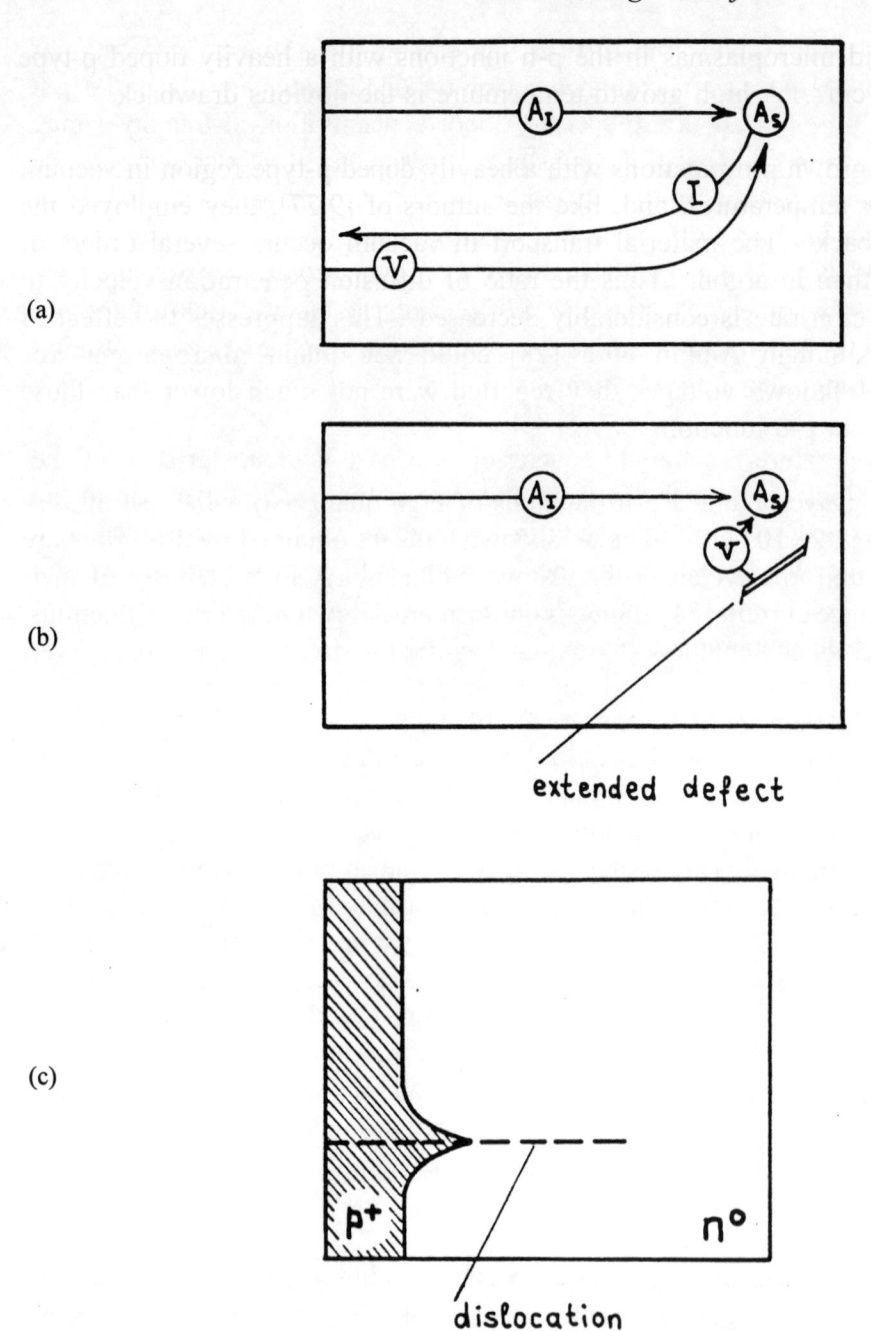

FIGURE 17 Effect of imperfections on interstitial-substitutional diffusion: (a) the interstitial to substitutional impurity conversion in a perfect crystal is limited by intrinsic point defect transport, (b) a crystal imperfection promotes the conversion. The subscripts I and S denote interstitial and substitutional positions. (c) Diffusion of acceptor along dislocation distorts p-n junction.

(iii) By decreasing the impurity concentration up to the level insufficient for breaking the point defect equilibrium.

All these three methods have been employed more or less successfully.

(i) The present author as well as Mokhov and co-workers have employed epitaxial growth at extremely high temperatures [9,71,74,77]. The advantage of this approach is the possibility to avoid microplasmas in the p-n junctions with a heavily doped p-type region, up to 10^{21} cm^{-3}. A high growth temperature is the obvious drawback.

(ii) Anikin et al have grown p-n junctions with a heavily doped p-type region in vacuum at sufficiently low temperatures, and, like the authors of [9,77], they employed the sublimation etch-back. The material transport in vacuum occurs several orders of magnitude faster than in argon. Thus the ratio of diffusion penetration velocity to growth or evaporation rate is considerably decreased. This suppresses the effect of imperfections. Although Anikin et al [78] could not obtain microplasma free breakdown, the breakdown voltages they reported were not much lower than those exhibited by uniform p-n junctions.

(iii) Edmond et al [73] have obtained p-n junctions of high quality with the use of very low acceptor doping, 2×10^{18} cm^{-3}. The breakdown voltages obtained by this group are very close to the uniform avalanche breakdown voltages. A high stability of their diodes to large reverse current [74] suggests uniform breakdown in their p-n junctions. The main drawback of this method is the high specific resistivity of the p-type layer, since the carrier concentration in non-degenerately doped p-type material is much lower than the dopant concentration.

The growth of high quality p-n junctions requires, therefore, a certain compromise between the acceptor concentration, growth temperature and the device perfection desired. The use of a higher acceptor doping suggests the use of a higher growth temperature or a reduction of p-n junction perfection.

J CONCLUSION

Sublimation methods are the most efficient and productive means for the growth of SiC. The 'sandwich' sublimation method, where the source and substrate are situated close to each other, permits faster growth under near-equilibrium conditions resulting in high quality material. Variations of these sublimation methods can be made to produce both bulk crystals and epitaxial layers. It is important to minimise carbide inclusions and this is done either by using tantalum furnace components or by etch-back at the growth temperature in a reverse temperature gradient. Both resistive and induction heating can be used, the advantage of the latter being the easier temperature field control. Sources can be SiC polycrystalline boules or CVD-grown polycrystalline layers or single crystal SiC (for epitaxial layer growth). Higher temperature growth is favoured due to a decrease in the Si:C ratio in the vapour. Control of equlibrium conditions is via sources enriched in silicon or via carbon gettering, usually using a tantalum getter. Growth rate varies linearly with temperature difference (source to substrate) except when impurities are present. These are commonly nitrogen (donor) and boron and aluminium (acceptors) which are also used as deliberate dopants. Low level nitrogen doping (10^{16} cm^{-3}) in epitaxial layers can be achieved in vacuum but growth rates must be limited to avoid the formation of donor defects. Nitrogen contamination can be decreased by using a tantalum crucible. Aluminium contamination can be removed via a vacuum anneal of the furnace but boron is not easily removed. Nitrogen is captured on the C face while the acceptors are captured by the Si face. When added as dopants, acceptor elements are in elemental form or as carbides while nitrogen is added as gaseous N_2. Pores and cubic inclusions tend to form at the early growth stage. Twinning and polytype formation are also problems which remain. Heating in argon prior to growth has alleviated the latter problem. High quality p-n junctions can be grown at high temperatures with aluminium as the acceptor. Etch-back at the growth temperature in a reverse temperature gradient and silicon excess both improve surface quality prior to growth. Reverse voltages of 1 kV for a 10 μm base width and $> 10^{16}$ cm^{-3} doping level have been achieved. Enhanced acceptor diffusion down dislocations can lead to breakdown but means to overcome this have been devised. A compromise still has to be made between acceptor concentration, growth temperature and device performance.

REFERENCES

[1] V.F. Knippenberg [*Philips Res. Rep. (Netherlands)* vol.18 no.1 (1965) p.161-274]
[2] J.A. Lely [*Ber. Dtsch. Keram. Ges. (Germany)* vol.32 (1955) p.229-34]
[3] A.R. Verma, P. Krishna [*Polymorphism and Polytypism in Crystals* (Wiley, New York, 1966)]
[4] R.B. Campbell, Y.C. Chang [*Semicond. Semimet. (USA)* vol.7b (1976) p.625-36]
[5] J. Smiltens [*Mater. Res. Bull. (USA)* vol.4 (1969) p.S85-96]
[6] Yu.A. Vodakov, E.N. Mokhov [Patent USSR No.403275 prior 5.11.1970, Patent USA No.4147572]
[7] Yu.M. Tairov, V.F. Tsvetkov [*J. Cryst. Growth (Netherlands)* vol.43 no.2 (1978) p.209-12]
[8] Yu.M. Tairov, V.F. Tsvetkov [*J. Cryst. Growth (Netherlands)* vol.52 no.1 (1981) p.146-50]
[9] A.O. Konstantinov, D.P. Litvin, V.I. Sankin [*Pis'ma Zh. Tekh. Fiz. (USSR)* vol.7 no.21 (1981) p.1335-9 (*Sov. Tech. Phys. Lett. (USA)* vol.7 no.11 (1981) p.572-3)]
[10] E.N. Mokhov et al [*Doped Semiconducting Materials* (Nauka, Moscow, 1990) p.45-52 (in Russian)]
[11] E.N. Mokhov, M.G. Ramm, A.D. Roenkov, A.A. Volfson, A.S. Tregubova, I.L. Shulpina [*Pis'ma Zh. Tekh. Fiz. (USSR)* vol.13 no.11 (1987) p.641-5]
[12] R.I. Scase, G.A. Slack [*Silicon Carbide - A High Temperature Semiconductor* (Pergamon Press, New York, 1960) p.23-8]
[13] W.F. Knippenberg, G. Verspui [*Mater. Res. Bull. (USA)* vol.4 (1969) p.S33-44]
[14] W.F. Knippenberg, G. Verspui [*Mater. Res. Bull. (USA)* vol.4 (1969) p.S45-56]
[15] C.J. Kapteyns, W.F. Knippenberg [*J. Cryst. Growth (Netherlands)* vol.7 no.1 (1970) p.20-8]
[16] R.M. Potter, J.H. Satelle [*J. Cryst. Growth (Netherlands)* vol.12 no.3 (1972) p.245-8]
[17] A.A. Glagovskii et al [*Problems of Physics and Technology of Wide Bandgap Semiconductors* (LIYaF, Leningrad, 1979) p.226-40 (in Russian)]
[18] J. Drowart, G. De Maria [*Silicon Carbide - A High Temperature Semiconductor* (Pergamon Press, New York, 1960) p.16-21]; see also J. Drowart, G. De Maria, M.G. Inghram [*J. Chem. Phys. (USA)* vol.41 no.5 (1958) p.1015-23]
[19] R.G. Behrens, G.H. Rinehart [*NBS Special Publication (USA)* no.561 *Proc. 10th Materials Research Symposium on High Temperature Vapours and Gases* (1979) p.125-42]
[20] H.J. Kim, D.W. Shin [*Springer Proc. Phys. (Germany)* vol.56 (1992) p.23-8]

[21] G. Ziegler, P. Lanig, D. Theis, C. Weirich [*IEEE Trans. Electron Devices (USA)* vol.ED-30 no.4 (1983) p.277-81]

[22] P.A. Glasow [*Springer Proc. Phys. (Germany)* vol.34 (1989) p.13-33]

[23] S. Nakata, K. Koga, Y. Ueda, T. Niina [*Springer Proc. Phys. (Germany)* vol.43 (1989) p.26-34]

[24] D.L. Barrett, R.G. Seidensticker, W. Gaida, R.H. Hopkins, W.J. Choyke [*Springer Proc. Phys. (Germany)* vol.56 (1992) p.33-9]

[25] S. Nishino, Y. Kojima, J. Saraie [*Springer Proc. Phys. (Germany)* vol.56 (1992) p.15-21]

[26] S.K. Lilov, Yu.M. Tairov, V.F. Tsvetkov, B.F. Yudin [*J. Cryst. Growth (Netherlands)* vol.32 no.2 (1976) p.170-6]

[27] S.K. Lilov, Yu.M. Tairov, V.F. Tsvetkov, B.F. Yudin [*J. Cryst. Growth (Netherlands)* vol.40 no.11 (1977) p.59-63]

[28] W.A. Chupka, J. Berkowits, C.F. Grieze, M.G. Inghram [*J. Phys. Chem. (USA)* vol.62 no.5 (1958) p.611-4]

[29] J. Drowart, G. De Maria, A.J.H. Boebrom, M.G. Inghram [*J. Chem. Phys. (USA)* vol.30 no.1 (1959) p.308-13]

[30] G.K. Safaraliev, Yu.M. Tairov, V.F. Tsvetkov [*Kristallografya (USSR)* vol.20 no.5 (1975) p.1004-6]

[31] Yu.A. Vodakov, E.N. Mokhov, M.G. Ramm, A.D. Roenkov [*Krist. Tech. (Germany)* vol.4 no.6 (1979) p.729-40]

[32] E.N. Mokhov, Yu.A. Vodakov, G.A. Lomakina [*Problems of Physics and Technology of Wide Bandgap Semiconductors* (LIYaF, Leningrad, 1979) p.136-49 (in Russian)]

[33] Yu.M. Tairov, V.F. Tsvetkov [*J. Cryst. Growth (Netherlands)* vol.46 no.3 (1979) p.403-9]

[34] Yu.M. Tairov, V.F. Tsvetkov, V.A. Taranets [*Neorg. Mater. (USSR)* vol.15 no.1 (1979) p.9-12]

[35] A.O. Konstantinov, E.N. Mokhov [*Pis'ma Zh. Tekh. Fiz. (USSR)* vol.7 no.21 (1981) p.1335-9]

[36] M.M. Anikin, A.A. Lebedev, S.N. Pyatko, A.M. Strel'chuk, A.L. Syrkin [*Mater. Sci. Eng. B (Switzerland)* vol.11 no.1 (1992) p.113-5]

[37] F. Reihel, Yu.M. Tairov, V.F. Tsvetkov [*Neorg. Mater. (USSR)* vol.19 no.1 (1983) p.67-71]

[38] M.G. Ramm, E.N. Mokhov, R.G. Verenchikova [*Neorg. Mater. (USSR)* vol.15 no.12 (1981) p.2233-4]

[39] E.N. Mokhov et al [*Doped Semiconducting Materials* (Nauka, Moscow, 1985) p.42-51 (in Russian)]

[40] Yu.A. Vodakov, E.N. Mokhov, M.G. Ramm, A.D. Roenkov [*Springer Proc. Phys. (Germany)* vol.56 (1992) p.329-34]

[41] Yu.A. Vodakov et al [*Properties of Doped Semiconductors* (Nauka, Moscow, 1977) p.48-52 (in Russian)]

[42] G.F. Yuldashev, M.M. Usmanova, Yu.A. Vodakov [*At. Energ. (Russia)* vol.18 no.1 (1972) p.592-3]

[43] E.I. Radovanova, R.G. Verenchikova, Yu.A. Vodakov [*Fiz. Tekh. Poluprovodn. (USSR)* vol.17 no.7 (1983) p.113-8]

[44] S.H. Hagen, A.W.C. van Kamenade, J.A.W. van der Das de Bye [*J. Lumin. (Netherlands)* vol.8 no.1 (1973) p.18-31]

[45] Yu.A. Vodakov et al [*Fiz. Tekh. Poluprovodn. (USSR)* vol.11 no.2 (1977) p.373-9 (*Sov. Phys.-Semicond. (USA)* vol.11 no.2 (1977) p.214)]

[46] R.M. Potter [*Mater. Res. Bull. (USA)* vol.4 (1969) p.S223-30]

[47] E.N. Mokhov, M.G. Ramm, A.D. Roenkov, M.I. Fedorov, R.G. Verenchikova [*Pis'ma Zh. Tekh. Fiz. (USSR)* vol.16 no.14 (1990) p.33-7]

[48] S.K. Lilov, Yu.M. Tairov, V.F. Tsvetkov [*Neorg. Mater. (USSR)* vol.13 no.3 (1977) p.448-9]

[49] H.J. Kim, R.F. Davis [*J. Electrochem. Soc. (USA)* vol.133 no.11 (1984) p.2350-7]

[50] Yu.A. Vodakov, G.A. Lomakina, E.N. Mokhov, B.S. Makhmudov, M.M. Usmanova, G.F. Yuldashev [*Doped Semiconducting Materials* (Nauka, Moscow, 1982) p.89-93 (in Russian)]

[51] Yu.A. Vodakov, G.A. Lomakina, E.N. Mokhov, E.I. Radovanova [*Fiz. Tverd. Tela (Russia)* vol.20 no.2 (1978) p.448-51]

[52] Yu.M. Tairov, V.F. Tsvetkov [*Crystal Growth* (Nauka, Moscow, 1991) vol.18 p.51-65 (in Russian)]; see also Yu.M. Tairov, V.F. Tsvetkov [*Prog. Cryst. Growth Charact. (UK)* vol.8 (1983) p.111-62]

[53] D. Pandey, P. Krishna [*Silicon Carbide-1973*, Proc. 3rd Int. Conf. (University of South Carolina Press, 1973) p.198-206]; see also D. Pandey, P. Krishna [*J. Cryst. Growth (Netherlands)* vol.31 no.1 (1971) p.66-71]

[54] Yu.M. Tairov, V.F. Tsvetkov, M.A. Chernov [*Kristallografya (USSR)* vol.24 no.4 (1979) p.772-7]

[55] W.S. Yoo, S. Nishino, S. Matsunami [*Springer Proc. Phys. (Germany)* vol.43 (1989) p.35-9]

[56] Yu.A. Vodakov, E.N. Mokhov, A.D. Roenkov, D.T. Saidbekov [*Phys. Status Solidi (Germany)* vol.51 no.1 (1979) p.209-15]

[57] D.R. Moskvina, I. Petzold, E.N. Potapov, Yu.M. Tairov [*Fiz. Tekh. Poluprovodn. (USSR)* vol.23 no.12 (1989) p.2240-3]

[58] L.S. Aivasova, L.G. Nikolaeva, V.G. Sidyakin, G.G. Shmatko [*Neorg. Mater. (USSR)* vol.9 no.1 (1973) p.149-50]

[59] A.O. Konstantinov [*Fiz. Tekh. Poluprovodn. (USSR)* vol.21 no.4 (1987) p.670-5 (*Sov. Phys.-Semicond. (USA)* vol.21 no.4 (1992) p.410-4)]

[60] Yu.A. Vodakov, G.A. Lomakina, E.N. Mokhov [*Fiz. Tverd. Tela (USSR)* vol.11 no.2 (1969) p.519-21]

[61] V.I. Levin, Yu.M. Tairov, M.G. Travajan, V.F. Tsvetkov, M.A. Chernov [*Neorg. Mater. (USSR)* vol.14 no.6 (1978) p.1062-6]

[62] G.K. Safaraliev, Yu.M. Tairov, V.F. Tsvetkov [*Pis'ma Zh. Tekh. Fiz. (USSR)* vol.2 no.15 (1976) p.699-701]

[63] H. Vakhner, Yu.M. Tairov [*Fiz. Tverd. Tela (USSR)* vol.11 (1969) p.2440]

[64] Yu.A. Vodakov, E.N. Mokhov, A.D. Roenkov, M.M. Anikin [*Pis'ma Zh. Tekh. Fiz. (USSR)* vol.5 no.6 (1979) p.367-70]

[65] E.N. Mokhov [*Proc. 3rd USSR Conf. Physics and Technology of Wide Bandgap Semiconductors* Mahachkala, 1986, p.4-5]

[66] A.S. Barash, Yu.A. Vodakov, E.N. Koltsova, A.A. Maltsev, E.N. Mokhov, A.D. Roenkov [*Pis'ma Zh. Tekh. Fiz. (USSR)* vol.14 no.24 (1988) p.2222-6]

[67] Yu.A. Vodakov, G.A. Lomakina, E.N. Mokhov, V.G. Oding [*Fiz. Tverd. Tela (USSR)* vol.19 no.9 (1969) p.1812-4]

[68] P.T.B. Shaffer [*Mater. Res. Bull. (USA)* vol.4 (1969) p.S13-24]

[69] R.G. Verenchikova et al [*Fiz. Tekh. Poluprovodn. (USSR)* vol.16 no.11 (1982) p.2029-32]

[70] C. van Opdorp [*Solid-State Electron. (UK)* vol.16 (1971) p.643]

[71] A.P. Dmitriev, A.O. Konstantinov, D.P. Litvin, V.I. Sankin [*Fiz. Tekh. Poluprovodn. (USSR)* vol.17 no.6 (1983) p.1093-109 (*Sov. Phys.-Semicond. (USA)* vol.17 no.6 (1984) p.686]

[72] W. von Muench, I. Pfaffeneder [*J. Appl. Phys. (USA)* vol.48 no.11 (1977) p.4831-3]

[73] J.A. Edmond, D.G. Waltz, S. Brueckner, H.-S. Kong, J.W. Palmour, C.H. Carter Jr. [*Proc. Conf. High Temperature Electronics*, Albuquerque, 1991]

[74] S.G. Konnikov, A.O. Konstantinov, V.I. Litmanovich [*Fiz. Tekh. Poluprovodn. (USSR)* vol.18 no.9 (1984) p.1556-60 (*Sov. Phys.-Semicond. (USA)* vol.18 no.9 (1984) p.975-7)]

[75] W. Frank, U. Gosele, H. Mehrer, A. Seeger [*Diffusion in Crystalline Solids* Eds G.E. Murch, A.S. Nowick (Academic Press, New York, 1984) p.63-96]

[76] A.O. Konstantinov [*Fiz. Tekh. Poluprovodn. (USSR)* vol.26 no.2 (1992) p.270-9 (*Sov. Phys.-Semicond. (USA)* vol.26 no.2 (1992) p.151-6)]

[77] Yu.A. Vodakov, D.P. Litvin, V.I. Sankin, E.N. Mokhov, A.D. Roenkov [*Fiz. Tekh. Poluprovodn. (USSR)* vol.19 no.5 (1985) p.814-9 (*Sov. Phys.-Semicond. (USA)* vol.19 no.5 (1985) p.502)]

[78] M.M. Anikin, A.A. Lebedev, A.L. Syrkin, B.V. Tsarenkov, V.E. Chelnokov [*Fiz. Tekh. Poluprovodn. (USSR)* vol.22 no.1 (1988) p.133-6]

8.3 Chemical vapour deposition of SiC

S. Nishino

June 1995

A   INTRODUCTION

Epitaxial growth of SiC by chemical vapour deposition (CVD) is essentially the same as the method used in Si epitaxial growth. Carrier gases such as Ar or hydrogen are passed through the reaction tube and reaction gases such as $SiCl_4 + CCl_4$, $HSiCl_3 + C_6H_{14}$, $SiH_4 + C_3H_8$, $SiCl_4 + C_6H_{14}$, and $SiC_{14} + C_3H_8$ are used. For device applications thin film formation by CVD is needed. However, usable sizes of single crystalline layer were not established for many years. Many researchers attempted to grow SiC on foreign substrates because of a lack of large SiC substrates [1-20]. For this reason, heteroepitaxial growth of SiC on Si has been frequently used. Si substrates have many merits such as crystalline perfection, orientation control and large area. However, the large thermal and lattice mismatches between the Si and SiC have limited the material quality. The use of other substrate materials, such as sapphire, was also reported [6]. However, thick single crystalline layers were not obtained for many years [1-20].

Homoepitaxial growth was also carried out using 6H-SiC substrates made by the Acheson method [153-167]. Initially the 6H-SiC basal plane was used as the substrate for homoepitaxial growth, and higher temperatures were required to obtain single crystalline layers [157,158]. As misoriented 6H-SiC substrates were employed in the homoepitaxial growth, the temperature was reduced to about 1500 °C. This innovative approach is called step-controlled epitaxy [166]. Recent SiC devices using α-SiC are made by the step-controlled epitaxial growth process [168-183]. There are many review articles concerning crystal growth of SiC and its properties: see for example [66,126,127].

B   3C-SiC

3C-SiC has some material property advantages over 6H-SiC, such as high electron mobility at low-field, which might be employed to produce superior devices and circuits for microwave power and other applications [135]. As for epitaxial growth of SiC, 3C-SiC has had a long history compared with α-SiC, because it is a stable phase at low temperature. However, good crystal growth was not obtained until a buffer layer technique was reported [19-22].

Heteroepitaxial growth of 3C-SiC on Si substrates has been known to be difficult due to the large lattice mismatch (20%) and thermal expansion coefficient mismatch (8%) between 3C-SiC and Si. To circumvent this lattice mismatch, buffer layer growth was employed just before the crystal growth. The temperature profile of crystal growth is shown in FIGURE 1. In this figure, a key point to reproduce good single crystalline cubic SiC is a carbonisation process [20,22]. Controlling parameters of the carbonisation process were a flow rate of propane, silane, temperature rise time etc. The silicon surface should be rapidly covered with a hard thin SiC layer to prevent Si vapour escaping from the substrate [47]. After this report, many 3C-SiC films on Si substrates were studied [23-152]. The nature of the carbonised layer was studied by MEIS (medium energy ion scattering), soft X-ray spectroscopy (SXS) and transmission electron microscopy (TEM). By these observations, the buffer layer was confirmed to be thin (few nm) single crystalline 3C-SiC [91-93].

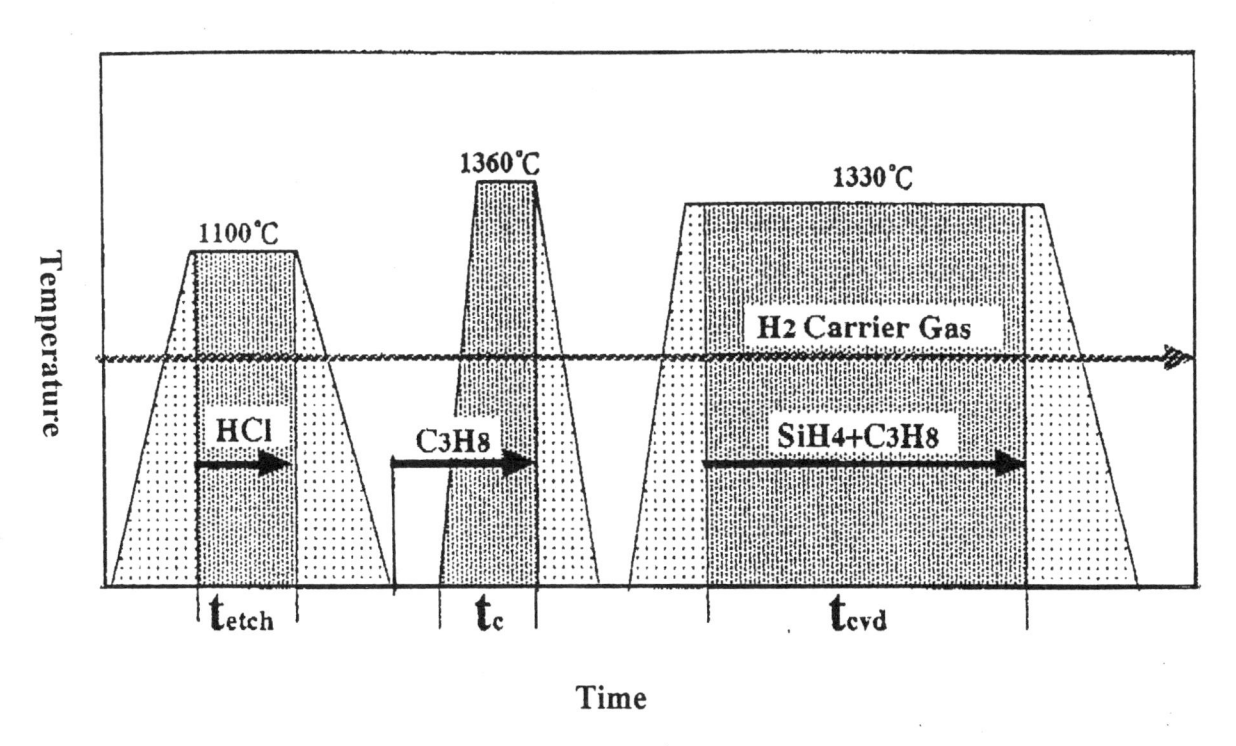

FIGURE 1 Temperature program of SiC single crystal growth: t_{etch}, t_c and t_{cvd} are the times of etching, carbonization and single crystal growth, respectively.

SiC films that were grown on on-axis Si(100) substrates exhibit a high density of interfacial twinning, stacking faults and antiphase disorder. The surface morphology of films grown on spherically polished substrates indicated that the tilt direction should be [110] to eliminate antiphase disorder [39,53]. Silicon substrates tilted toward the (100) plane were not effective in eliminating antiphase disorder. With respect to surface morphology, the on-axis films are less sensitive to the Si/C ratio of the input process gases than the misoriented films. The surface roughness for the on-axis films was nominally three times greater than for the misoriented films [75]. Micro-cracks were observed in epitaxial 3C-SiC films grown on Si(111), and thermal stress in the 3C-SiC(111) films was calculated to be approximately twice as large as in the 3C-SiC(100) films [75].

Despite continuing progress in the crystal growth, 3C-SiC films still contain many lattice defects. In particular, twins, stacking faults and antiphase boundaries (APBs) have been reported [64,65]. APBs occur as a common defect when a polar film, SiC in this case, is heteroepitaxially grown on a non-planar substrate. To eliminate this particular defect in 3C-SiC films, Si substrates misoriented from the (100) plane have been used, as stated above [39,53].

Low temperature growth has been attempted in many forms such as plasma-CVD, low pressure-CVD etc. [57,80,84,85,91,103,106,110,118,125,142,151]. Heterojunction devices are also reported using those low temperature growth methods [38,57,80,98,104].

C 6H-SiC

In early studies of homoepitaxial growth of 6H-SiC, researchers used basal plane 6H-SiC made by the Acheson method. In this case, high epitaxial temperatures were employed [153-165]. As SiC substrates can now be obtained, studies are focusing on the growth of SiC homoepitaxially at lower temperatures. Single crystals of 6H-SiC can be homoepitaxially grown at 1500 °C, which is 300 °C lower than previously reported, by the introduction of misorientation into the 6H-SiC(0001) substrates [166]. This misorientation of the substrate produces increased surface steps and enables growth of single crystalline 6H-SiC with a very smooth surface at lower temperatures. The direction of misorientation was also an important parameter to obtain smooth surfaces. This method is termed step controlled epitaxial [166,167]. The growth mechanism was studied experimentally and was also explained based on the BCF (Burton, Cabrera, Frank) theory [173,174,181,183].

In general, background nitrogen (N) causes unintentionally doped crystals to be n-type. In the best 6H-SiC samples, the most developed of the polytypes, background carrier concentrations in the mid-10^{15} cm^{-3} range have been achieved. A 'site-competition' epitaxy technique based on the use of the Si/C ratio for dopant control has been developed for SiC CVD epitaxial layers [180]. In this novel technique, very low net carrier concentrations of about 1×10^{14} cm^{-3} are reported. When N-doping is introduced during the growth, carrier concentrations as high as 10^{18} cm^{-3} have been realised.

p-Type doping is a problem in SiC, although considerable progress has been made. All of the acceptor impurities such as Al, B, Ga and Sc produce deep levels and are difficult to activate, generally requiring a high-temperature anneal. Al is somewhat difficult to incorporate into the SiC lattice, and high carrier concentrations are difficult to achieve. p-Type carrier concentrations in the 10^{19} - 10^{20} cm^{-3} range using tri-methylaluminium, TMA, in a CVD process on the Si face of 6H-SiC were obtained; however, a hole concentration of 2×10^{18} cm^{-3} was obtained on the C face. The carrier concentration was easily controllable down to low 10^{16} cm^{-3}.

D CONCLUSION

Higher quality 3C-SiC is obtained when a buffer layer is used between the silicon substrate and the epitaxial layer. Layers on (100) silicon contain twins, stacking faults and antiphase regions while those on (111) silicon exhibit microcracks. Use of misoriented (100) substrates eliminates the antiphase regions. Low temperature CVD processes have produced material which has been used in heterojunction devices. Step-controlled epitaxy using misoriented 6H-SiC(0001) substrates is used to produce smooth single crystal layers of the 6H-SiC polytype. Controlling the Si/C ratio in the so-called 'site-competition' growth process produces very low background donor levels. Nitrogen doping up to 10^{18} cm^{-3} has been achieved. Acceptor doping is more problematic but aluminium doping up to 10^{20} cm^{-3} (Si face) and 2×10^{18} cm^{-3} (C face) and down to 10^{16} cm^{-3} can be accomplished in 6H-SiC material.

REFERENCES

[1] D.M. Jackson Jr., R.W. Howard [*Trans. Metall. Soc. AIME (USA)* vol.233 (1965) p.468]

[2] N.C. Tombs, J.J. Comer, J.F. Fitzgerald [*Solid-State Electron. (UK)* vol.8 (1956) p.839-42]

[3] H. Nakashima, T. Sugano, H. Yanai [*Jpn. J. Appl. Phys. (Japan)* vol.5 (1966) p.374]

[4] J.J. Rohan, J.L. Sampson [*J. Phys. Chem. Solids (UK)* suppl. no.1 (1967) p.523]

[5] K.E. Bean, P.S. Gleim [*J. Electrochem. Soc. (USA)* vol.114 (1987) p.1158]

[6] R.B. Bartlett, R.A. Mueller [*Mater. Res. Bull. (USA)* vol.4 (1969) p.S341-54]

[7] P. Rai-Chaudhury, N.P. Formigoni [*J. Electrochem. Soc. (USA)* vol.116 (1969) p.1440]

[8] K.A. Jacobson [*J. Electrochem. Soc. (USA)* vol.118 (1971) p.1001]

[9] Y. Avigal, M. Schieber [*J. Cryst. Growth (Netherlands)* vol.9 (1971) p.127]

[10] J. Graul, E. Wagner [*Appl. Phys. Lett. (USA)* vol.21 (1972) p.67]

[11] G. Gramberg, M. Koniger [*Solid-State Electron. (UK)* vol.15 (1972) p.285]

[12] K. Kuroiwa, T. Sugano [*J. Electrochem. Soc. (USA)* vol.120 (1973) p.138]

[13] W.F. Knippenberg, G. Vesuui, A.W.C. van Kemenade [*Silicon Carbide 1973* Eds R.C. Marshall, J.W. Faust Jr., C.E. Ryan (University of South Carolina Press, 1974) p.92-107]

[14] W.F. Knippenberg, G. Verspui [*Silicon Carbide 1973* Eds R.C. Marshall, J.W. Faust Jr., C.E. Ryan (University of South Carolina Press, 1974) p.108-22]

[15] C.J. Mogab, H.J. Leamy [*Silicon Carbide 1973* Eds R.C. Marshall, J.W. Faust Jr., C.E. Ryan (University of South Carolina Press, 1974) p.58-63]

[16] H.J. Leamy, C.J. Mogab [*Silicon Carbide 1973* Eds R.C. Marshall, J.W. Faust Jr., C.E. Ryan (University of South Carolina Press, 1974) p.64-71]

[17] H. Matsunami, S. Nishino, T. Tanaka [*J. Cryst. Growth (Netherlands)* vol.45 (1978) p.138-43]

[18] A. Leonhardt, D. Selbmann, E. Wolf, M. Schonherr, C. Herrmann [*Krist. Tech. (Germany)* vol.13 (1978) p.523]

[19] S. Nishino, Y. Hazuki, H. Matsunami, T. Tanaka [*J. Electrochem. Soc. (USA)* vol.127 (1980) p.2674]

[20] H. Matsunami, S. Nishino, H. Ono [*IEEE Trans. Electron Devices (USA)* vol.ED-38 (1981) p.1235]

[21] S. Nishino, H. Suhara, H. Matsunami [*Extended Abstracts 16th Conf. Solid State Devices and Materials* Tokyo (Business Center for Academic Societies Japan, Tokyo, 1983) p.317]

[22] S. Nishino, J.A. Powell, W. Will [*Appl. Phys. Lett. (USA)* vol.42 (1983) p.460]

[23] A. Adamiano, J.A. Sprague [*Appl. Phys. Lett. (USA)* vol.44 (1984) p.525]

[24] A. Addamiano, P.H. Klein [*J. Cryst. Growth (Netherlands)* vol.70 (1984) p.291]

[25] K. Sasaki, E. Sakuma, S. Misawa, S. Yoshida, S. Gonda [*Appl. Phys. Lett. (USA)* vol.45 (1984) p.72]

[26] H.P. Liaw, R.F. Davis [*J. Electrochem. Soc. (USA)* vol.131 (1984) p.3014-8]

[27] A. Suzuki, K. Furukawa, J. Higashigaki, S. Harada, S. Nakajima, T. Inouguchi [*J. Cryst. Growth (Netherlands)* vol.70 (1984) p.287]

[28] R. Kaplan [*J. Appl. Phys. (USA)* vol.56 (1984) p.1636-41]

[29] P. Liaw, R.F. Davis [*J. Electrochem. Soc. (USA)* vol.132 (1985) p.642]

[30] S. Yoshida, K. Sasaki, K. Sakuma, S. Misawa, S. Gonda [*Appl. Phys. Lett. (USA)* vol.46 (1985) p.766]

[31] C.H. Carter Jr., J.A. Edmond, J.W. Palmour, J. Rye, H.J. Kim, R.F. Davis [MRS Meeting, San Francisco, USA, April 1985]

[32] H. Ryu, H.J. Kim, R.F. Davis [*Appl. Phys. Lett. (USA)* vol.17 (1985) p.850]

[33] Y. Kondo, T. Takahashi, K. Ishii, Y. Yoshida [*IEEE Electron Device Lett. (USA)* vol.EDL-7 (1986) p.404-6]

[34] K. Furukawa, A. Uemoto, M. Shigeta, A. Suzuki, S. Nakajima [*Appl. Phys. Lett. (USA)* vol.48 (1986) p.1536-7]

[35] Y. Fujiwara, E. Sakuma, S. Misawa, K. Endo, S. Yoshida [*Appl. Phys. Lett. (USA)* vol.49 (1986) p.388]

[36] H.J. Kim, R.F. Davis [*J. Appl. Phys. (USA)* vol.60 (1986) p.2897]

[37] H.J. Kim, R.F. Davis [*J. Electrochem. Soc. (USA)* vol.133 (1986) p.2350-7]

[38] Y. Furumura, M. Doki, F. Mieno, M. Maeda [*Trans. Inst. Electron. Commun. Eng. Jpn. (Japan)* vol.69 (1986) p.705]

[39] K. Shibahara, S. Nishino, H. Matsunami [*J. Cryst. Growth (Netherlands)* vol.78 (1986) p.538]

[40] C.H. Carter Jr., R.F. Davis, S.R. Nutt [*J. Mater. Res. (USA)* vol.1 (1986) p.811]

[41] K. Shibahara, T. Saito, S. Nishino, H. Matsunami [*IEEE Electron Device Lett. (USA)* vol.EDL-7 (1986) p.692-3]

[42] K. Shibahara, T. Saitoh, S. Nishino, H. Matsunami [*IEEE Electron Device Lett. (USA)* vol.EDL-7 (1986) p.692]

[43] K. Shibahara, T. Saito, S. Nishino, H. Matsunami [*Ext. Abstracts 18th Conf. Solid State Devices and Materials* Tokyo (Business Center for Academic Societies Japan, Tokyo, 1986) p.717]

[44] H.S. Kong, J.T. Glass, R.F. Davis [*Appl. Phys. Lett. (USA)* vol.49 (1986) p.1074-6]

[45] C.M. Chorey, P. Pirouz, J.A. Powell, T.E. Mitchell [*Semiconductor-Base Heterojunctions: Interfacial Structure and Stability* Eds M.L. Green, J.E. Gablin, G.Y. Chin, H.W. Deckman, W. Mayo, D. Narasnham (The Metallurgical Society Inc, Warrendale, PA, 1986) p.115-25]

[46] A. Suzuki, A. Uemoto, M. Shigeta, K. Furukawa, S. Nakajima [*Ext. Abstracts 18th Int. Conf. Solid State Devices and Materials* 1989 p.101-4]

[47] S. Nishino, H. Suhara, H. Ono, H. Matsunami [*J. Appl. Phys. (USA)* vol.61 (1987) p.4889-93]

[48] C.D. Stinespring, J.C. Wormhoudt [*J. Appl. Phys. (USA)* vol.65 (1989) p.1733-42]

[49] G.L. Harris et al [*Mater. Res. Soc. Symp. Proc. (USA)* vol.97 (1987) p.201-6]

[50] H. Matsunami [*Optoelectronics-Device and Technology* vol.2 (Mita Press, Tokyo, Japan, 1987) p.29-44]

[51] J.A. Powell, L.G. Matus, A. Kaczmarski [*J. Electrochem. Soc. (USA)* vol.134 (1987) p.15]

[52] J.A. Powell, L.G. Matus, M.A. Kuczmarski, C.M. Chorey, T.T. Cheng, P. Pirouz [*Appl. Phys. Lett. (USA)* vol.51 (1987) p.823]

[53] K. Shibahara, S. Nishino, H. Matsunami [*Appl. Phys. Lett. (USA)* vol.50 (1987) p.1888]

[54] K. Furukawa, A. Uemoto, Y. Fujii, M. Shigeta, A. Suzuki, S. Nakajima [*Ext. Abstracts 19th Conf. Solid State Devices and Materials* Tokyo (Business Center for Academic Societies Japan, Tokyo, 1987) p.231]

[55] K. Shibahara, N. Kuroda, S. Nishino, H. Matsunami [*Jpn. J. Appl. Phys. (Japan)* vol.26 (1987) p.L1815-7]
[56] J.W. Palmour, H.S. Kong, R.F. Davis [*Appl. Phys. Lett. (USA)* vol.51 (1987) p.2028-30]
[57] T. Sugii, T. Itoh, Y. Furumura, M. Doki, F. Mieno, M. Maeda [*J. Electrochem. Soc. (USA)* vol.134 (1987) p.2545-9]
[58] M. Yamanaka, H. Daimon, E. Sakuma, S. Misawa, S. Yoshida [*J. Appl. Phys. (USA)* vol.61 (1987) p.599]
[59] H. Mukida et al [*J. Appl. Phys. (USA)* vol.62 (1987) p.254]
[60] D.E. Ioannou, N.A. Papanicolaou, P.E. Nordquist [*IEEE Trans. Electron Devices (USA)* vol.ED-34 (1987) p.1697]
[61] S. Yoshida, E. Sakuma, H. Okumura, S. Misawa, K. Endo [*J. Appl. Phys. (USA)* vol.62 (1987) p.303]
[62] H.J. Kim, R.F. Davis, X.B. Cox, R.W. Linton [*J. Electrochem. Soc. (USA)* vol.134 (1987) p.2269]
[63] H. Daimon et al [*Appl. Phys. Lett. (USA)* vol.51 (1987) p.2106]
[64] S.R. Nutt, D.J. Smith, H.J. Kim, R.F. Davis [*Appl. Phys. Lett. (USA)* vol.50 (1987) p.203-5]
[65] P. Pirouz, C.M. Chorey, J.A. Powell [*Appl. Phys. Lett. (USA)* vol.50 (1987) p.221-3]
[66] H. Matsunami [*Mater. Res. Soc. Symp. Proc. (USA)* vol.97 (1987) p.171-82]
[67] K. Shibahara, T. Takeuchi, T. Saito, S. Nishino, H. Matsunami [*Mater. Res. Soc. Symp. Proc. (USA)* vol.97 (1987) p.247-52]
[68] S. Yoshida et al [*Mater. Res. Soc. Symp. Proc. (USA)* vol.97 (1987) p.259-64]
[69] H. Kong et al [*Mater. Res. Soc. Symp. Proc. (USA)* vol.97 (1987) p.223-45]
[70] H.S. Kong, J.W. Palmour, J.T. Glass, R.F. Davis [*Appl. Phys. Lett. (USA)* vol.51 (1987) p.442-4]
[71] G. Kelner, S. Bibari, K. Sleger, H. Hong [*Mater. Res. Soc. Symp. Proc. (USA)* vol.97 (1987) p.227-31]
[72] K. Furukawa et al [*IEEE Electron Device Lett. (USA)* vol.EDL-8 (1987) p.48-9]
[73] H.S. Kong, J.T. Glass, R.F. Davis, S.R. Nutt [*Mater. Res. Soc. Symp. Proc. (USA)* vol.97 (1987) p.405-10]
[74] H. Matsunami, K. Shibahara, N. Kuroda, W. Yoo, S. Nishino [*Springer Proc. Phys. (Germany)* vol.34 (1988) p.34-9]
[75] L.G. Matus, J.A. Powell [*Springer Proc. Phys. (Germany)* vol.34 (1988) p.40-4]
[76] S. Nishino, J. Saraie [*Springer Proc. Phys. (Germany)* vol.34 (1988) p.45-50]
[77] P. Pirouz, T.T. Cheng, J.A. Powell [*Springer Proc. Phys. (Germany)* vol.34 (1988) p.56]
[78] J.A. Edmond, J. Ryu, J.T. Glass, R.F. Davis [*J. Electrochem. Soc. (USA)* vol.135 (1988) p.359-62]
[79] J.A. Edmond, K. Das, R.F. Davis [*J. Appl. Phys. (USA)* vol.63 (1988) p.922-9]
[80] Y. Furumura, M. Doki, F. Mieno, T. Eshita, T. Sugii, M. Maeda [*J. Electrochem. Soc. (USA)* vol.135 (1988) p.1255]
[81] J.W. Palmour, H.S. Kong, R.F. Davis [*J. Appl. Phys. (USA)* vol.64 (1988) p.2168-77]
[82] H.S. Kong, J.T. Glass, R.F. Davis [*J. Appl. Phys. (USA)* vol.64 (1988) p.2672]
[83] J.W. Palmour, H.S. Kong, R.F. Davis [*J. Appl. Phys. (USA)* vol.64 (1988) p.2168]
[84] Y. Ohshita, A. Ishitani [*J. Appl. Phys. (USA)* vol.66 (1989) p.4635]
[85] Y. Ohshita, A. Ishitani [*J. Appl. Phys. (USA)* vol.66 (1989) p.4535]

[86] S. Nishino, J. Saraie [*Springer Proc. Phys. (Germany)* vol.34 (1989) p.186-90]
[87] K. Shibahara, T. Takeuchi, S. Nishino, H. Matsunami [*Jpn. J. Appl. Phys. (Japan)* vol.28 (1989) p.1341]
[88] H. Matsunami [*Springer Proc. Phys. (Germany)* vol.43 (1989) p.2-7]
[89] S. Nishino, J. Saraie [*Springer Proc. Phys. (Germany)* vol.43 (1989) p.8-13]
[90] J.A. Powell, L.G. Matus [*Springer Proc. Phys. (Germany)* vol.43 (1989) p.14-19]
[91] J.C. Liao, J.L. Crowley, P.H. Klein [*Springer Proc. Phys. (Germany)* vol.43 (1989) p.20-5]
[92] M. Iwami, M. Kusaka, M. Hirai, H. Nakamura, K. Shibahara, H. Matsunami [*Surf. Sci. (Netherlands)* vol.199 (1988) p.467]
[93] M. Iwami et al [*Nucl. Instrum. Methods B (Netherlands)* vol.33 (1988) p.615]
[94] M. Iwami et al [*Springer Proc. Phys. (Germany)* vol.43 (1989) p.146]
[95] M. Shigeta, Y. Fujii, K. Furukawa, A. Suzuki, S. Nakajima [*Appl. Phys. Lett. (USA)* vol.55 (1989) p.1522]
[96] H.S. Kong, J.T. Glass, R.F. Davis [*J. Mater. Res. (USA)* vol.4 (1989) p.204]
[97] J.A. Freitas Jr., S.G. Bishop [*Appl. Phys. Lett. (USA)* vol.55 (1989) p.2757-9]
[98] T. Sugii, T. Yamazaki, T. Itoh [*IEEE Trans. Electron Devices (USA)* vol.37 (1990) p.2331]
[99] H. Matsunami, T. Ueda, H. Nishino [*Mater. Res. Soc. Symp. Proc. (USA)* vol.162 (1990) p.397]
[100] T. Tachibana, H.S. Kong, Y.C. Wang, R.F. Davis [*J. Appl. Phys. (USA)* vol.67 (1990) p.6375-81]
[101] J.A. Powell et al [*Appl. Phys. Lett. (USA)* vol.56 (1990) p.1353]
[102] K. Takahashi, S. Nishino, J. Saraie [*J. Cryst. Growth (Netherlands)* vol.115 (1991) p.617]
[103] K. Ikoma, M. Yamanaka, H. Yamaguchi, Y. Shichi [*J. Electrochem. Soc. (USA)* vol.138 (1991) p.3028]
[104] T. Sugii, T. Yamazaki, T. Itoh [*Jpn. J. Appl. Phys. (Japan)* vol.30 (1991) p.L970]
[105] M.I. Chaudhry, R.L. Write [*Appl. Phys. Lett. (USA)* vol.59 (1991) p.51-3]
[106] J.C. Pazik, G. Kelner, N. Bottka [*Appl. Phys. Lett. (USA)* vol.58 (1991) p.1419]
[107] H. Nagasawa, Y. Yamaguchi [*J. Cryst. Growth (Netherlands)* vol.115 (1991) p.612]
[108] J.A. Powell et al [*Appl. Phys. Lett. (USA)* vol.59 (1991) p.333-5]
[109] K. Ikoma, M. Yamanaka, H. Yamaguchi, Y. Shichi [*Amorphous and Crystalline Silicon Carbide IV* Eds C.Y. Yang, M.M. Rahman, G.L. Harris (Springer-Verlag, Berlin, 1992) p.60]
[110] A.J. Steckl, J.P. Li [*Appl. Phys. Lett. (USA)* vol.60 (1992) p.2107]
[111] I. Golecki, F. Reidinger, J. Marti [*Appl. Phys. Lett. (USA)* vol.60 (1992) p.1703]
[112] K.G. Irvine, M.G. Spencer, M. Aluko [*Springer Proc. Phys. (Germany)* vol.56 (1992) p.291]
[113] S. Nishino, K. Takahashi, H. Ishida, J. Saraie [*Springer Proc. Phys. (Germany)* vol.56 (1992) p.295]
[114] J.A. Powell, J.B. Petit, L.G. Matus, S.E. Lempner [*Springer Proc. Phys. (Germany)* vol.56 (1992) p.313]
[115] A. Suzuki, K. Furukawa, Y. Fujii, M. Shigeta, S. Nakashima [*Springer Proc. Phys. (Germany)* vol.56 (1992) p.101-3]
[116] H. Matsunami [*Springer Proc. Phys. (Germany)* vol.71 (1992) p.3]
[117] J.A. Powell, D.J. Larkin, J.B. Petit, J.H. Edgar [*Springer Proc. Phys. (Germany)* vol.71 (1992) p.23]

[118] H. Nagasawa, Y. Yamaguchi [*Springer Proc. Phys. (Germany)* vol.71 (1992) p.40]
[119] A.J. Steckl, J.P. Li [*Springer Proc. Phys. (Germany)* vol.71 (1992) p.49]
[120] K. Ikoma, M. Yamanaka, H. Yamaguchi, Y. Shichi [*Springer Proc. Phys. (Germany)* vol.71 (1992) p.60]
[121] W.S. Yoo, H. Matsunami [*Springer Proc. Phys. (Germany)* vol.71 (1992) p.66]
[122] K. Takahashi, S. Nishino, J. Saraie, K. Harada [*Springer Proc. Phys. (Germany)* vol.71 (1992) p.78]
[123] K. Takahashi, S. Nishino, J. Saraie [*Appl. Phys. Lett. (USA)* vol.61 (1992) p.2081-3]
[124] L.M. Baranov, V.A. Dmitriev, I.P. Nikitina, T.S. Kondrateva [*Springer Proc. Phys. (Germany)* vol.71 (1992) p.116]
[125] H. Shimizu, K. Naito, S. Ishino [*Springer Proc. Phys. (Germany)* vol.71 (1992) p.119]
[126] P.A. Ivanov, V.E. Chelnokov [*Semicond. Sci. Technol. (UK)* vol.7 (1992) p.1-10]
[127] H. Morkoc, S. Strite, G.B.B. Gao, M.E. Lin, B. Sverdlov, M. Burns [*J. Appl. Phys. (USA)* vol.76 (1994) p.1363-98]
[128] B. Bahavar, M.I. Chaudhry, R.J. McCluskey [*Appl. Phys. Lett. (USA)* vol.63 (1993) p.914-6]
[129] M. Yamanaka, K. Ikoma [*Physica B (Netherlands)* vol.185 (1983) p.308]
[130] J.D. Parsons, R.F. Bunshahand, O.M. Stafsudd [*J. Electrochem. Soc. (USA)* vol.140 (1993) p.1756-62]
[131] A.K. Chaddha, J.D. Parson, J. Wu, H.S. Chen, D.A. Roberts, H. Hockenhull [*Appl. Phys. Lett. (USA)* vol.62 (1993) p.3097]
[132] J.P. Li et al [*Appl. Phys. Lett. (USA)* vol.62 (1993) p.3135-7]
[133] M. Yamanaka, I. Ikoma, M. Ohtsuka, T. Ishizawa, Y. Shichi [*Jpn. J. Appl. Phys. (Japan)* vol.33 (1994) p.997]
[134] N. Nordell, S. Nishino, J.W. Yang, C. Jacob, P. Pirouz [*Appl. Phys. Lett. (USA)* vol.64 (1994) p.1647]
[135] F.R. Chien, S.R. Nutt, J.M. Carulli Jr., N. Buchan, C.P. Beetz Jr., W.S. Yoo [*J. Mater. Res. (USA)* vol.9 (1994) p.2086]
[136] P.G. Neudeck, D.J. Larkin, J.E. Starr, J.A. Powell, C.S. Salupo, L.G. Matus [*IEEE Trans. Electron Devices (USA)* vol.41 (1994) p.826-35]
[137] B.S. Sywe, Z.J. Yu, S. Burckhard, J.H. Edgar [*J. Electrochem. Soc. (USA)* vol.141 (1994) p.510-3]
[138] J. Takahashi, M. Kanaya, T. Hoshino [*Inst. Phys. Conf. Ser. (UK)* no.137 (1994)]
[139] H. Matsunami [*Inst. Phys. Conf. Ser. (UK)* no.137 (1994) p.45-50]
[140] S. Nishino, K. Takahashi, H. Tanaka, J. Saraie [*Inst. Phys. Conf. Ser. (UK)* no.137 (1994) p.63-6]
[141] V.A. Dmitriev, K.G. Irvine, M.G. Spencer, I.P. Nikitina [*Inst. Phys. Conf. Ser. (UK)* no.137 (1994) p.67-70]
[142] H. Nagasawa, Y. Yamaguchi [*Inst. Phys. Conf. Ser. (UK)* no.137 (1994) p.71-4]
[143] A.J. Steckl, C. Yuan, J.P. Li, M.J. Loboda [*Inst. Phys. Conf. Ser. (UK)* no.137 (1994) p.79-82]
[144] C.W. Liu, J.C. Strum [*Inst. Phys. Conf. Ser. (UK)* no.137 (1994) p.87-9]
[145] H. Shimizu, K. Naito [*Inst. Phys. Conf. Ser. (UK)* no.137 (1994) p.83-6]
[146] W.S. Yoo et al [*Inst. Phys. Conf. Ser. (UK)* no.137 (1994) p.259-62]
[147] L.S. Hong, S. Misawa, H. Okumura, S. Yoshida [*Inst. Phys. Conf. Ser. (UK)* no.137 (1994) p.239-42]

[148] A. Henry, O. Kordina, C. Hallin, R.C. Glass, E. Janzen [*Inst. Phys. Conf. Ser. (UK)* no.137 (1994) p.305-8]

[149] K. Kamimura, H. Tanaka, S. Miyazaki, T. Homma, S. Yonekubo, Y. Onuma [*Inst. Phys. Conf. Ser. (UK)* no.137 (1994) p.109-12]

[150] V.V. Luchinin, S.V. Kostromin, A.V. Korlyakov, O.L. Komarov, A.A. Petrov [*Trans. 2nd Int. High Temperature Electronics Conf. (HiTEC)* Charlotte, NC, USA 5-10 June 1994 p.187-90]

[151] Y. Ohshita [*J. Cryst. Growth (Netherlands)* vol.147 (1995) p.111-6]

[152] N. Nordell, S. Nishino, J.-W. Wang, C. Jacob, P. Pirouz [*J. Electrochem. Soc. (USA)* vol.142 (1995) p.565-71]

[153] R.B. Cambell, T.L. Chu [*J. Electrochem. Soc. (USA)* vol.113 (1966) p.825]

[154] V.L. Jennings, A. Sommer, H.C. Chang [*J. Electrochem. Soc. (USA)* vol.113 (1966) p.728]

[155] S. Minagawa, H.C. Gatos [*Jpn. J. Appl. Phys. (Japan)* vol.10 (1971) p.1680]

[156] B. Wessels, H.C. Gatos, A.F. Witt [*Silicon Carbide 1973* Eds R.C. Marshall, J.W. Faust Jr., C.E. Ryan (University of South Carolina Press, 1974) p.25-32]

[157] R.F. Rutz, J.J. Cuomo [*Silicon Carbide 1973* Eds R.C. Marshall, J.W. Faust Jr., C.E. Ryan (University of South Carolina Press, 1974) p.72-9]

[158] R.W. Brander [*Silicon Carbide 1973* Eds R.C. Marshall, J.W. Faust Jr., C.E. Ryan (University of South Carolina Press, 1974) p.8-24]

[159] H. Matsunami, S. Nishino, M. Odaka, T. Tanaka [*J. Cryst. Growth (Netherlands)* vol.31 (1975) p.72]

[160] W. von Muench, I. Pfaffeneder [*Thin Solid Films (Switzerland)* vol.31 (1976) p.39]

[161] W. von Muench, W. Kuerzinger, I. Pfaffeneder [*Solid-State Electron. (UK)* vol.19 (1976) p.871]

[162] W. von Muench, I. Pfaffeneder [*J. Appl. Phys. (USA)* vol.48 (1977) p.4831]

[163] S. Nishino, H. Matsunami, T. Tanaka [*J. Cryst. Growth (Netherlands)* vol.45 (1978) p.144-9]

[164] S. Nishino, A. Ibaraki, H. Matsunami, T. Tanaka [*Jpn. J. Appl. Phys. (Japan)* vol.19 (1980) p.L353-56]

[165] G. Ziegler, D. Theis [*IEEE Trans. Electron Devices (USA)* vol.ED-28 (1981) p.425]

[166] N. Kuroda, K. Shibahara, W.S. Yoo, S. Nishino, H. Matsunami [*Ext. Abstracts 19th Conf. Solid State Devices and Materials* Tokyo (Business Center for Academic Societies Japan, Tokyo, 1987) p.227]

[167] T. Ueda, H. Nishino, H. Matsunami [*J. Cryst. Growth (Netherlands)* vol.104 (1990) p.695]

[168] J.A. Powell et al [*Appl. Phys. Lett. (USA)* vol.56 (1990) p.1442]

[169] S. Karmann et al [*Appl. Phys. Lett. (USA)* vol.72 (1992) p.5437]

[170] S. Nishino, K. Takahashi, Y. Kojima, J. Saraie [*Springer Proc. Phys. (Germany)* vol.56 (1992) p.363]

[171] H. Matsunami, H. Nishino, T. Ueda [*Springer Proc. Phys. (Germany)* vol.56 (1992) p.161]

[172] T. Kimoto, H. Nishino, A. Yamashita, W.S. Yoo, H. Matsunami [*Springer Proc. Phys. (Germany)* vol.71 (1992) p.31]

[173] T. Kimoto, H. Nishino, W.S. Yoo, H. Matsunami [*J. Appl. Phys. (USA)* vol.73 (1993) p.726]

[174] A. Itoh, H. Akita, T. Kimoto, H. Matsunami [*Appl. Phys. Lett. (USA)* vol.65 (1994) p.1400]

8.3 Chemical vapour deposition of SiC

[175] H.M. Hobgood, P.J. McHugh, J. Greggi, R.H. Hopkins, M. Skowronski [*Inst. Phys. Conf. Ser. (UK)* no.137 (1994)]

[176] A.A. Burk Jr. et al [*Inst. Phys. Conf. Ser. (UK)* no.137 (1994) p.29-32]

[177] K. Nishino, T. Kimoto, H. Matsunami [*Inst. Phys. Conf. Ser. (UK)* no.137 (1994) p.33-6]

[178] O. Kordina, C. Hallin, R.C. Glass, A. Henry, E. Janzen [*Inst. Phys. Conf. Ser. (UK)* no.137 (1994) p.41-4]

[179] D.J. Larkin, P.G. Neudeck, J.A. Powell, L.G. Matsud [*Inst. Phys. Conf. Ser. (UK)* no.137 (1994) p.51-4]

[180] T. Kimoto, H. Matsunami [*Inst. Phys. Conf. Ser. (UK)* no.137 (1994) p.55-8]

[181] A. Itoh, A. Akita, T. Kimoto, H. Matsunami [*Inst. Phys. Conf. Ser. (UK)* no.137 (1994) p.59-62]

[182] T. Kimoto, H. Matsunami [*Inst. Phys. Conf. Ser. (UK)* no.137 (1994) p.95-8]

[183] L.L. Clemen et al [*Inst. Phys. Conf. Ser. (UK)* no.137 (1994) p.147-50]

8.4 LPE of SiC and SiC-AlN

V.A. Dmitriev

March 1994

A   INTRODUCTION

Liquid Phase Epitaxy (LPE) is a fruitful method of SiC p-n junction fabrication. Single crystal 6H-SiC layers with thicknesses from 0.2 to 100 µm have been grown by LPE from the silicon melt with the growth rate of 0.01 - 2 µm min^{-1}. Nitrogen is used as a donor impurity and aluminium, gallium and boron are used as acceptors. The concentration $N_d - N_a$ in the n-layers may be controlled from 8×10^{15} to 10^{20} cm^{-3} and can be changed during the growth. The concentration of the acceptors (Al) in the p-layers varied from 10^{17} to 10^{20} cm^{-3}, depending on Al concentration in the melt. LPE grown material displays high carrier mobility and low deep centre concentration.

High quality 6H-SiC and 4H-SiC p-n junctions have been made by LPE. During a single epitaxial run, SiC p-n structures of different types have been grown: n(substrate)-n$^+$-n$^-$-p$^+$, n(substrate)-n$^-$-p-n$^+$, and n(substrate)-n-p$^+$-n$^-$. For the first time, SiC normally-off JFETs, dynistors, violet LEDs and ultraviolet LEDs have been made based on p-n structures grown by LPE.

New directions of SiC LPE will be discussed: (1) heteropolytype growth from the silicon melt, (2) SiC p-n structures growth from alternative melts at 1100 - 1200 °C, and (3) growth of the SiC-AlN solid solution.

B   PRINCIPLES OF SiC LPE

LPE growth is a version of crystallisation from a solution [1-3]. The growth of the semiconductor material takes place from the supersaturated solution of the semiconductor in a solvent. The main feature of the LPE growth is that the growing layers are in equilibrium with the liquid phase. The growth process is determined by the phase diagram for SiC and the material which is used as a solvent. The usual melt for SiC LPE is silicon [4-12]. The phase diagram for the Si-C system was studied by Scace and Slack [13] and is shown in FIGURE 1. Alternative materials like Sn, Ge, Ga and their mixtures are also used for SiC LPE [14-16]. Unfortunately, triple phase diagrams Si-C-M, where M is a material of alternative melts, are not well established and this limits the development of the SiC LPE technique.

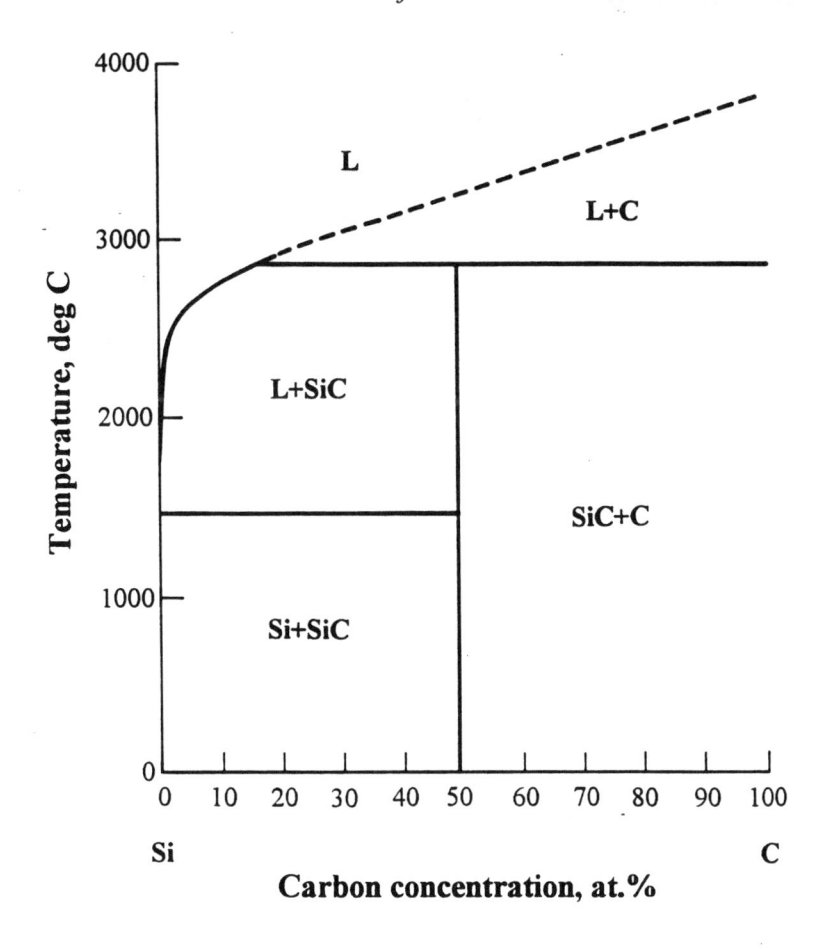

FIGURE 1 Si - C phase diagram [13].

C SILICON CARBIDE LPE GROWTH FROM THE SILICON MELT

C1 LPE Growth With the Substrate Fixed in the Crucible

The epitaxial technique of SiC growth from carbon-saturated silicon solution was developed more than 25 years ago by Brander and Sutton [4]. The silicon was contained in a crucible (FIGURE 2) made of dense, high-purity graphite. The substrates, α-SiC, were held between graphite nuts threaded onto the central stem. Silicon was placed at the top of the stem. When the system was heated, Si melted and ran down to surround the substrates. Carbon was dissolved from the crucible walls during the growth. Growth took place in a temperature gradient. Carbon transport in the melt occurs by diffusion and convection.

At a temperature of 1650 °C SiC single crystal layers up to 100 μm thick were reproducibly grown. X-ray analysis showed that layers were single crystal of the same polytype as the substrate. With this technique, SiC p-n junctions have been prepared using nitrogen, aluminium and boron impurities. Diode properties have been examined [4].

About 10 years later, the same ideas of using Si as a solvent and fixing the substrate in the crucible were used by Munch and Kurzinger [5]. The p-type SiC substrate was clamped to the bottom of a graphite crucible containing about 1 cm^3 of silicon. The system operated in

8.4 LPE of SiC and SiC-AlN

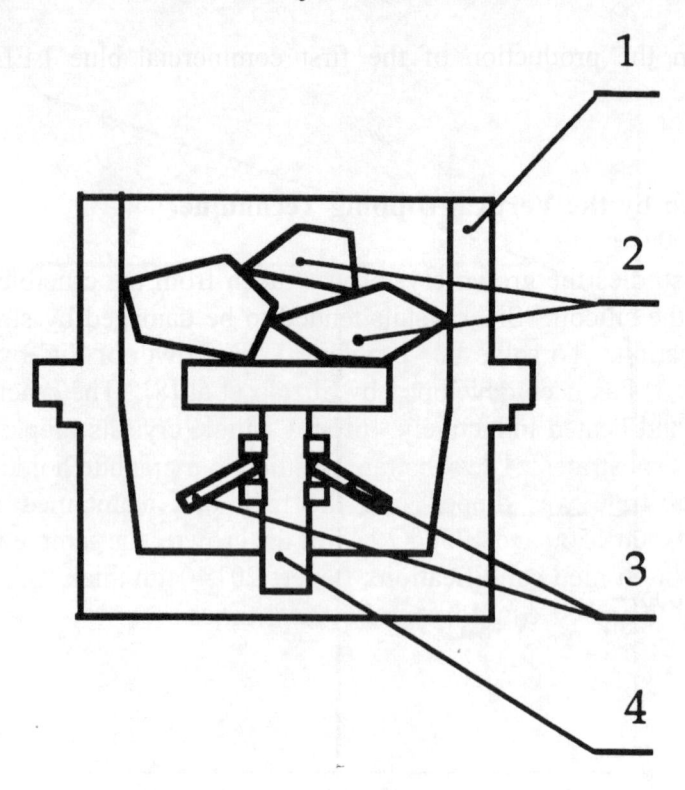

FIGURE 2 Schematic cross-section of LPE growth cell [4]: 1 - crucible, 2 - silicon, 3 - SiC substrates, 4 - graphite stem.

1 atm of high-purity argon, except for the nitrogen-doping period. Growth rates up to 1 µm min^{-1} have been obtained at the melt temperature of 1800 °C and with a temperature gradient of 30 °C cm^{-1}. The cool-down process after the growth was carefully controlled to avoid substrate cracking during the melt solidification.

To make SiC p-n junctions, the authors in [5] employed two methods:

(1)   An Al-doped p-layer was grown on the p-type substrate. The p-n junction was fabricated by a second LPE process with nitrogen as a donor dopant (two-step epitaxy).

(2)   The growth process was started with a certain Al concentration in the melt to grow the p-layer. Nitrogen gas was added after a growth period of approximately 10 min, thus giving rise to conversion to n-type during the growth (overcompensation method).

Both methods were employed by these researchers to fabricate SiC LEDs. The efficiencies of the blue LEDs produced by the overcompensation technique were nearly ten times higher than those of the double-epitaxy devices. The highest quantum efficiency that has been obtained was 4×10^{-5} at a current density of 2 A cm^{-2}.

More recently, Hoffmann et al [6] have achieved significant success in growing p-n junctions using the same growth equipment. The quality of p-n junctions made by the overcompensation method and LED efficiency have been improved by using a special temperature-time program. Thicknesses of the n- and p-layer were around 8 µm. This research resulted in the production of the first commercial blue LED by Siemens AG (Germany) [7].

C2   LPE Growth by the Vertical Dipping Technique

In all of the above studies the grown crystal was taken from the crucible after the epitaxial process by etching the silicon. SiC crystals tended to be damaged by stresses induced with the Si melt solidification. To solve this problem, LPE growth of SiC by a vertical dipping technique (FIGURE 3) has been developed by Suzuki et al [8]. The crucible was filled with a 26 g charge of Si and heated inductively. 6H-SiC single crystals prepared by the Acheson method were used as substrates. The substrate was tied to a graphite holder. After the silicon was melted, the substrate was dipped into the melt and maintained for 5 hr. Growth temperature was between 1500 and 1750°C. After the growth, the sample was pulled up from the crucible before the Si melt solidification. Layers 20 - 40 µm thick were obtained. By this technique, the crucible and Si could be used several times.

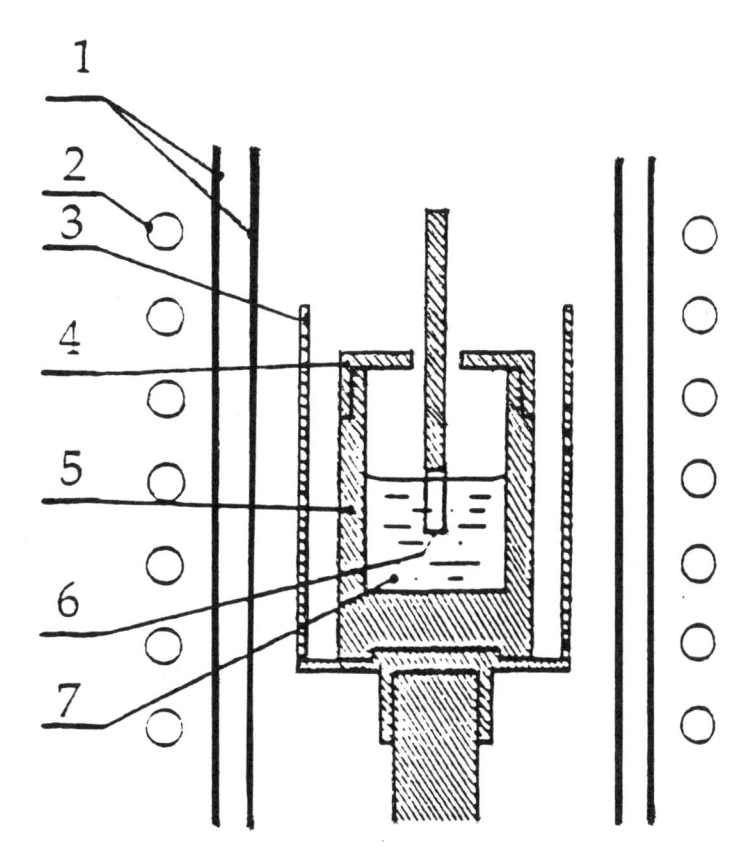

FIGURE 3   Schematic cross-section of the growth arrangement [8]: 1 - quartz, 2 - RF coil, 3 - radiation shield, 4 - graphite lid, 5 - crucible, 6 - substrate, 7 - silicon melt.

The dipping method has been employed for growth of SiC p-n junctions and multilayer structures by Ikeda et al [9]. This group used the growth system consisting of three crucibles with the melts for the growth of p-layers, n-layers, and for rinsing. The substrate holder could be rotated and translated vertically. The crucibles could be exchanged by the rotation. The growth temperature was varied from 1600 to 1700°C. The undoped 6H-SiC layers were n-type with an electron concentration, n_e, of $\sim 5 \times 10^{17}$ cm^{-3} and electron mobility (μ_e) of 174 cm^2 V^{-1} s^{-1}. Aluminium, Ga, B and N were used as dopants. A detailed study of electroluminescence (EL) of 6H-SiC p-n junctions has been made in the temperature range from 100 to 400 K. Some data on EL of 3C-SiC and 15R-SiC p-n junctions have also been reported. The maximum EL external quantum efficiency of 1.2×10^{-5} was obtained for 6H-SiC p-n structures with an Al-compensated n-layer [9].

More recently, the dipping technique has been employed by Nakata et al [10] for blue LED fabrication using off-oriented 6H-SiC substrates. The dependence of surface morphology on off-orientation angle has been studied. It was found that material quality and EL characteristics may be significantly improved by using off-angle substrates.

C3 LPE Growth Without a Crucible; Container-Free LPE

Recently, Dmitriev et al [11,12] developed a new version of LPE, container-free LPE (CFLPE), based on the electromagnetic crucible technique. In this method, liquid metal is suspended in a high frequency electromagnetic field. Si is a metal in the liquid state, so it is possible to keep Si melt in the electromagnetic crucible.

The epitaxy was carried out in a metal chamber (diameter ~ 30 cm, height ~ 40 cm) in the He ambient. The pressure in the chamber was varied from 10^{-5} to 760 torr. To produce the electromagnetic field a two-turn copper inductor was used (FIGURE 4 [17]). The substrates were n-type 6H-SiC and 4H-SiC crystals grown by the Lely method, usually with {0001} basal-plane orientation. Silicon with a resistivity of 100 - 200 Ω cm (300 K) was used as a solvent. The melt volume was varied from 4 to 10 cm^3. Aluminium was used as an acceptor impurity by introducing Al metal pellets into the melt. Nitrogen served as a donor by filling the reactor with nitrogen gas.

Both non-isothermal and isothermal conditions were used for CFLPE growth [17-19]. The isothermal growth was possible due to a temperature gradient which exists across the suspended Si drop. The top part of the melt is always cooler than the bottom part. The time-temperature plot of the isothermal growth of an n(substrate)-p-n$^+$ structure is shown in FIGURE 5. First, a solid piece of Si was heated to ~ 1000 °C by radiation heating. This heating is necessary to give Si a metallic conductivity (time $t_0 - t_1$ in FIGURE 5). When the conductivity is sufficient, silicon reaches a suspended state. Further Si heating was done by the electromagnetic field ($t_1 - t_6$). Silicon carbide crystals were put on the Si melt at the temperature of ~ 1450 °C ($t_2 - t_3$). SiC crystals which served as source material were placed in the bottom of the Si melt. SiC crystals which were to be used as substrates were placed on the top of the melt. The crystals were kept on the melt surface by surface tension forces. To grow a p-layer, Al was introduced into the melt ($t_2 - t_3$). Si-Al melt with SiC crystals was heated up to the temperature of 1500 - 1650 °C. The p-layer was grown during the $t_4 - t_5$ stage of the process. Nitrogen was then introduced into the chamber (t_5) and an n$^+$-SiC layer was grown ($t_5 - t_6$). The advantage of this method is that the substrate temperature is constant during the growth.

The time-temperature plot of the growth under non-isothermal conditions is shown in FIGURE 6. The first few stages are the same as for the growth under isothermal conditions

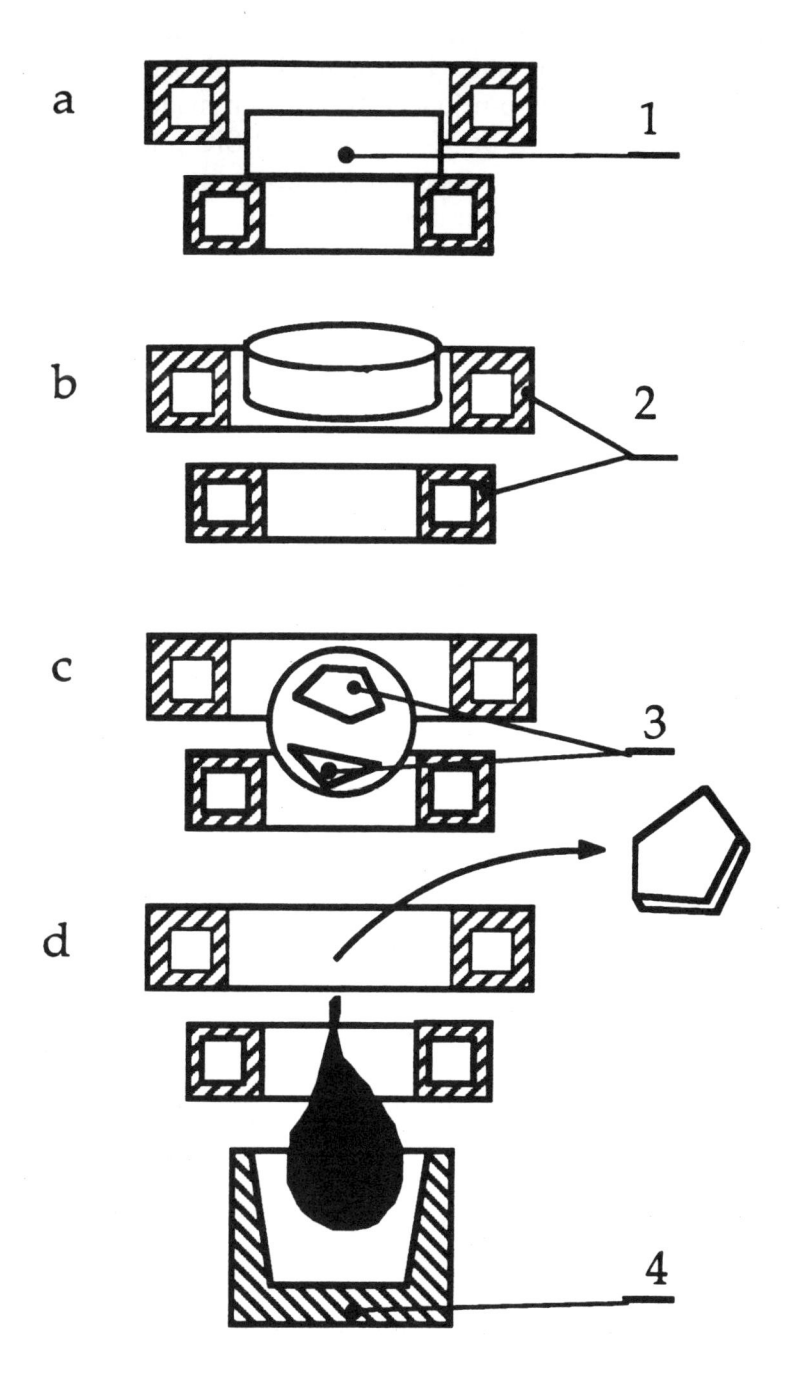

FIGURE 4 Schematic of CFLPE [17]: 1 - silicon, 2 - inductor, 3 - SiC substrates, 4 - copper container; a - solid Si in copper inductor, b - solid Si suspended in the inductor, c - liquid Si with SiC crystals suspended in electromagnetic field, d - silicon is dropped in container.

(t_0 - t_4). SiC crystals were placed only on the top of the Si drop. The silicon was then saturated with carbon by partially dissolving the SiC substrates (t_4 - t_5) in the temperature range of 1500 to 1700°C. The SiC epitaxial layer was grown by cooling the Si-C solution (t_5 - t_6). To grow multilayer p-n structures, during the t_5 - t_6 stage, impurities were introduced into or removed from the melt. After the growth, SiC samples were removed from the melt (t_6).

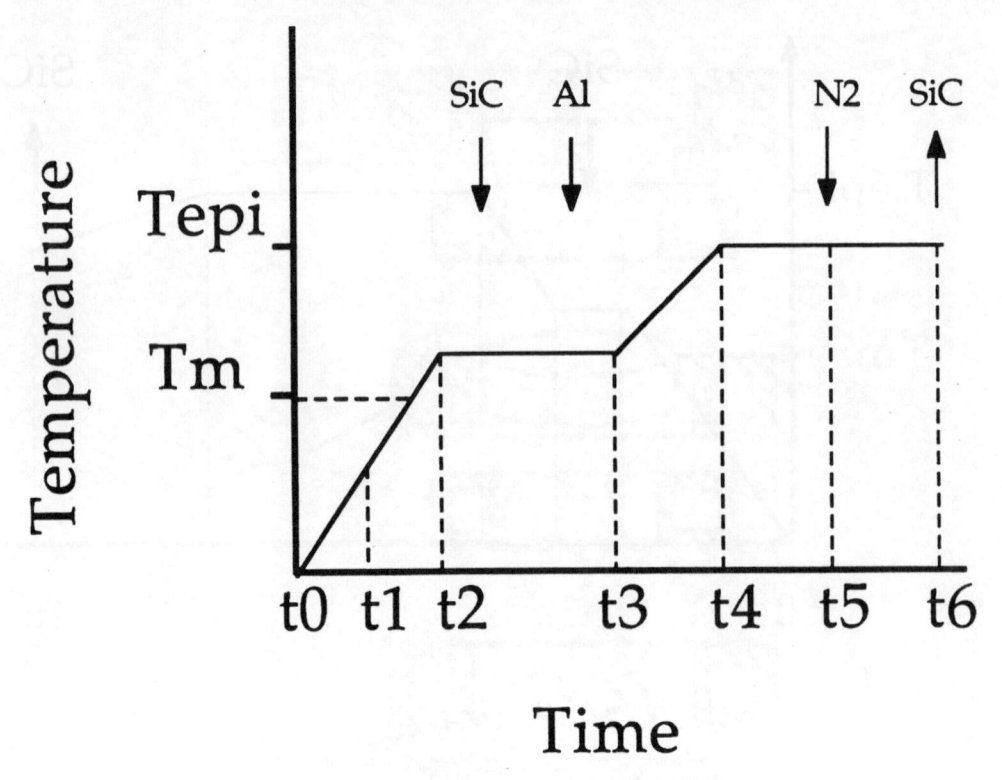

FIGURE 5 Time-temperature plot of CFLPE under isothermal conditions: T_m - silicon melting point, T_{epi} - substrate temperature during the growth (see text).

C4 Silicon Carbide Layers Grown by LPE from Si Melt

Layer thicknesses ranged from 0.2 to 100 µm [19]. Growth rate was controlled from 0.02 to 2 µm min^{-1}. At doping levels lower than 10^{19} cm^{-3}, the dislocation density revealed by chemical etching was less than 10^3 cm^{-2}. Britun et al [18] state that according to TEM data, regions of the layers with a size larger than 100 mm were completely free of planar and linear defects. The half-width of the X-ray rocking curve, $W_{1/2}$, of the undoped 6H-SiC layers was from 10 to 25 arcsec. Dislocation density increases with increasing doping concentration. At N_{Al} of 10^{20} cm^{-3} a dislocation density of 10^6 cm^{-2} has been measured.

Undoped 6H-SiC layers grown by CFLPE had the uncompensated electrically active donor concentration $N_d - N_a$ of 8×10^{15} cm^{-3} [19]. Deep level concentrations in the undoped layers were less than 2×10^{14} cm^{-3}. For nitrogen-doped n-type layers the $N_d - N_a$ concentration was controlled from 10^{16} to 10^{20} cm^{-3} [12]. The maximum Hall mobility for the n-layer was 252 cm^2 V^{-1} s^{-1} [10] (FIGURE 7). The nitrogen concentration for the layer grown on the (0001)C substrate face was higher than for that on the (0001)Si face. This resulted in a different layer-substrate lattice mismatch for the layers grown in the same epitaxial run on the different faces. The hole diffusion length was ~ 0.9 µm for n-type material with $N_d - N_a = 5 \times 10^{17}$ cm^{-3}.

8.4 LPE of SiC and SiC-AlN

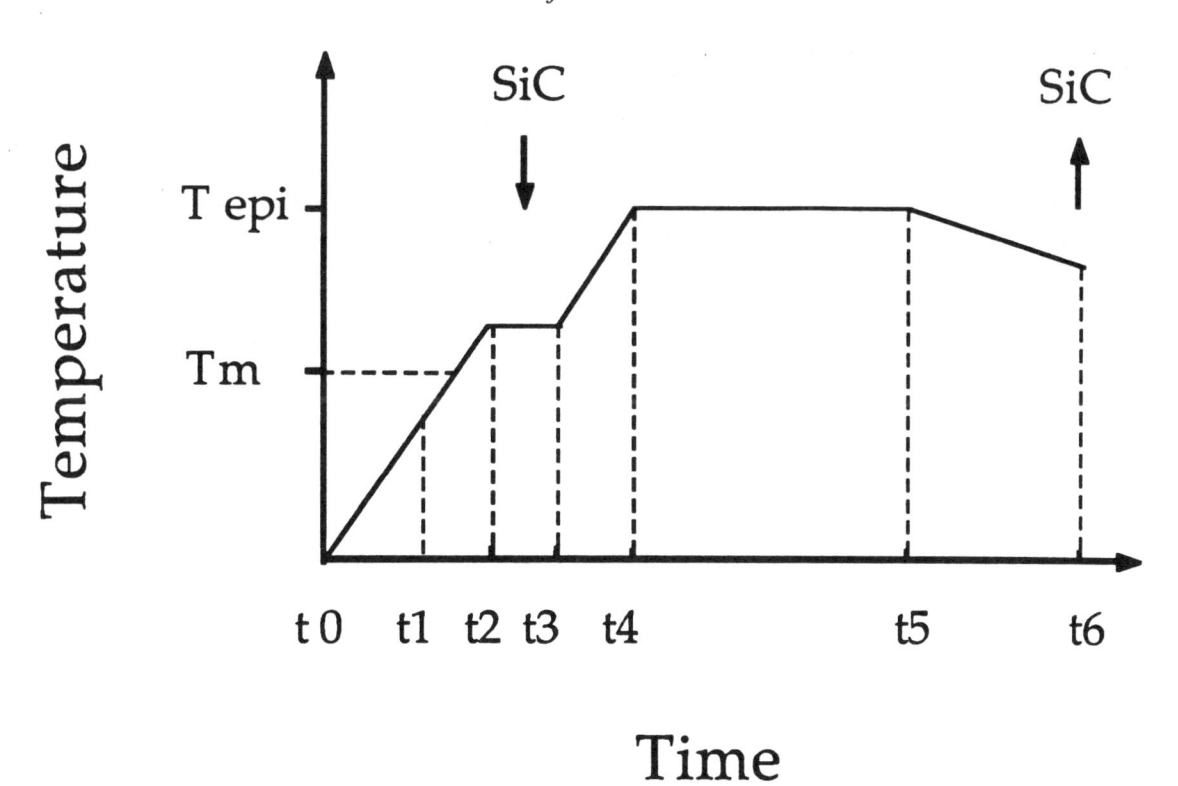

FIGURE 6 Time-temperature plot of CFLPE growth under non-isothermal conditions (T_m - Si melting point).

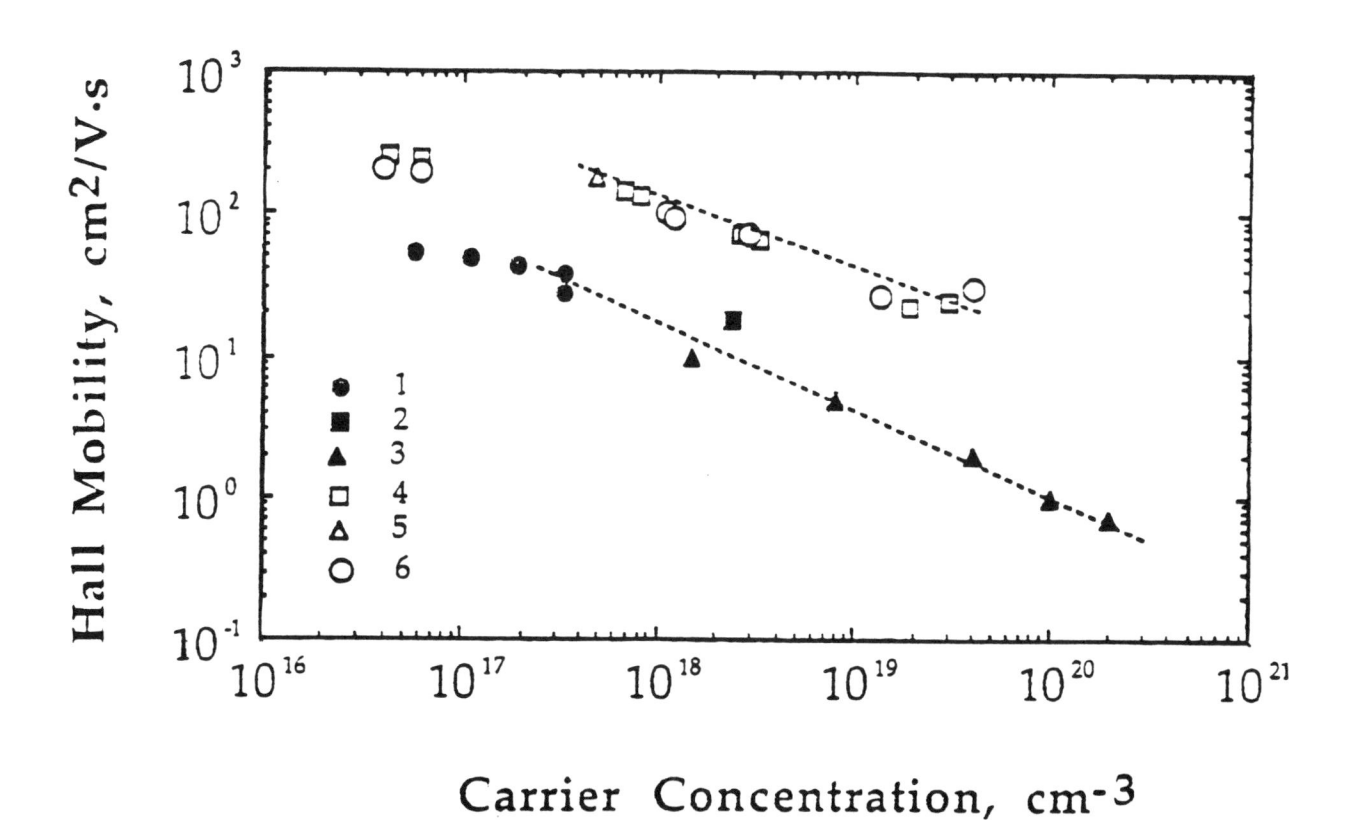

FIGURE 7 Hall mobility for holes (1, 2, 3) and for electrons (4, 5, 6) as a function of carrier concentration (300 K): 1 - [10], 2 - [9], 3 - [17], 4 - [10], 5 - [9], 6 - [10].

- 221 -

p-Type 6H-SiC layers with Al concentration N_{Al} from 10^{17} to 2×10^{20} cm^{-3} have been obtained by controlling the Al concentration in the melt [20]. The Al effective distribution coefficient k_{eff} was estimated at 2×10^{-2} for the substrate temperature of 1600 °C and growth rate of 1 μm min^{-1}. The value of hole concentration measured by the Hall method was in the range 5×10^{16} - 2×10^{20} cm^{-3} (300 K). The hole mobility ranged from 1 to 54 cm^2 V^{-1} s^{-1} (FIGURE 7). An electron diffusion length of 0.4 μm was measured for material with $N_a - N_d = 2 \times 10^{18}$ cm^{-3}. Electron lifetime for the same material was estimated at 3 ns [21].

C5 SiC p-n Structures Grown by LPE from Si Melt

The LPE technique has been employed to fabricate epitaxial n$^+$(substrate)-n$^-$-p$^+$ structures, in which n- and p-type regions were grown in the same epitaxial run [22]. The C-V characteristics were straight lines when plotted using the coordinated C^{-2} and V, which indicated that the p-n junctions were abrupt. The capacitance cut-off voltage was 2.65 V for 6H-SiC and 2.85 V for 4H-SiC (300 K).

The forward branches of the I-V characteristics at the current density $j < 1$ A cm^{-2} usually consisted of two exponential parts [23]. Each part was described by the expression:

$$j = j_o \exp \frac{eV}{nkT} \quad (1)$$

For the lower part, at $j < 10^{-4}$ A cm^{-2}, the ideality factor n was 2 and was independent of temperature in the range 300 to 600 K. It was shown by Strelchuk [23] that for this region the current is described in terms of the Sah-Noyce-Shockley model for recombination via a deep level close to the middle of the forbidden gap. According to the model, the pre-exponential factor j_o is determined by the equation:

$$j_o = \frac{e n_i \alpha}{\tau_{st}} \quad (2)$$

where n_i is the intrinsic carrier concentration in SiC ($n_i = 10^{-6}$ cm^{-3}, 300 K), $\alpha = kT/eE$ is the effective width of the recombination region, E is the electric field strength in the p-n junction and τ_{st} is the stationary lifetime of minority carriers. The j_o values for 6H-SiC structures grown by CFLPE are four orders of magnitude lower than for the structures grown by sublimation epitaxy [24]. This means that τ_{st} in LPE structures is about 10 times higher. Anikin et al [25] show that this is due to a lower concentration of deep centres in LPE structures.

For the upper exponential part of the I-V characteristic ($10^{-4} < j < 1$ A cm^{-2}) the factor n for different p-n structures was equal to 6/5 or 4/3. The n was independent of temperature. It was found that for p-n structures with n = 6/5 the temperature dependence of j_o was exponential with the activation energy of 2.7 eV; at room temperature the j_o value for this part of the I-V characteristic was 10^{-37} A cm^{-2} [24]. This research stated that the current flow mechanism for this part of the I-V characteristic corresponds to a modified Sah-Noyce-Shockley model for recombination via a multiply charged centre in the space charge region.

The breakdown of the 6H-SiC p-n structures with {0001} basal plane orientation was observed at voltages of 30 - 500 V. If $N_d - N_a$ in the base region was higher than 10^{17} cm^{-3}, the breakdown had a microplasma avalanche nature. The maximum electric field strength in the space-charge region, E_m, was about 4.5×10^6 V cm^{-1}. At a lower $N_d - N_a$ value, breakdown usually took place on the mesa surface. The reverse leakage current outside the breakdown region did not exceed 100 nA at room temperature and 0.1 mA at 500 °C. Detailed investigation of microplasma breakdown of these structures has been carried out by Kondratiev et al [26]. The temperature coefficient of the breakdown voltage (K_T) was negative in the temperature range from 300 to 360 K, and positive at temperatures between 360 and 700 K. For p-n structures grown on the (1120) face of 6H-SiC crystals E_m was 1.5 - 2 times lower than for the structure on the (0001) face [27]. The K_T was positive at $2 - 3 \times 10^{-4}$ deg^{-1} at 300 - 600 K.

The EL properties of the p-n structures grown by LPE have been described in several publications [4-6,9,27-31]. The EL spectrum of the 6H-SiC n(substrate)-n-p$^+$ structure with an undoped n-layer includes a violet peak ($\lambda_{max} \sim 425$ nm) which dominates at current densities higher than 7 A cm^{-2} (300 K). The peak has been interpreted in [9] as due to free exciton recombination. The EL spectrum also contains a blue peak from donor-acceptor recombination, and a green peak of defect luminescence. A dependence of EL quantum yield, h_{EL}, on p-layer doping has been studied by Kogan and Morozenko [28]. The EL of the 4H-SiC p-n structures grown by CFLPE have been described by Dmitriev et al [27]. For the first time, ultraviolet ($\lambda_{max} \sim 395$ nm) and violet ($\lambda_{max} \sim 430$ nm) peaks have been detected.

Based on the p-n structure grown by LPE, a variety of SiC devices have been made. Devices for (1) high temperature electronics including diodes, FETs, bipolar transistors and dynistors operating up to 500 °C, and for (2) optoelectronics including green, blue, violet and UV LEDs have been recently developed (see [22]).

D SiC LPE GROWTH FROM ALTERNATIVE SOLUTIONS

The principal limitation of SiC LPE growth from Si melt is the high melting point of silicon. To grow SiC at temperatures below 1400 °C alternative melts are required. In FIGURE 8 we summarise data on SiC solubility in Sn, Ge, Ga and Si melts [13,14,32]. We also show carbon solubility data for Pb and Al melts [33]. It is clear that the same value of solubility for Si melt may be obtained for alternative melts at a much lower temperature.

Tairov et al [14] were the first to propose Sn and Ga melts as solvents for SiC low temperature LPE (LTLPE). They used the usual graphite sliding boat to deposit SiC layers at temperatures between 1100 and 1400 °C. Growth in a temperature gradient of 10 °C cm^{-1} has been employed. Layers had single crystal structure and thicknesses from 0.5 to 4 µm. More recently, Dmitriev et al [16] reported 6H-SiC p-n junction growth and SiC selective growth at 1100 - 1200 °C by LTLPE. This is the minimum temperature ever achieved for SiC

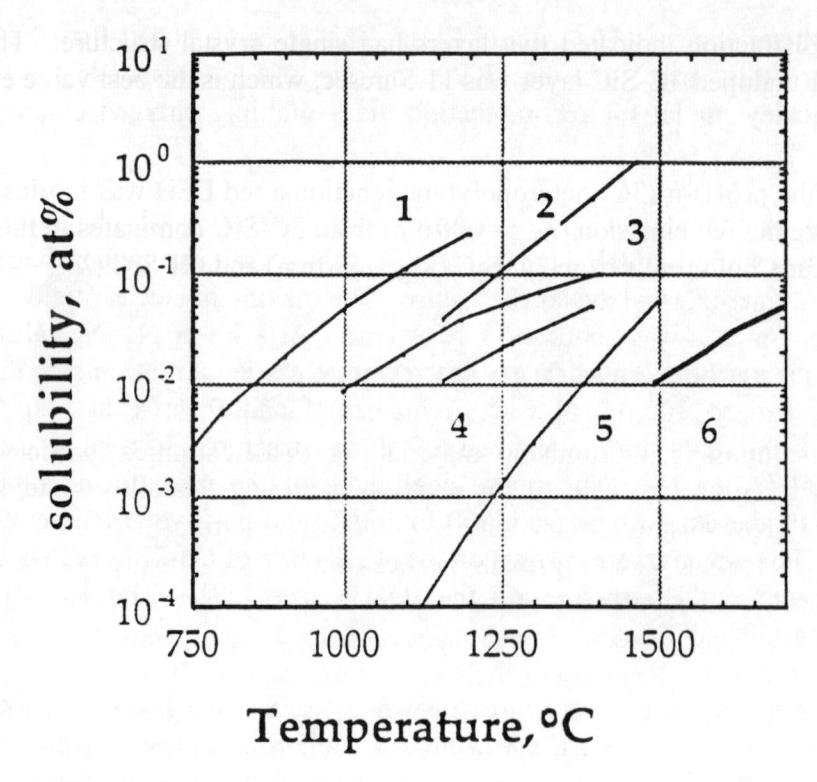

FIGURE 8 Solubility of SiC in: 2 - Sn [32], 3 - Ge [32], 4 - Ga [32], 6 - Si [13] and carbon solubility in: 1 - Pb [33], 5 - Al [33].

p-n junction growth. LPE growth was carried out under isothermal conditions (temperature gradient ~ 30 °C cm^{-1}). Ga melt and molten mixtures of Sn-Al-Si, Ga-Al-Si, and Ge-Si were used as solvents. Layers were grown on the (0001)Si face of n-type 6H-SiC Lely crystals, at a rate of ~ 0.1 μm hr^{-1}. To grow p-layers, melts containing Al or Ga have been used.

For structures with a p-layer grown from Ga-Al-Si melt, C-V characteristics indicated an abrupt p-n junction. The V_c value of 2.7 V was measured which is typical for 6H-SiC p-n junctions. Forward I-V characteristics at voltages below 2.4 V had an exponential part described by Eqn (1). At 300 K j_o ~ 3 x 10^{-35} A cm^{-2} and n = 1.25. Breakdown voltage was 40 - 63 V (E_m ~ 3.5 x 10^6 V cm^{-1}).

For structures with a p-layer grown from Ga melt on 6H-SiC Lely crystals uniform avalanche breakdown was obtained. The description of the breakdown was given by Vainshtein et al [34]. It was the first observation of the uniform breakdown of SiC p-n junctions grown by LPE. The E_m was ~ 5 x 10^6 V cm^{-1}.

The EL from p-n junctions grown by LTLPE has not been studied in detail. It was mentioned in [16] that the EL spectrum was similar to that for structures grown from the Si melt.

E   HETEROPOLYTYPE LPE GROWTH

Epitaxial layers of 3C-SiC have been grown on the (0001)Si face of 6H-SiC substrates at temperatures of 1500 °C and lower. The layers had (111) orientation and thicknesses of 3 to

30 μm. X-ray diffraction indicated that layers had single crystal structure. The minimum value of $W_{1/2}$ for undoped 3C-SiC layer was 11.5 arcsec, which is the best value ever obtained for 3C-SiC.

On the basis of the p(6H)-n$^+$(3C) heteropolytype junction a red LED was fabricated [19]. At room temperature the red emission ($\lambda_{max} \sim 600$ nm) from 3C-SiC dominates in the EL spectra. At low temperature, both peaks from 3C-SiC ($\lambda_{max} \sim 600$ nm) and 6H-SiC ($\lambda_{max} \sim 480$ nm) exist.

F GROWTH OF SiC-AlN SOLID SOLUTION

SiC-AlN solid solution is a promising material for optoelectronics, particularly for the ultraviolet spectra region (see [35,36] for reviews describing this alloy system). SiC-AlN layers have been grown using CFLPE on 6H-SiC substrates from Si-Al-N-C solution [37]. The same set-up as for SiC CFLPE growth has been employed. The growth temperature was 1450 - 1500°C. SiC-AlN layers had a thickness up to 40 μm. The AlN concentration in the layer measured by Auger spectroscopy changed from 0 to 20 mol% depending on Al concentration in the melt. Layers with AlN concentration up to 10 mol% had single crystal structure. Solid solutions with both types of conductivity were obtained. n-type layers were grown from nitrogen-rich solution, and p-layers were grown from Al-rich solution. By changing the melt composition in a single epitaxial run a two-layer p-n structure has been grown [38].

G CONCLUSION

To date, 6H-SiC layers with thicknesses from 0.2 to 100 μm have been grown by LPE from the Si melt on 6H-SiC substrates with the growth rate of 0.01 - 2 μm min^{-1}. Epitaxial growth has been carried out at temperatures of 1450 - 1700°C in both isothermal and non-isothermal conditions. The concentration $N_d - N_a$ in the n-layers was controlled from 8×10^{15} to 10^{20} cm^{-3} and can be changed during the growth. For the first time, SiC epitaxial structures with doping profile control have been grown by LPE. The concentration of the acceptors (Al) in the p-layers varied from 10^{17} to 10^{20} cm^{-3} depending on Al concentration in the melt. LPE grown material displays high Hall carrier mobility and low deep centre concentration. High quality 6H-SiC and 4H-SiC p-n junctions have been made by LPE from Si melt. During a single epitaxial run, SiC p-n structures of different types may be grown: n(substrate)-n$^+$-n$^-$-p$^+$, n(substrate)-n$^-$-p-n$^+$ and n(substrate)-n-p$^+$-n$^-$.

New directions of SiC LPE include: (1) multilayer structure growth with the layers thinner than 0.1 mm, (2) growth of epitaxial layers with the $N_d - N_a$ concentration less than 10^{15} cm^{-3}, (3) low temperature (1000 - 1200 °C) LPE from alternative melts, (4) selective epitaxial growth, (5) heteropolytype epitaxy, and (6) SiC-AlN p-n structure growth.

REFERENCES

[1] H. Nelson [*RCA Rev. (USA)* vol.24 (1963) p.603]

[2] L.R. Dawson [*Progress in Solid State Chemistry* vol.7, Eds H. Reiss, J.O. McCaldin (Pergamon Press, 1972) p.117]

[3] H. Kressel, H. Nelson [*Physics of Thin Films* vol.7 (Academic Press, New York and London, 1973) p.115]

[4] R.W. Brander, R.P. Sutton [*Br. J. Appl. Phys. (UK)* vol.2 (1969) p.309]

[5] W. Munch, W. Kurzinger [*Solid-State Electron. (UK)* vol.21 (1978) p.1129]

[6] L. Hoffmann, D. Ziegler, D. Theis, C. Weyrich [*J. Appl. Phys. (USA)* vol.53 (1982) p.6962]

[7] Light Emitting Diodes, Short Form Catalog 1986/87, Siemens, Munich (1986)

[8] A. Suzuki, M. Ikeda, H. Matsunami, T. Tanaka [*J. Electrochem. Soc. (USA)* vol.122 (1975) p.1741]

[9] M. Ikeda, T. Hayakawa, S. Yamagiwa, S. Matsunami. T. Tanaka [*J. Appl. Phys. (USA)* vol.50 (1979) p.8215]

[10] T. Nakata, K. Koga, Y. Matsushita, Y. Ueda, T. Niina [*Springer Proc. Phys. (Germany)* vol.43 (1989) p.26]

[11] V.A. Dmitriev et al [*Sov. Tech. Phys. Lett. (USA)* vol.11 (1985) p.98]

[12] V.A. Dmitriev, A.E. Cherenkov [*Pis'ma Zh. Tekh. Fiz. (Russia)* vol.17 (1991) p.43]

[13] R.I. Scace, G.A. Slack [*J. Chem. Phys. (USA)* vol.30 (1959) p.1551]

[14] Yu.M. Tairov, F.I. Raikhel, V.F. Tsvetkov [*Izv. Akad. Nauk SSSR Neorg. Mater. (Russia)* vol.18 (1982) p.1390]

[15] S.V. Rendakova [*Physics and Technology of Wide-Gap Semiconductors* Abstracts of 3rd All-Union Meeting (Makhachkala, 1986) p.28 (in Russian)]

[16] V.A. Dmitriev, L.B. Elfimov, N.D. Il'inskaya, S.V. Rendakova [*Springer Proc. Phys. (Germany)* vol.56 (1992) p.307]

[17] V.A. Dmitriev, I.V. Popov, V.E. Chelnokov [*Growth and Synthesis of Semiconductor Crystals and Films* Ed. F.A. Kuznetsov (Nauka, Novosibirsk, 1986) p.74 (in Russian)]

[18] V.F. Britun et al [*Sov. Phys.-Tech. Phys. (USA)* vol.31 (1986) p.129]

[19] V.A. Dmitriev, A.E. Cherenkov [*J. Cryst. Growth (Netherlands)* in print]

[20] V.A. Dmitriev, S.V. Kazakov, V.V. Tretjakov, V.G. Ohding, M.A. Chernov [*J. Electrochem. Soc. (USA)* vol.137 (1989) p.711]

[21] V.A. Dmitriev, M.E. Levinshtein, S.N. Vainshtein, V.E. Chelnokov [*Electron. Lett. (UK)* vol.24 (1988) p.1031]

[22] V.A. Dmitriev, P.A. Ivanov, Ya.V. Morozenko, V.E. Chelnokov, A.E. Cherenkov [*Applications of Diamond Films and Related Materials* Eds Y. Tzeng, M. Yoshikawa, M. Murakawa, A. Felman (Elsevier, Amsterdam, 1991) p.769]

[23] A.M. Strelchuk [PhD Thesis, A.F. Ioffe Institute (1992)]

[24] M.M. Anikin, V.V. Evstropov, I.V. Popov, A.M. Strelchuk, A.L. Syrkin [*Fiz. Tekh. Poluprovodn. (Russia)* vol.23 (1989) p.1813-8]

[25] M.M. Anikin, A.S. Zubrilov, A.A. Lebedev, A.M. Strelchuk, A.E. Cherenkov [*Fiz. Tekh. Poluprovodn. (Russia)* vol.25 (1991) p.479]

[26] B.S. Kondratiev, I.V. Popov, A.M. Strelchuk, M.L. Tiranov [*Fiz. Tekh. Poluprovodn. (Russia)* vol.24 (1990) p.647]

[27] V.A. Dmitriev et al [*Sov. Tech. Phys. Lett. (USA)* vol.13 (1987) p.489]

[28] L.M. Kogan, Ya.V. Morozenko [*Springer Proc. Phys. (Germany)* vol.56 (1992) p.131]

8.4 LPE of SiC and SiC-AlN

[29] B.I. Vishnevskaya et al [*Sov. Phys.-Semicond. (USA)* vol.22 (1988) p.414]

[30] V.A. Dmitriev, L.M. Kogan, Ya.V. Morozenko, B.V. Tsarenkov, V.E. Chelnokov, A.E. Cherenkov [*Sov. Phys.-Semicond. (USA)* vol.23 (1989) p.23]

[31] V.A. Dmitriev, I.Yu. Linkov, Ya.V. Morozenko, V.E. Chelnokov [*Pis'ma Zh. Tekh. Fiz. (Russia)* vol.18 (1992) p.19]

[32] Yu.M. Tairov, V.F. Tsvetkov [*Handbook on Electrotechnical Materials* Eds Yu. Koritskiy, V.V. Pasinov, B.M. Tareev (Energoatomizdat, Leningrad, 1988) p.446 (in Russian)]

[33] I.S. Kulikov [*Thermodynamics of Carbides and Nitrides* (Metallurgiya, Chelyabinsk, 1988) (in Russian)]

[34] S.N. Vainshtein, V.A. Dmitriev, M.E. Levinshtein, S.V. Rendakova [*Sov. Tech. Phys. Lett. (USA)* vol.13 (1987) p.308]

[35] G.K. Safaraliev [*Wide-Gap Semiconductors*, Makhachkala, 1988, p.34 (in Russian)]

[36] V.A. Dmitriev [*Springer Proc. Phys. (Germany)* vol.56 (1992) p.3]

[37] V.A. Dmitriev et al [*Sov. Tech. Phys. Lett. (USA)* vol.17 (1991) p.214]

[38] V.A. Dmitriev et al [*Amorphous and Crystalline Silicon Carbide IV, Proc. 4th Int. Conf. on Amorphous and Crystalline Silicon Carbide*, Santa Clara, USA, 1991 (Springer-Verlag, Berlin, Germany, 1992]

CHAPTER 9

SiC DEVICES AND OHMIC CONTACTS

9.1   Ohmic contacts to SiC
9.2   Overview of SiC devices
9.3   SiC p-n junction and Schottky barrier diodes
9.4   SiC field effect transistors
9.5   SiC bipolar junction transistors and thyristors
9.6   SiC optoelectronic devices
9.7   Potential performance/applications of SiC devices and integrated circuits

9.1 Ohmic contacts to SiC

G.L. Harris, G. Kelner and M. Shur

November 1993

A INTRODUCTION

Low resistance, reliable, temperature-stable ohmic contacts are a prerequisite for the commercialization of SiC device technology. Still, these contacts are not yet satisfactory for a variety of reasons: the annealing temperature is too high, the contacts penetrate too deep, they deteriorate when devices operate at elevated temperature, and the contact resistance is quite high. The problem of stable contacts to SiC may be resolved by using refractory metals. Refractory metals can be used to form both carbides and silicides. Silicides appear to provide a stable resistance if carbides are not present [1]. Contact systems based on such metals as nickel, chromium, titanium, cobalt and tungsten have been demonstrated for n-type SiC. Most contact systems for p-type SiC are Al-based and this imposes a limitation on the operating temperature of SiC devices with p-type contacts.

B OHMIC CONTACTS

Recently, Kelner and co-workers [2] obtained a specific ohmic contact resistance of $3.5 \times 10^{-6} \, \Omega \, cm^2$ with a carrier concentration of $4 \times 10^{18} \, cm^{-3}$ to 6H-SiC. The contacts were resistively evaporated nickel that was rapidly thermally annealed at $1000\,°C$, for 30 s, in a forming gas atmosphere. Crofton et al [3] recently reported a contact resistance as low as $10^{-5} \, \Omega \, cm^2$ to p-type SiC. Another problem to be solved is the fabrication of low resistivity Au or Al overlays for device interconnects. These overlays have to be separated from contacts by a diffusion barrier layer (for example, W, Cr, Ti, or conducting nitrides).

All in all, SiC technology is progressing. A summary of contacts to the various polytypes is listed in TABLES 1 to 3. As mentioned above, the problem of reliable high temperature low resistance ohmic contacts is the most important problem to be solved. A second issue is the development of semi-insulating SiC for microwave and/or integrated circuits. SiC is now the prime candidate for commercial applications in high temperature devices and circuits and for devices and circuits operating in a harsh environment.

TABLE 1 Ohmic contacts to 4H- and 15R-SiC.

Type of contact	Specific contact resistance r_c ($\Omega \, cm^2$)	Annealing conditions	Comments	Ref
Ta	Ohmic	Annealed	n-type, 4H-SiC Ar ion sputtered	[8]
Al/Ti	Ohmic	Annealed	p-type, 4H-SiC	[8,21]
Ni	Ohmic	1200 °C 10 min.	n-type, 4H-SiC	[18,19]
Ti	Ohmic	900 °C	n-type, 15R-SiC	[20]

9.1 Ohmic contacts to SiC

TABLE 2 Ohmic contacts to 6H-SiC.

Type of contact	Specific contact resistance r_c ($\Omega\ cm^2$)	Annealing conditions	Comments	Ref
Al-W	$2 - 5 \times 10^{-4}$	1000 °C 120 s	p-type	[4]
Am W-W	$2 - 5 \times 10^{-4}$	1800 °C 120 s	p-type	
W	5×10^{-4}	~1600 °C 120 s	n-type, KOH etch, $(N_D - N_A) \sim 10^{19}\ cm^{-3}$, (0001)C face	
W	1×10^{-4}	1200 - 1400 °C 120 s	n-type, KOH etch, $(N_D - N_A) \sim 10^{19}\ cm^{-3}$, (0001)Si face	
Mo	1×10^{-4}	as-deposited	n-type, $(N_D - N_A) > 10^{19}\ cm^{-3}$, RF magnetron sputtered	[5]
	2×10^{-4}	as-deposited	p-type	
Ta	$\sim 1 \times 10^{-4}$	as-deposited	n-type, $(N_D - N_A) > 10^{19}\ cm^{-3}$, RF magnetron sputtered	
	7×10^{-4}	as-deposited	p-type	
Ti	$\sim 1 \times 10^{-4}$	as-deposited	n-type, $(N_D - N_A) > 10^{19}\ cm^{-3}$, RF magnetron sputtered	
	7×10^{-4}	as-deposited	p-type	
Ti/W 10/90	1×10^{-4} at RT 5×10^{-7} at 600 °C	750 °C 5 min.	n-type, Si face sputter	[6]
Ti/Au (200 Å /2000 Å)	2×10^{-5}	700 °C 5 min.	n^{++} capping layer	[7]
Ni	3.5×10^{-6}	1000 °C 30 s RTA	n-type SiC in forming gas	[1]
Ti/Al	Ohmic	950 °C 5 min.	p-type, $(N_D - N_A) \sim 10^{17} - 10^{18}\ cm^{-3}$	[15]
Al/Si	Ohmic	950 °C 5 min.	p-type, $(N_D - N_A) \sim 10^{17} - 10^{18}\ cm^{-3}$	
Ni/Ti/W	10^{-4}	1050 °C 5 min.	n-type, N_2 atmosphere	[16]
Al/Pt/W	$\sim 10^{-3}$	850 °C 10 min.	p-type, N_2 atmosphere anneal	
Al	$\sim 10^{-3}$	800 °C 10 min.	p-type, N_2 atmosphere anneal	
Ni	7×10^{-7}	950 °C 2 min.	n-type, Si-face, $(N_D - N_A) \sim 10^{19}\ cm^{-3}$	[17]
Ni	4×10^{-6}	925 °C 2 min.	n-type, C-face, $(N_D - N_A) \sim 10^{19}\ cm^{-3}$	

9.1 Ohmic contacts to SiC

TABLE 3 Ohmic contacts to β-SiC.

Type of contact	Specific contact resistance r_c (Ω cm^2)	Annealing conditions	Comments	Ref
W	0.24 at 23 °C 0.08 at 900 °C	as-deposited	n-type, ion beam deposited, SiC ion cleaned	[10]
Ta/W W/Am Ti/W/Am C/Ti/W/Am	20×10^{-4} 11×10^{-4} 80×10^{-4} 90×10^{-4}	800 °C 1h 800 °C 1h 800 °C 1h 800 °C 1h	n-type, sputtered n-type, sputtered n-type, sputtered n-type, sputtered	[11]
Mo Ni	6.2×10^{-2} $4.1 - 3.2 \times 10^{-2}$	1150 °C 15 min. 580 °C 15 min.	n-type, Ar ambient p-type, Ar ambient	[12]
Ni	Ohmic	900 °C 3 - 5 min.	n-type, SiC sputtered	[13]
Al/Si Ti/P$^+$ MoNi	2×10^{-1} 1×10^{-6} 1.4×10^{-4}	700 °C 30 s 950 °C 30 s 700 °C 30 s	p-type, SiC e-beam evaporated n-type, SiC e-beam evaporated n-type, SiC e-beam evaporated	[14]

C   CONCLUSION

Contacts formed with a variety of metals, and combinations of metals, to 4H-, 15R-, 6H- and β-SiC polytypes have been described. Reliable low resistance contacts which can survive high temperatures are still elusive. Low resistivity overlays for use as interconnects in devices are also required.

REFERENCES

[1]   M. Östling [personal communication, Royal Institute of Technology, Kista, Sweden]
[2]   G. Kelner [*Proc. 16th Nordic Semiconductor Meeting*, Hanmeenlinna, Finland, 1992, p.5-12]
[3]   J. Crofton, P.A. Barnes, J.R. Williams, J.A. Edmonds [*Appl. Phys. Lett. (USA)* vol.62 (1993) p.384-6]
[4]   M.M. Anikin, M.G. Rastegaeva, A.L. Syrkin, I.V. Chuiko [*Springer Proc. Phys. (Germany)* vol.56 (1992) p.183-7]
[5]   J.B. Petit, R.G. Neudeck, G.S. Salupo, D.J. Larkin, J.A. Powell [*Inst. Phys. Conf. Ser. (UK)* no.137 (1994) p.679-82]
[6]   J. Crofton, J.R. Williams, M.J. Bozack, P.A. Barnes [*Inst. Phys. Conf. Ser. (UK)* no.137 (1994) p.719-22]
[7]   P.G. Young, N.C. Varaljay, P.G. Neudeck, E.J. Haugland, D.J. Larkin, J.A. Powell [*Inst. Phys. Conf. Ser. (UK)* no.137 (1994) p.667-70]
[8]   R.A. Ivanov, N.S. Savkina, T.P. Samsonova, V.N. Panteleev, V.E. Chelnokov [*Inst. Phys. Conf. Ser. (UK)* no.137 (1994) p.593-5]

[9] J. Crofton, P.A. Barnes, J.R. Williams, J.A. Edmond [*Appl. Phys. Lett. (USA)* vol.62 (1993) p.384-6]

[10] K.M. Geib, J.E. Mahan, C.W. Wilmsen [*Springer Proc. Phys. (Germany)* vol.42 (1989) p.224-8]

[11] P.G. Mullin, J.A. Spitznagel, J.R. Szedon, J.A. Costello [*Springer Proc. Phys. (Germany)* vol.56 (1992) p.275-81]

[12] H.J. Cho, C.S. Hwang, W. Bank, H.J. Kim [*Inst. Phys. Conf. Ser. (UK)* no.137 (1994) p.663-6]

[13] A.J. Steckl, J.N. Su, P.H. Yih, C. Yuan, J.P. Li [*Inst. Phys. Conf. Ser. (UK)* no.137 (1994) p.653-6]

[14] D. O. Arugu [Ph.D. Thesis, Howard University, 1992]

[15] T. Nakata, K. Koga, Y. Matsushita, Y. Veda, T. Niina [*Springer Proc. Phys. (Germany)* vol.43 (1989) p.26-33]

[16] S. Adams, C. Severt, J. Leonard, S. Liu, S.R. Smith [*Trans. 2nd Int. High Temp. Electronics Conf.* vol.I, 1994, (HiTEC) p.XIII-9-14]

[17] J. Crofton, P.G. McMullin, J.R. Williams, M.J. Bozack [*Trans. 2nd Int. High Temp. Electronics Conf.* vol.I, 1994, (HiTEC) p.XIII-15-25]

[18] A. Itoh, H. Akita, T. Kimoto, H. Matsunami [*Inst. Phys. Conf. Ser. (UK)* no.137 (1994) p.59-62]

[19] W.J. Schaffer, H.S. Kong, G.H. Negley, J.W. Palmour [*Inst. Phys. Conf. Ser. (UK)* no.137 (1994) p.155-9]

[20] T. Troffer et al [*Inst. Phys. Conf. Ser. (UK)* no.137 (1994) p.173-6]

[21] J.W. Palmour, J.A. Edmond, H.S. Kong, C.H. Carter Jr. [*Inst. Phys. Conf. Ser. (UK)* no.137 (1994) p.499-522]

9.2 Overview of SiC devices

G. Kelner and M. Shur

November 1993

The application of modern epitaxial techniques (see Chapter 8) has led to a rapid improvement in the material quality of SiC and made practical SiC devices a reality. The applications of SiC include high-temperature, high-power devices, microwave devices (both avalanche diodes and field effect transistors), and optoelectronic devices, such as photodiodes and light-emitting diodes which emit throughout the visible spectrum into the ultraviolet.

SiC exists in more than 170 polytypes (see Chapter 1). The properties of the polytypes are so different that, in fact, SiC represents a group of closely related materials. The two most important polytypes are hexagonal 6H-SiC (also called α-SiC) and cubic 3C-SiC (also called β-SiC). The relative material quality of these polytypes can be crudely evaluated by comparing the values of electron mobility at room temperature. These values are approximately 300, 600, 800 and 1000 $cm^2 V^{-1} s^{-1}$ for 6H, 4H, 15R and 3C SiC polytypes. Interest in 6H α-SiC for device applications has been recently renewed due to the availability of 2 inch diameter 6H α-SiC substrates. At the present time, no comparable substrates exist for β-SiC material. 6H-SiC homoepitaxial films have been grown with a compensation ratio as low as 2% [1]. β-SiC epilayers are grown on Si substrates and have compensation ratios approaching 100%. This is the primary reason why 6H-SiC devices have demonstrated better performance. However, Ivanova et al [2,3] have recently grown and characterised bulk 3C-SiC with remarkable electrical properties. The mobility of the material grown in [2,3] was estimated to be as high as 70,000 $cm^2 V^{-1} s^{-1}$ at 28 K, with an n-type carrier concentration of $7 \times 10^{13} cm^{-3}$. The compensation of the background n-type impurity was on the order of 10%. This material was grown by the decomposition of methyl trichlorosilane in hydrogen on graphite electrodes. The sample size of 3C-SiC grown by this technique is very small (approximately 3×3 mm). Nevertheless, these results show the potential superiority of β-SiC over all other polytypes.

Recent developments in solid state solutions of AlN/SiC/InN/GaN open up the possibility of a new generation of heterostructure devices based on SiC. Single crystal epitaxial layers of AlN/SiC/InN have been recently demonstrated by Dmitriev [4]. A whole range of solid state solutions has been grown. Recently Dmitriev et al [5] reported on an $(AlN_x-SiC_{1-x})-(AlN_y SiC_{1-y})$ p-n junction. Solid state solutions of AlN-SiC [6,7] are also expected to lead to direct gap ternary materials for UV and deep blue optoelectronics, including the development of visible lasers. The direct to indirect bandgap transition is predicted to occur at between 70 and 80% of AlN in SiC.

The device potential of a semiconductor material is often estimated in terms of figures of merit. The Johnson's and Keyes' figures of merit are frequently used. Johnson's Figure of Merit (JFM) addresses the potential of a material for high-frequency and high-power applications [8]:

$$JFM = (E_B^2 v_s^2)/4\pi^2 \qquad (1)$$

9.2 Overview of SiC devices

where E_B is the breakdown electric field and v_s is the electron saturation velocity. In terms of this figure of merit, SiC is 260 times better than Si and is inferior only to diamond (see FIGURE 1).

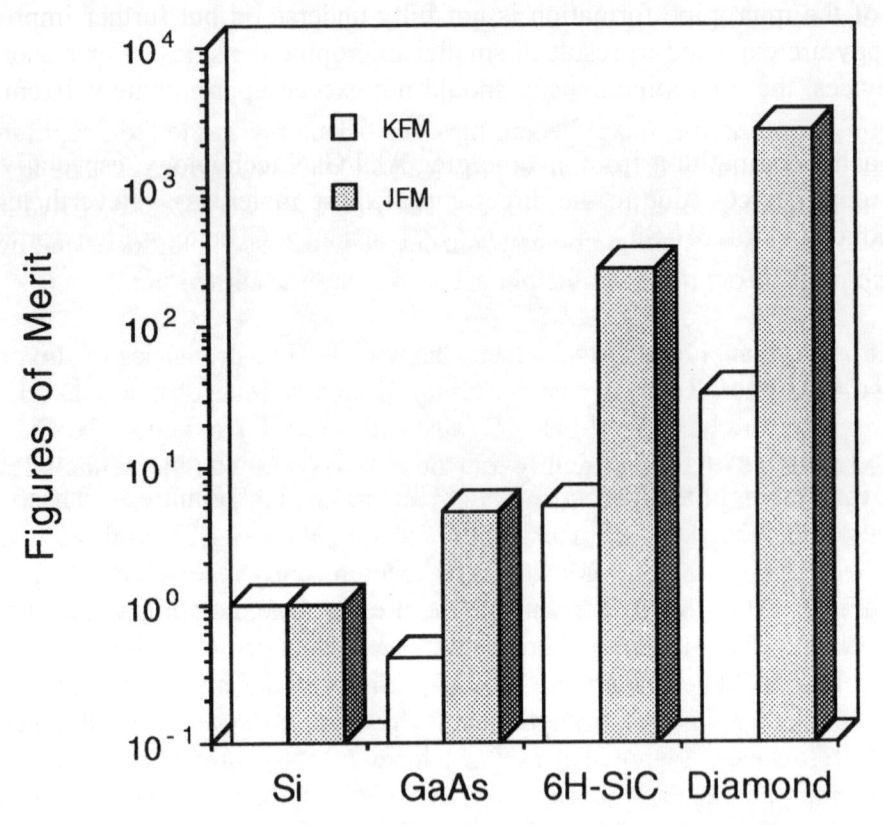

FIGURE 1 Figures of merit for Si, GaAs and SiC. KFM - Keyes' figure of merit, JFM - Johnson's figure of merit.

Keyes' Figure of Merit (KFM) [9] is relevant to integrated circuits applications:

$$KFM = \chi[(cv_s)/(4\pi\varepsilon_0)]^{1/2} \qquad (2)$$

where c is the velocity of light, ε_o is the static dielectric constant, and χ is the thermal conductivity. In terms of this figure of merit, SiC is 5.1 times better than Si and, again, is inferior only to diamond. These figures of merit illustrate the SiC device potential for power and high voltage devices which are a prime target area for SiC devices.

The only commercially available SiC devices are blue LEDs which are used in large area advertising displays. These devices are produced by Cree Research, Siemens, Sanyo and other manufacturers.

SiC power devices are expected to be superior to Si devices even at room temperature. However, the practical difficulties are related to a limited yield of large area devices due to the existence of so-called micropipes (or voids) [10]. The density of these micron size defects has steadily decreased over the years reaching approximately 100 to 400 cm^{-2} in the state-of-the-art 6H-SiC substrates [11]. These defects are formed during the substrate growth. They have a hexagonal cross-section and propagate through the length of the boule grown along the c-axis and through the epitaxial layer. The micropipes are less of a problem in the 4H-SiC polytype (where the micropipe density is presently close to $50 - 200 \, cm^{-2}$, see [11]). The mechanism of the micropipe formation is not fully understood but further improvements in SiC technology are expected to result in smaller micropipe densities. For reasonable yields of power devices, the micropipe density should not exceed approximately $10 \, cm^{-2}$.

SiC is also facing competition from an emerging AlN/GaN technology, especially in the area of optoelectronic devices (due to the direct gap of these materials). Nevertheless, the high thermal conductivity of 6H-SiC ($4.9 \, W \, cm^{-1} \, K^{-1}$) makes SiC better suited for most power devices.

ACKNOWLEDGEMENT

The work on Datareviews 9.2 - 9.7 at the Naval Research Laboratory was supported by the Office of Naval Research, and the work at the University of Virginia was partially supported by the Office of Naval Research under the contract # NOOO 14-92-1580.

The authors are grateful to J.W. Palmour of Cree Research, Inc. and D.M. Brown of General Electric for valuable comments on Datareviews 9.2-9.7.

REFERENCES

[1] T.Tachibana, H.S. Kong, Y.C. Wang, R.F. Davis [*J. Appl. Phys. (USA)* vol.67, no.10 (1990) p.6375-81]

[2] W.J. Moore, P.J. Lin-Chung, J.A. Freitas, Y.M. Altaiski, V.L. Zuev, L.M. Ivanova [*Proc. 5th Int. Conf. SiC and Related Materials*, Washington, DC, USA, 1993, p.181-4]

[3] R. Kaplan, W.J. Moore, J.A. Freitas, Y.M. Altaiski, V.L. Zuev, L.M. Ivanova [*Proc. 5th Int. Conf. SiC and Related Materials*, Washington, DC, USA, 1993, p.207-10]

[4] V. Dmitriev [AlN/SiC/GaN Solid Solutions, presented at the Panel Discussion of ISDRS-93, Charlottesville, Va, USA (1993)]

[5] V.A. Dmitriev et al [*Proc. 4th Int. Conf. on Amorph. and Cryst. SiC*, Santa Clara, CA, 1991 (Springer-Verlag, Berlin, 1992) p.101-4]

[6] V.A. Dmitriev [*Proc. Electrochem. Soc. Meeting*, Washington, DC, vol.138 (1991) p.214-5]

[7] G.K. Safaraliev, Yu.M. Tairov, V.F. Tsvetkov [*Sov. Phys.-Semicond. (USA)* vol.25, no.8 (1991) p.865-8]

[8] E.O. Johnson [*RCA Rev. (USA)* vol.26 (1965) p.163-77]

[9] R.W. Keyes [*Proc. IEEE (USA)* vol.60 (1972) p.225-6]

[10] R.F. Davis [*Proc. 5th Int. Conf. on SiC and Related Materials*, Washington, DC, USA, 1993, p.1-6]

[11] J.W. Palmour [Cree Research Inc., private communication, 1993]

9.3 SiC p-n junction and Schottky barrier diodes

G. Kelner and M. Shur

November 1993

A INTRODUCTION

This Datareview covers the state of diodes in SiC polytypes both made from p-n junctions and using Schottky barriers. After defining the necessary theoretical background to junction diodes the various device structures used and the performances obtained are detailed. A range of metals has been used to produce Schottky barriers on three of the SiC polytypes and the performances achieved in this type of device are listed.

B JUNCTION DIODES

SiC p-n diodes clearly illustrate the advantages of wide bandgap semiconductors in general and SiC in particular. The elementary theory of p-n junctions yields the following expression for the reverse saturation current density, J_R, for a p^+-n junction [1]:

$$J = q\sqrt{D_p/\tau_p}\,\frac{n_i^2}{N_D} + \frac{qn_i W}{\tau_e} \qquad (1)$$

where q is the electronic charge, n_i is the intrinsic carrier density, N_D is the doping density of the n-type region, D_p is the hole diffusion coefficient, τ_e is the effective lifetime and W is the thickness of the depletion region:

$$W = \sqrt{2\varepsilon_0\varepsilon(V_{bi} - V)/qN_D} \qquad (2)$$

where N_D is the donor concentration and V_{bi} is the built-in voltage.

In wide bandgap materials, n_i is very small and therefore the second term in the right hand part of Eqn (1) is dominant. In α-SiC at room temperature, the theoretical value of n_i is as low as 1 carrier per m^3! For an estimate, let us assume certain values for m_n and m_p, independent of an energy gap, since this assumption will not change the order of magnitude of the resulting number. Let us take $m_n = 0.3\,m_e$ and $m_p = 0.6\,m_e$ to be specific. Then we find:

$$n_i = 1.34 \times 10^{21} \times T^{3/2} \exp(-E_g/2k_B T) \qquad (3)$$

where T is the temperature in K. Choosing the transistor volume to be $1 \times 0.1 \times 10\,\mu m^3$ and assuming the generation time of 1 ns, we obtain:

9.3 SiC p-n junction and Schottky barrier diodes

$$I_{leakage} = 2.14 \times 10^{-7} \times T^{3/2} \exp(-E_g/2k_BT) \tag{4}$$

where $I_{leakage}$ is the leakage current in A. The resulting dependence of the minimum leakage current on the energy gap for room temperature is shown in FIGURE 1 [2]. However, experimentally, such low values of the saturation current density have not yet been observed (see below), and certainly the conventional theory of a p-n junction is not valid when n_i is very small.

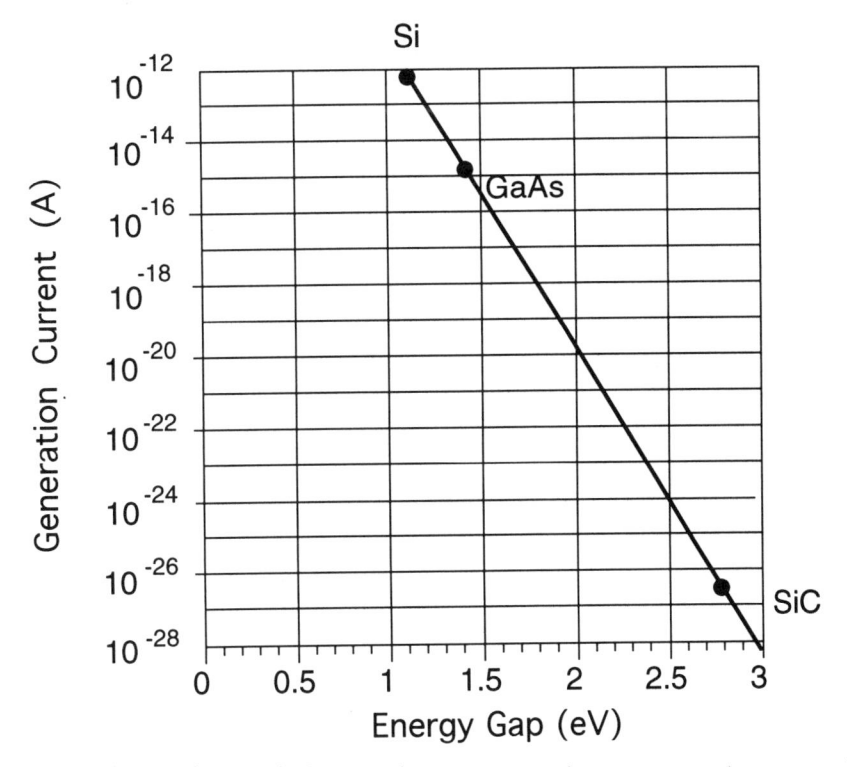

FIGURE 1 Dependence of theoretical generation current on the energy gap (at room temperature) [2].

Under forward bias, the diode current voltage characteristic is described by the following equation [2]:

$$J_F = q\sqrt{D_p/\tau_p}\frac{n_i^2}{N_D}\exp\left[\frac{qV}{k_BT}\right] + \alpha q n_i W \sigma v_{th} N_t \exp\left[\frac{qV}{2k_BT}\right] \tag{5}$$

where τ_p is the minority carrier lifetime in an n-type semiconductor, v_{th} is the carrier thermal velocity, N_t is the trap density, σ is the capture cross-section, and α is a numerical factor on the order of unity. Once again, in wide band materials, the second term in the right hand part of Eqn (5) may be dominant because of small values of n_i. In this case, J_F is proportional to $\exp(qV/nk_BT)$ with $n=2$. This is, indeed, often the case for SiC p-n diodes [3-5]. However, as was pointed out by Edmond et al [6], recombination through multielectron centres may lead to ideality factors close to 1.5 which are often observed in SiC p-n junctions.

9.3 SiC p-n junction and Schottky barrier diodes

SiC p$^+$-n junction diodes have been fabricated using both α-SiC and β-SiC. High voltage diodes have been fabricated in 6H-SiC [4,7-12]. FIGURE 2 illustrates a state-of-the-art result achieved for 6H SiC p-n diodes fabricated in SiC material homoepitaxially grown on off-axis substrates. As can be seen from the figure, breakdown voltages in excess of 1100 V have been achieved with a leakage current of less than 20 nA. Edmond et al [7], Matus and Powell [9] and Neudeck et al [12] fabricated SiC p-n junction diodes by epitaxial growth using nitrogen as the n-type and aluminium as the p-type dopants. The diode structures are shown in FIGURE 3. Low current forward and reverse bias tests were performed on Cree diodes which had a breakdown voltage of 710 V (V_B) at temperatures up to 623 K [7]. (Even higher breakdown voltages in SiC p-n junction diodes, up to 2000 V, have been reported by Neudeck et al [12].) Ref [6] shows the forward I-V characteristics on a semilog scale. The diode ideality factor, n, varied between 1.5 and 2 and was close to 2 at relatively low current densities, illustrating the dominant role of the recombination current. Ref [6] also shows the effect of temperature on the reverse bias current density. The reverse bias current density is very low even at 623 K. The series resistance of these rectifier diodes was on the order of $1 \times 10^{-3} \, \Omega \, cm^2$, illustrating the need for improving the ohmic contact resistance.

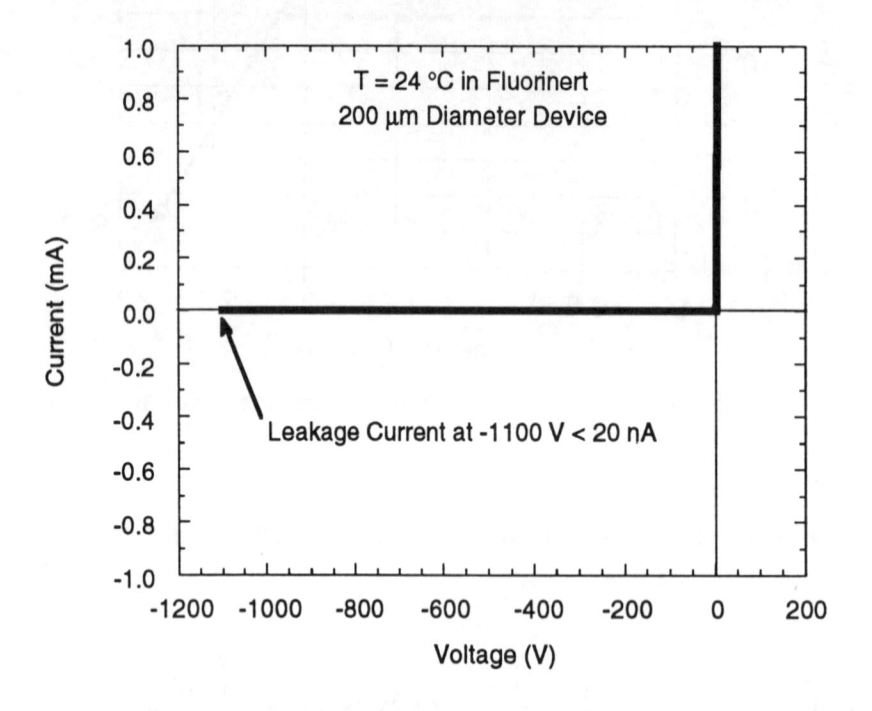

FIGURE 2 Current-voltage characteristics of 6H-SiC p-n diodes [4].

Reverse recovery times, T_{rr}, were measured by passing 100 mA of forward current through the device and switching the device to the 20 V reverse bias. The value of T_{rr} was 14 ns at 300 K, increasing to 14.5 ns at 623 K [7].

Ref [8] shows the dependence of the reverse breakdown voltage of 6H-SiC and depletion width on the background carrier concentration measured in n-type layers. The dependence of the breakdown voltage on the carrier concentration is approximately linear on a log-log

9.3 SiC p-n junction and Schottky barrier diodes

scale. This is similar to the observation of Anikin et al [10,11]. An increase in the breakdown field, E_B, with an increase in the carrier concentration can be approximated as:

$$E_B = A_B + B_B \lg(N_D/10^{15}) \tag{6}$$

where $A_B = 2 \times 10^6 \, V \, cm^{-1}$ and $B_B = 10^6 \, V \, cm^{-1}$. These diodes had a built-in voltage of 2.76 V at room temperature which decreased with temperature to approximately 2.31 V at 623 K. These values are consistent with the energy gap of 6H-SiC.

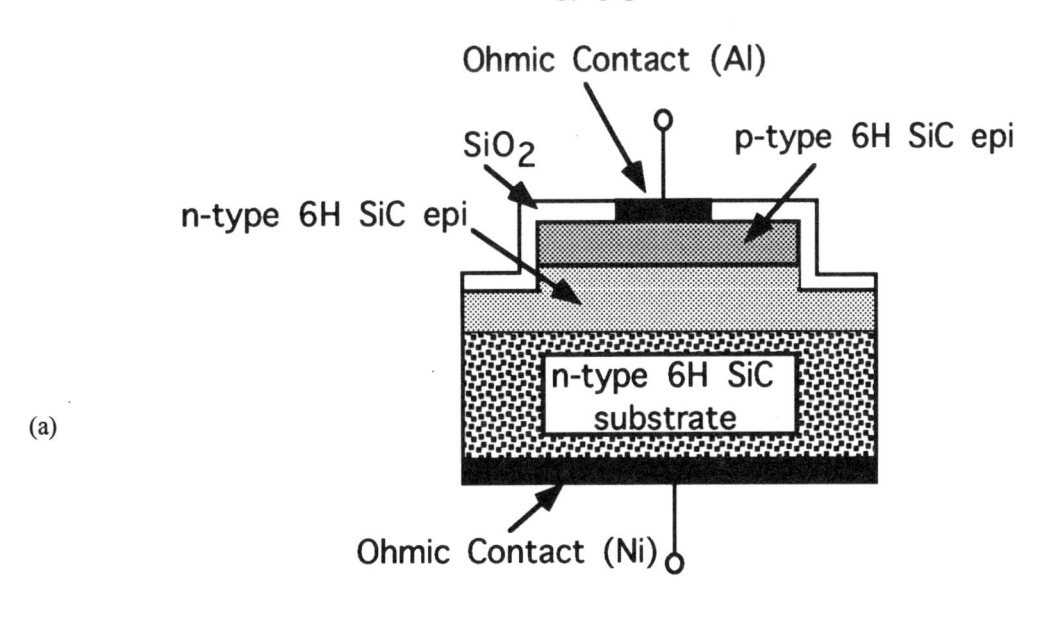

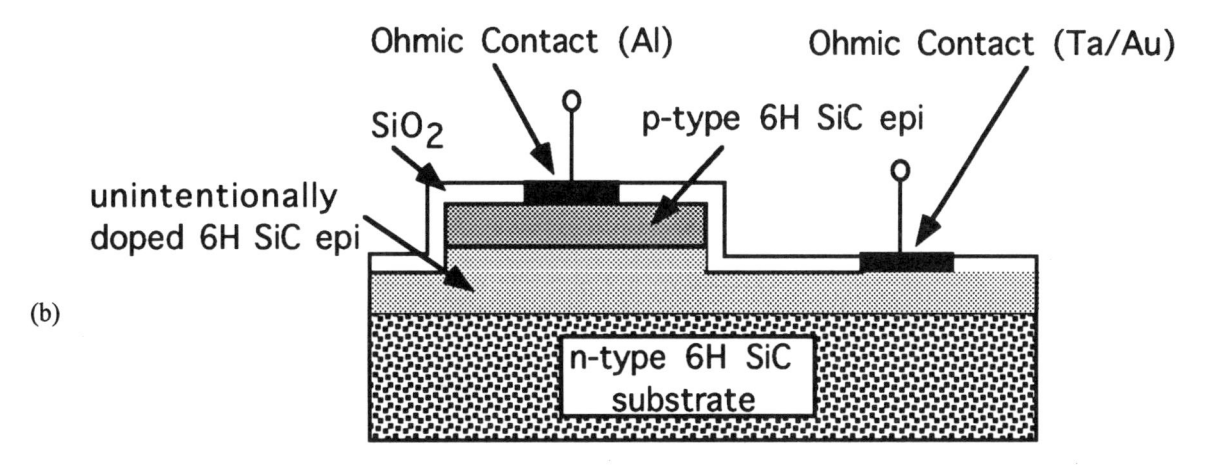

FIGURE 3 Cross-section of 6H-SiC p-n diodes - (a) from [7] and (b) from [9].

Ghezzo et al [13] reported on ion implanted planar p-n junction 6H-SiC diodes fabricated by ion implantation of B into n-type SiC with the donor concentration of $9 \times 10^{15} \, cm^{-3}$. The implantation was performed at 25 °C and 1000 °C followed by a 1300 °C post-implant furnace anneal. The diodes had an ideality factor of 1.77 at room temperature, the reverse leakage current of $10^{-10} \, A \, cm^{-2}$ at -10 V and the reverse breakdown voltage of -650 V.

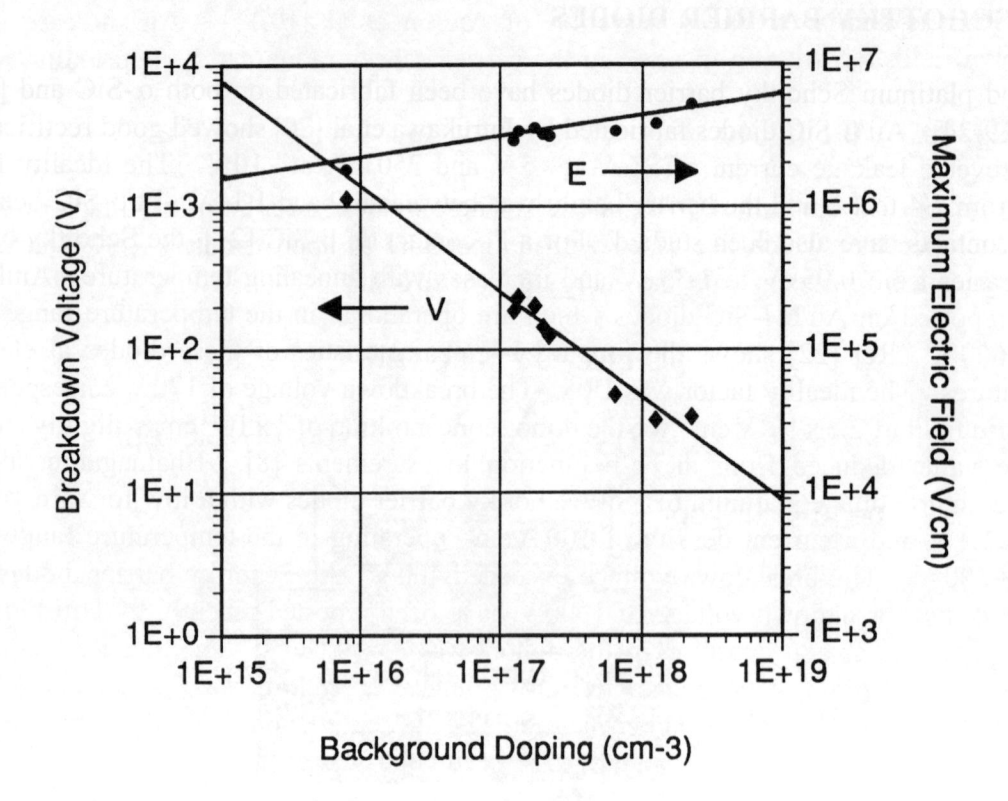

FIGURE 4 Measured dependence of the reverse breakdown voltage and maximum electric field on background carrier concentration for 4H-SiC [18].

Suzuki and co-workers [14,15] fabricated β-SiC p-n junction diodes using a mesa structure. The turn-on voltage at room temperature was 1.2 V and a reverse leakage current of 5 μA was measured at -5 V. This leakage current is much greater than that for α-SiC diodes. Both parameters were severely degraded at elevated temperatures. Edmond et al [16] and Avila et al [17] have also produced p-n junction diodes in β-SiC. These diodes were made by ion implantation and operated up to 673 K.

Larkin et al [5] recently obtained excellent results for 3C-SiC p-n diodes fabricated using epilayers grown on 6H-SiC substrates. The epilayers were practically free of double positioned boundaries (DPB) [5]. The material was grown on a slightly off-axis 6H-SiC (0001) substrate (the off-axis angle was approximately 0.3 degrees). These 3C-SiC p-n junction diodes exhibited a room temperature breakdown voltage greater than 200 V, a record for 3C-SiC material. An analysis of the data for devices with different areas showed that the breakdown was caused by processes at the sample periphery.

4H-SiC p^+-n and n^+-p junction diodes also exhibited excellent characteristics [18] with the breakdown voltages up to 1130 V (corresponding to the breakdown electric field of 2.2×10^6 V cm^{-1}) and low reverse leakage current (see FIGURE 4). Hence, Palmour et al [18] demonstrated that the breakdown field in 4H-SiC is only slightly smaller than that in 6H-SiC: see FIGURE 4.

C SCHOTTKY BARRIER DIODES

Gold and platinum Schottky barrier diodes have been fabricated on both α-SiC and β-SiC [16,17,19-24]. Au/β-SiC diodes fabricated by Furukawa et al [20] showed good rectification, with a reverse leakage current of 5 µA at -5 V and 250 mA at -10 V. The ideality factor varied from 1.4 to 1.6 and the barrier height was between 0.9 and 1.1 eV. Pt/β-SiC Schottky barrier contacts have also been studied. For a Pt contact on β-SiC [21], the Schottky barrier height varies from 0.95 eV to 1.35 eV and increases with annealing temperature. Anikin et al [22] reported on Au/6H-SiC diodes which are operational in the temperature range from 300 to 600 K. Ref [22] shows the forward I-V characteristics of these diodes at elevated temperatures. The ideality factor was 1.05. The breakdown voltage of 170 V corresponds to an electric field of 2.3×10^6 V cm^{-1} for the donor concentration of 1×10^{17} cm^{-3}, slightly smaller than the value deduced from the p-n junction measurements [8]. Bhatnagar et al [23] fabricated high voltage platinum 6H-SiC Schottky barrier diodes with a low forward voltage drop of 1.1 V and a current density of 100 A cm^{-2} operating in the temperature range from 25 °C to 200 °C. The breakdown voltage exceeded 400 V. Au Schottky barrier diodes with the record high breakdown voltage of 1000 V have been reported recently by Urushidani et al [24]. The carrier concentration of the donors was 3×10^{15} cm^{-3}. The ideality factor was 1.15. Prior to the onset of avalanche breakdown the leakage current was 2×10^{-10} A at -100 V and 2×10^{-9} A at -500 V. Porter and co-workers [25,26] reported on Schottky barrier 6H-SiC diodes using Hf, Ti and Co. All metals have been deposited by electron beam evaporation in ultra high vacuum. Hf forms excellent Schottky barrier contacts with the barrier height of 0.97 eV and the ideality factor <1.07. Co and Ti deposited on 6H-SiC had barrier heights of 1.14 and 0.97 eV respectively and ideality factors close to 1. Zhao et al [27] fabricated Pd/6H-SiC Schottky barrier diodes with near to a 1000 V breakdown voltage. At the reverse bias near to 1000 V the leakage current was on the order of 10^{-6} A.

State-of-the-art Schottky diodes operate up to 400 °C [28]: see FIGURE 5. The stability of the Schottky diodes limits the maximum temperature of operation of SiC MESFETs since SiC itself can withstand much higher temperatures. State-of-the-art ohmic contacts can operate at significantly higher temperatures than Schottky contacts.

The parameters of SiC diodes are summarized in TABLE 1.

D CONCLUSION

Junction diodes have been fabricated in both cubic and hexagonal SiC polytypes, usually with epitaxial material. Ohmic contact resistance is still a limiting factor in these devices. Ion implantation planar devices and ones using a mesa structure have been produced in β-SiC. Leakage currents are greater than in α-SiC diodes. Diodes with breakdown voltages exceeding 1000 V have been fabricated using Au as the Schottky barrier metal. Operating temperatures of 300-400 °C are routinely achieved in both junction and Schottky diodes made in SiC.

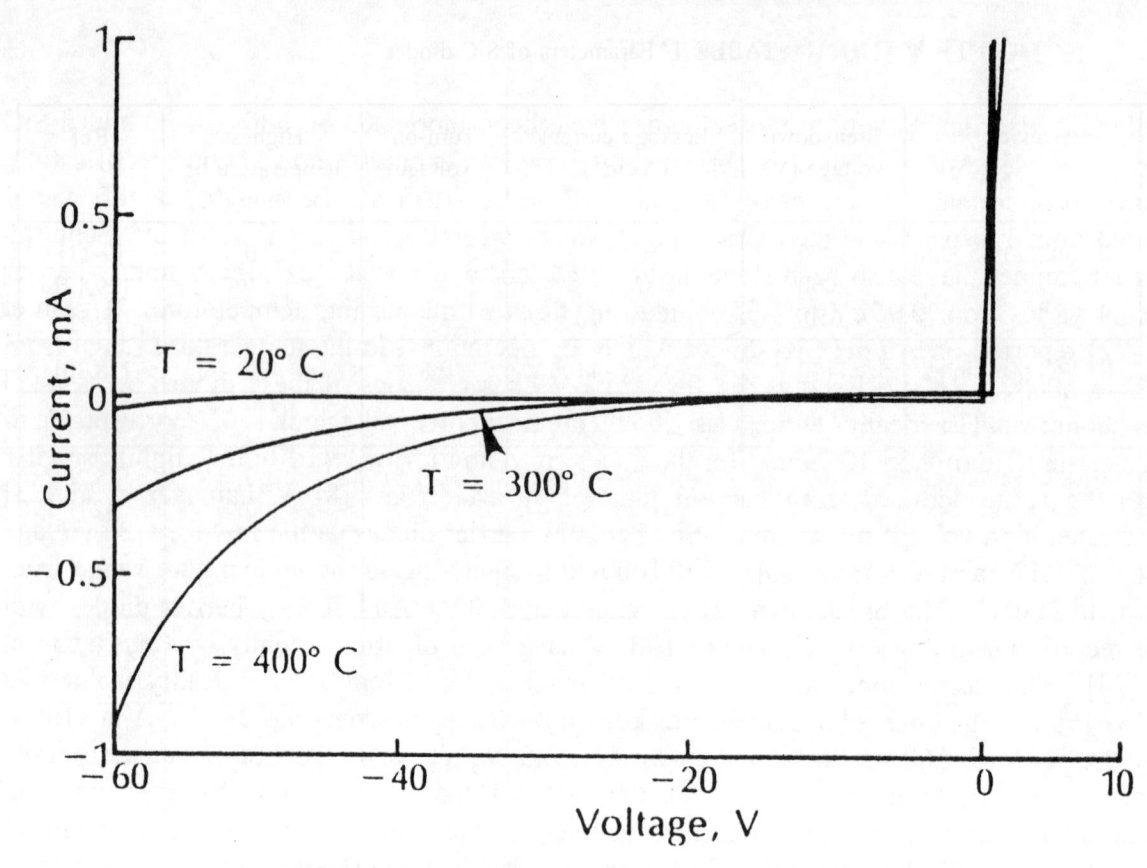

FIGURE 5 High temperature operation of SiC Schottky diodes [28].

REFERENCES

[1] S.M. Sze [*Physics of Semiconductor Devices* (John Wiley and Sons, New York, 1981)]
[2] M. Shur [*Introduction to Physics of Semiconductor Devices* to be published]
[3] V.A. Dmitriev [*Proc. Electrochem. Soc. Meeting*, Washington, DC, vol.138 (1991) p.214]
[4] P.G. Neudeck, D.J. Larkin, J.E. Starr, J.A. Powell, C.S. Salupo, L.G. Matus [*Abstracts of Workshop on SiC Materials and Devices* University of Virginia, Sept. 10-11, 1992 (Office of Naval Research, USA, 1992) p.38]
[5] D.J. Larkin et al [*Abstracts of Workshop on SiC Materials and Devices* University of Virginia, Sept. 10-11, 1992 (Office of Naval Research, USA, 1992) p.37]
[6] J.A. Edmond, J.W. Palmour, C.H. Carter Jr. [*Proc. Int. Semicond. Device Research Symposium* Charlottesville, Va, (1991) p.487-90]
[7] J.A. Edmond, H.S. Kong, C.H. Carter Jr. [*Proc. 4th Int. Conf. Amorphous and Crystalline Silicon Carbide*, Santa Clara, CA, USA, 9 - 11 Oct. 1991, Eds C.Y. Yang, M.M. Rahman, G.L. Harris (Springer-Verlag, Berlin, Germany, 1992) p.344-51]
[8] J.A. Edmond, D.G. Waltz, S. Brueckner, H.S. Kong [*Proc. 1st Int. High Temperature Electronics Conf.*, Albuquerque, NM, (1991) p.207-12]
[9] L.G. Matus, J.A. Powell [*Appl. Phys. Lett. (USA)* vol.59 (1991) p.1770-2]

9.3 SiC p-n junction and Schottky barrier diodes

TABLE 1 Parameters of SiC diodes.

Device	Breakdown voltage (V)	Leakage current (A cm^{-2})	Turn-on voltage (V)	Highest temperature of operation (°C)	Ref
6H-SiC p-n diode	1400 2000 650	5 x 10^{-4} 10^{-10} at -10 V	2.5 - 2.6	350	[9-11] [29,12] [13]
4H-SiC p-n diode	1130	3 x 10^{-4} at -1100 V	2.8	400	[18,30]
3C-SiC p-n diode	220		1.2		[5,14-17]
6H-SiC Schottky diode (Pt)	400	2 x 10^{-10} at -100 V	1.0 - 1.5	700 200	[25,26] [23]
6H-SiC Schottky diode (Ti)		5 x 10^{-8} at -10 V		700	[25,26]
6H-SiC Schottky diode (Co)		5 x 10^{-8} at -10 V		300	[25,26]
6H-SiC Schottky diode (Hf)		5 x 10^{-8} at -10 V		700	[25,26]
6H-SiC Schottky diode (Au)	1100	350 µA at -60 V	1.0 - 1.5	400	[22,24]
6H-SiC Schottky diode (Pd)	1000	1 x 10^{-6} at -900 V		200	[27]
3C-SiC Schottky diode (Au, Pt)	16	500 µA	1.2		[16,17, 19-21]

[10] M.M. Anikin et al [*Sov. Phys.-Semicond. (USA)* vol.20 (1989) p.844-5]

[11] M.M. Anikin, A.A. Lebedev, A.L. Syrkin, A.V. Suvorov, A.M. Strelchuk [*Ext. Abstr. Electrochem. Soc. (USA)* (1990) p.706]

[12] P.G. Neudeck, D.J. Larkin, C.S. Salupo, J.A. Powell, L.G. Matus [*Inst. Phys. Conf. Ser. (UK)* no.137 (1994) p.51-4]

[13] M. Ghezzo, D.M. Brown, E. Downey, J. Kretchmer, J.J. Kopansky [*Appl. Phys. Lett. (USA)* vol.63, no.9 (1993) p.1206-8]

[14] A. Suzuki, A. Uemeto, M. Shigota, K. Furukawa, S. Makajima [*Extended Abstracts 18th Conf. Solid State Devices and Materials*, Tokyo, Japan, 20 - 22 Aug. 1986 (Bus. Center Acad. Socs. Japan, Tokyo, 1986) p.101]

[15] K. Furukawa, A. Uemeto, Y. Fujii, M. Shigeta, A. Suzuki, S. Makajima [*Appl. Phys. Lett. (USA)* vol.48 (1986) p.1536-7]

[16] J.A. Edmond, K. Das, R.F. Davis [*J. Appl. Phys. (USA)* vol.63 (1988) p.922-9]

[17] R.E. Avila, J.J. Kopanski, C.D. Fung [*J. Appl. Phys. (USA)* vol.62 (1987) p.3469-71]

[18] J.W. Palmour, J.A Edmond, H.S Kong, C.H. Carter [*Proc. 5th Int. Conf. SiC and Related Materials*, Washington, DC, USA, 1 - 3 Nov. 1993, Eds M.G. Spencer et al (IOP Publishing, Bristol, UK, 1994) p.499-502]

[19] S. Yoshida, H. Daimon, M. Yamanaka, E. Sakuma, S. Misawa, K. Endo [*J. Appl. Phys. (USA)* vol.60 (1986) p.2989-92]

[20] K. Furukawa, A. Uemeto, Y. Fujii, M. Shigeta, A. Suzuki, S. Nakajima [*19th Conf. on Solid State Devices and Materials*, Tokyo, Japan, 25 - 27 Aug. 1987 (Bus. Center Acad. Socs. Japan, Tokyo, 1987) p.231-4]

[21] N. Papanicolaou, A. Christou, M. Gipe [*J. Appl. Phys.(USA)* vol.65 (1989) p.3526-30]

[22] M.M. Anikin, A.N. Andreev, A.A. Lebedev, S.N. Pyatko [*Sov. Phys.-Semicond. (USA)* vol.25 (1991) p.328-30]

[23] M. Bhathnagar, P.K. McLarty, B.J. Baliga [*IEEE Electron Device Lett. (USA)* vol.13, no.10 (1992) p.501-3]

[24] T. Urushidani, T. Kimoto, H. Matsunami [*Proc. 5th Int. Conf. SiC and Related Materials*, Washington, DC, USA, 1 - 3 Nov. 1993, Eds M.G. Spencer et al (IOP Publishing, Bristol, UK, 1994) p.471-4]

[25] L.M. Porter, R.F. Davis [*Tri-Service SiC Workshop*, NRL, Washington, DC, USA, 1993]

[26] L.M. Porter, R.F. Davis, J.S. Bow, M.J. Kim, R.W. Carpenter [*Proc. 5th Int. Conf. SiC and Related Materials*, Washington, DC, USA, 1 - 3 Nov. 1993, Eds M.G. Spencer et al (IOP Publishing, Bristol, UK, 1994) p.581-4]

[27] J.H. Zhao, K.Z. Xie, W. Buchwald, J. Flemish [*Proc. 5th Int. Conf. SiC and Related Materials*, Washington, DC, USA, 1 - 3 Nov. 1993, Eds M.G. Spencer et al (IOP Publishing, Bristol, UK, 1994) Abstr. 11-60]

[28] J.W. Palmour, J.A. Edmond [*Powertech. Mag. (USA)* (August 1989) p.18-21]

[29] J.W. Palmour, J.A. Edmond, H.S. Kong, C.H. Carter Jr. [*Abstracts of Workshop on SiC Materials and Devices*, University of Virginia, (USA) (1992) p.32]

[30] J.W. Palmour [Cree Research Inc., private communication]

9.4 SiC field effect transistors

G. Kelner and M. Shur

November 1993

A INTRODUCTION

Different types of SiC Field Effect Transistors, Metal Oxide Semiconductor Transistors (MOSFETs), Metal Semiconductor Field Effect Transistors (MESFETs), and Junction Field Effect Transistors (JFETs) compete for future applications in high temperature and harsh environment electronics. This Datareview details these various types of FETs, the structures used and the performances obtained. Interesting recent developments and potential applications, such as FET integrated circuits, a hybrid operational amplifier and an inverter circuit are also outlined.

B FETs

The first SiC MOSFETs were fabricated by Suzuki et al [1] in 3C-SiC. At the present time, SiC MOSFET research is carried out by several groups, including Cree Research [2-4] and Westinghouse [5]. Enhancement mode and depletion mode β-SiC MOSFETs have been fabricated by Palmour et al [3]. Enhancement mode devices with 5 μm gate lengths had a maximum transconductance (g_m) of 0.46 mS mm^{-1} at room temperature (see FIGURES 1 and 2). The devices were operational up to 823 K.

Palmour et al [2,6] have reported on an enhancement mode 6H-SiC MOSFET. The device structure is shown in FIGURE 3. This MOSFET exhibited a better performance than β-SiC MOSFETs and was operational up to 923 K (see FIGURE 4). The maximum g_m in the saturation regime was 2.8 mS mm^{-1} at room temperature for devices with a gate length of 7 μm. This value corresponds to a field effect mobility of ≈ 46 cm^2 V^{-1} s^{-1}. The transconductance of both α- and β-SiC MOSFETs increased somewhat with temperature. Depletion mode β-SiC MOSFETs, fabricated by Palmour et al [3] (see FIGURE 5), had a transconductance of 10 mS mm^{-1} for a 2.4 μm gate length device (see FIGURE 6). The device transconductance of the depletion mode β-SiC MOSFETs also increased with temperature to a maximum value of about 12 mS mm^{-1} at 673 K. These devices operated up to 923 K. The device transconductance in the saturation regime for transistors with the gate length of 7.2 μm was 5.32 mS mm^{-1}. This value corresponds to the field effect mobility of 37 cm^2 V^{-1} s^{-1}.

Depletion mode 6H-SiC MOSFETs have been fabricated by Westinghouse [5], Cree Research [7] and General Electric [8]. FIGURE 7 shows the room temperature I-V characteristics of the device fabricated by the Cree Research group. The maximum transconductance at $V_g = 0$ V for devices with the gate length of 5 μm was 2.3 mS mm^{-1}. The effective field effect mobility of these devices was estimated to be 21.4 cm^2 V^{-1} s^{-1}. All mobility values were estimated using a standard charge control model [9]. Such low mobility values for SiC MOSFETs can only be explained by a poor quality SiC-SiO$_2$ interface, leading to a high trap density. For p-type MOS capacitors, the surface state density of the SiC-SiO$_2$ interface was estimated to be on the order of 7 x 10^{11} cm^{-2} eV^{-1} [10] and 5 x 10^{12} cm^{-2} eV^{-1} [11]. As Palmour et al [2]

9.4 SiC field effect transistors

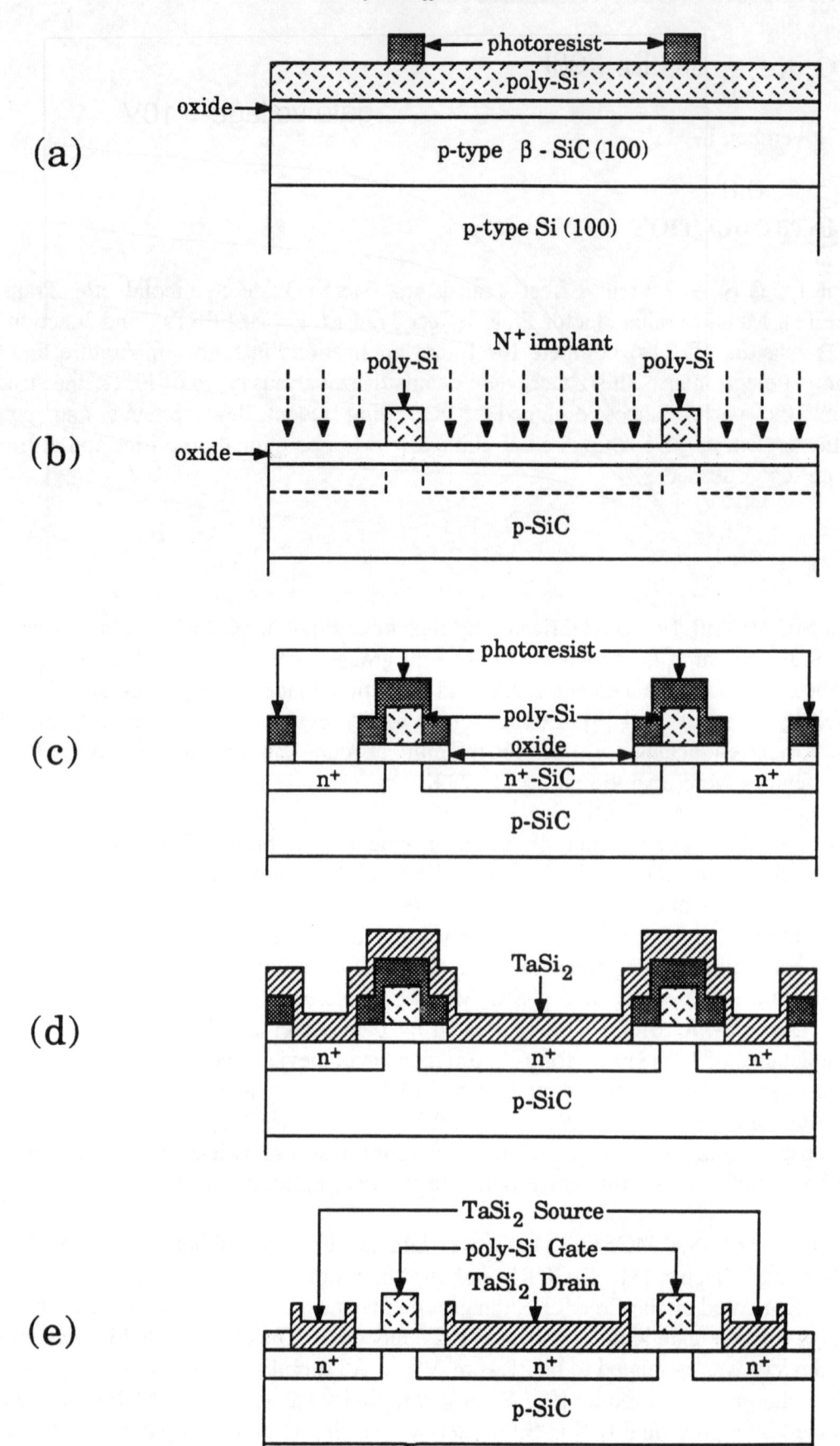

FIGURE 1 Fabrication steps for an enhancement mode β-SiC MOSFET [3].

9.4 *SiC field effect transistors*

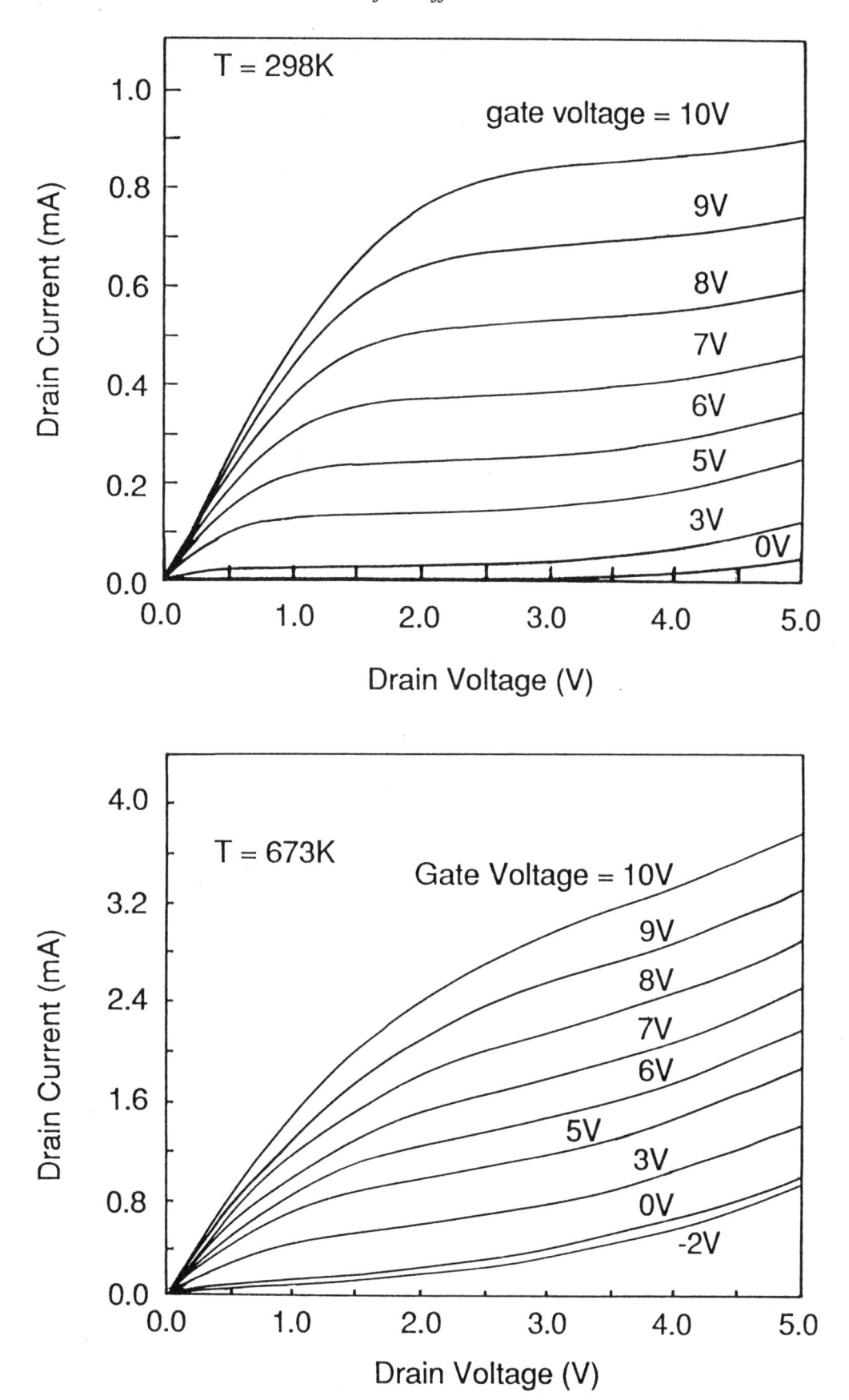

FIGURE 2 Current-voltage characteristics of enhancement mode β-SiC MOSFETs at room temperature and 673 K [3].

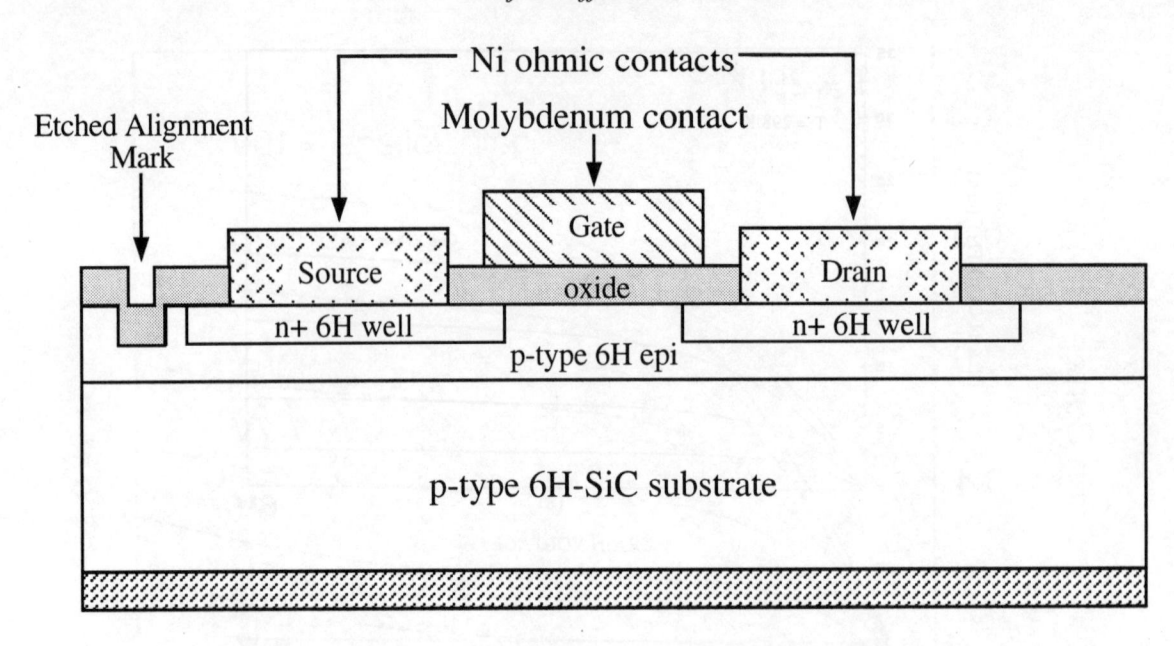

FIGURE 3 Cross-section of n-channel enhancement mode α-SiC MOSFET [6].

emphasized, an improved SiC-SiO$_2$ interface is a prerequisite for the improvement of SiC MOSFETs. If such an improvement is achieved, the performance of an optimized MOSFET at elevated temperatures is expected to be worse than at room temperature since the mobility should decrease at high temperatures [2].

The depletion mode 6H-SiC MOSFETs fabricated by the General Electric group [8] exhibited a fairly low leakage current of 5×10^{-11} A at 23 °C at -10 V gate bias and drain bias of 8 V with the on-current as high as 1.6 mA. However, the subthreshold was quite low. These devices operated up to 350 °C.

Recently, Cree Research Inc. reported on the first p-channel 6H-SiC MOSFET [6]. The device structure and output characteristics are shown in FIGURES 8 and 9, respectively. The device current and transconductance are very small (76 µA mm^{-1} at 40 V of drain bias and 16 µS mm^{-1}, respectively, for a 7 µm gate length device). The performance of this device was limited by a large parasitic series resistance. Nevertheless, even these preliminary results show the feasibility of SiC CMOS technology, capable of operating at elevated temperatures.

An extremely interesting application of the SiC technology is the development of power vertical n-channel α-SiC MOSFETs. In power Si MOSFETs, the drain layer must have a very low doping in order to ensure a large enough depletion width which can sustain a large voltage drop for power operation. This leads to a high device resistance in the linear region. The high breakdown voltage of SiC allows an increase in the doping level in the drain layer which leads to lowering of this resistance while maintaining high voltage and high power operation. For this application, SiC may be superior to Si even at room temperature. FIGURE 10 shows the output characteristics of the first power SiC MOSFET developed by Cree [6,12] which exhibited the maximum transconductance of 6.75 mS mm^{-1} and on-current of 32 mA mm^{-1} at the gate voltage of 12 V. The device threshold voltage was 3.7 V. The maximum drain-to-source voltage was 60 V. The specific on-resistance was 38 mΩ cm^2. The transistors withstood current densities of 190 A cm^{-2} (0.32 A cm^{-1} of gate periphery). These

9.4 SiC field effect transistors

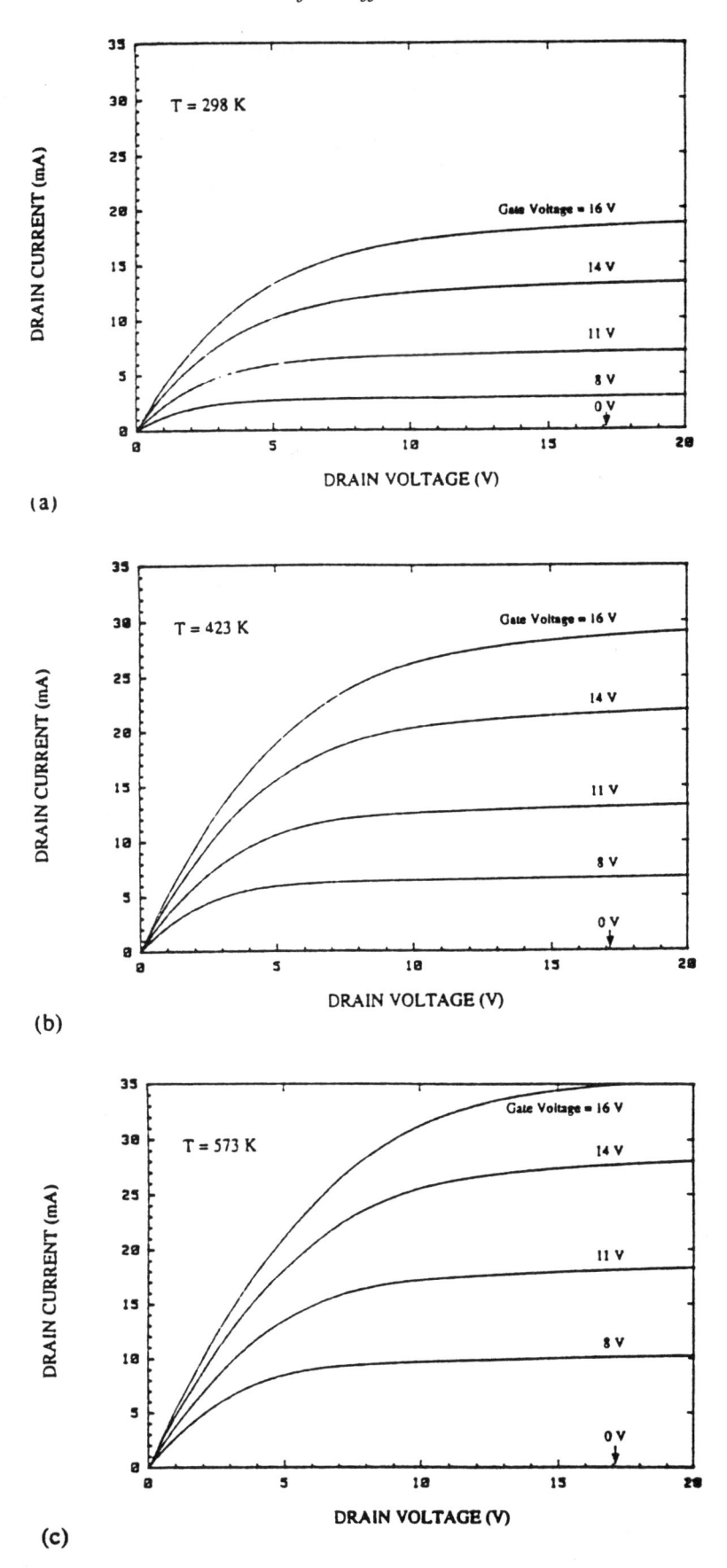

FIGURE 4 Current-voltage characteristics of n-channel enhancement α-SiC MOSFETs [6].

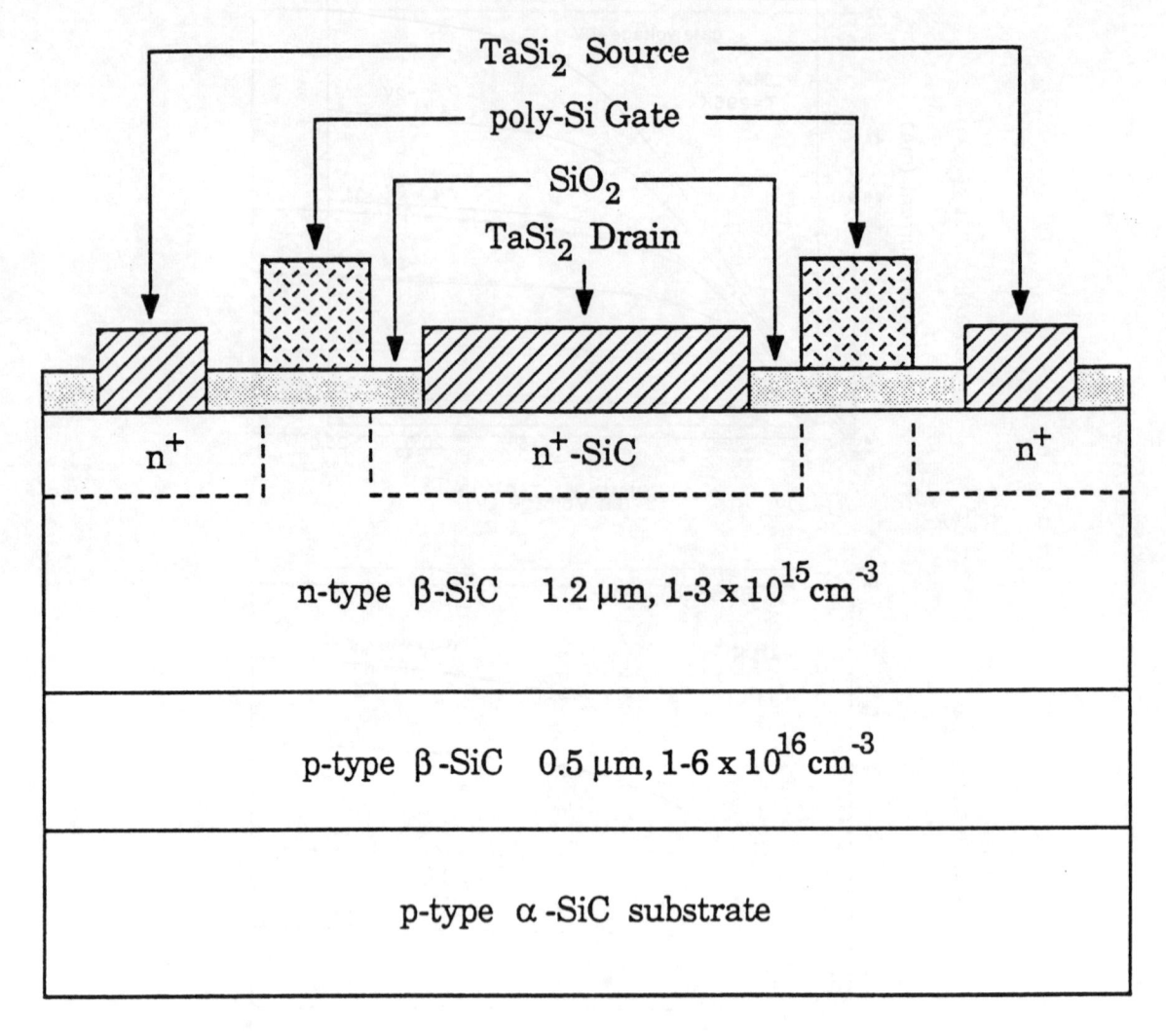

FIGURE 5 Cross-section of a depletion mode n-channel β-SiC MOSFET [3].

devices also operated at 300 °C with an increase of the on-state resistance of only 21 %. Vertical 4H-SiC MOSFETs have also been reported [12,13]. These devices exhibited a better performance than 6H devices. For the voltage of 180 V, the on-resistance was 17.5 mΩ cm^2, and the device withstood current densities of 550 A cm^{-2} (0.92 A cm^{-1} of gate periphery).

SiC MESFETs [2,5,14,15] and JFETs [2,16-21], have exhibited higher field effect mobilities than SiC MOSFETs (see TABLE 1). α-SiC MESFETs reported by Palmour et al [2] operated up to 773 K. The device structure is shown in FIGURE 11. The room temperature transconductance of a 6H-SiC MESFET with a 600 nm channel, 6.5×10^{16} cm^{-3} doping, and 24 µm gate length fabricated at Cree was approximately 4 mS mm^{-1}. At elevated temperatures, the device transconductance decreases due to the decrease in mobility. MESFETs did not exhibit a breakdown up to 100 V on the drain (see FIGURE 12). Using the square law MESFET model [9], we estimate that the field effect mobility in these MESFETs was approximately 300 cm^2 V^{-1}s^{-1}. This is nearly an order of magnitude higher than that for SiC MOSFETs. β-SiC MESFETs have also been fabricated [23] but, once again, α-SiC MESFETs exhibited a better performance that β-SiC MESFETs.

9.4 SiC field effect transistors

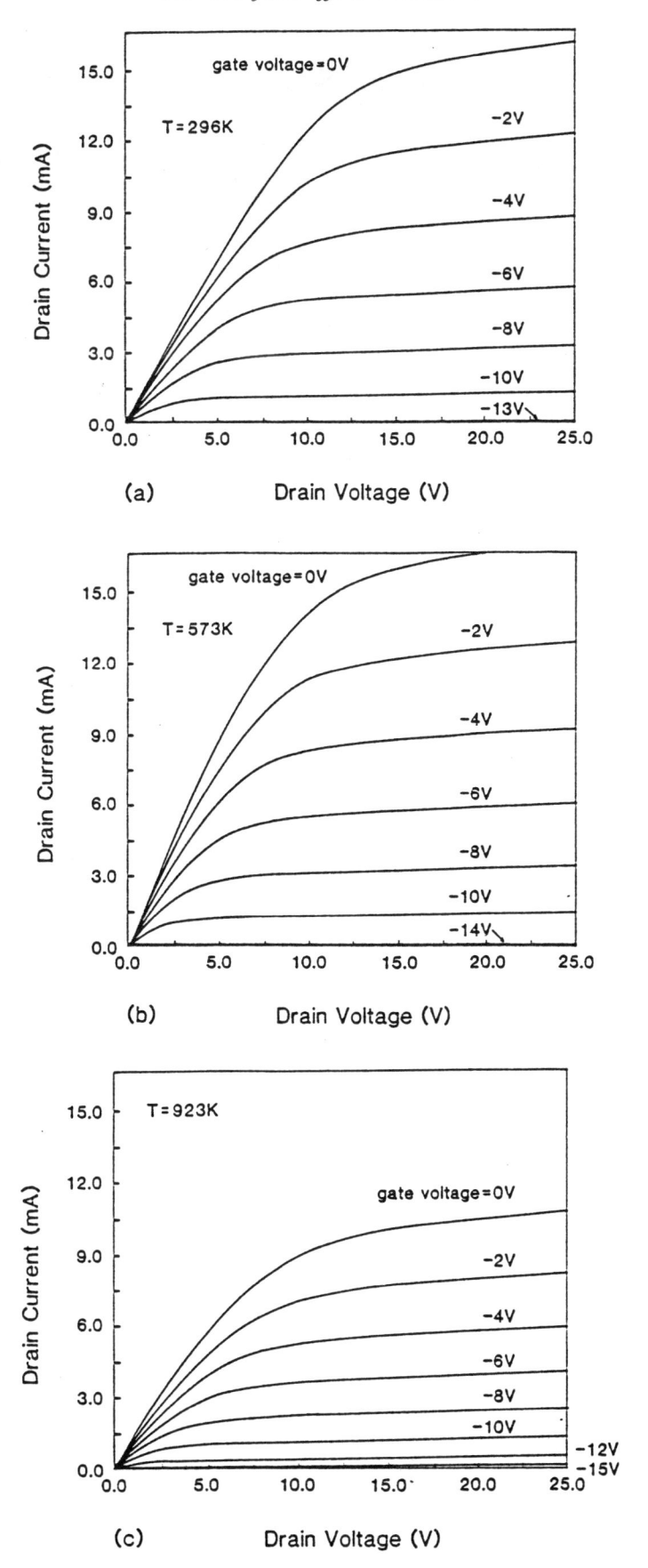

FIGURE 6 Current-voltage characteristics of a depletion mode n-channel β-SiC MOSFET [3].

9.4 SiC field effect transistors

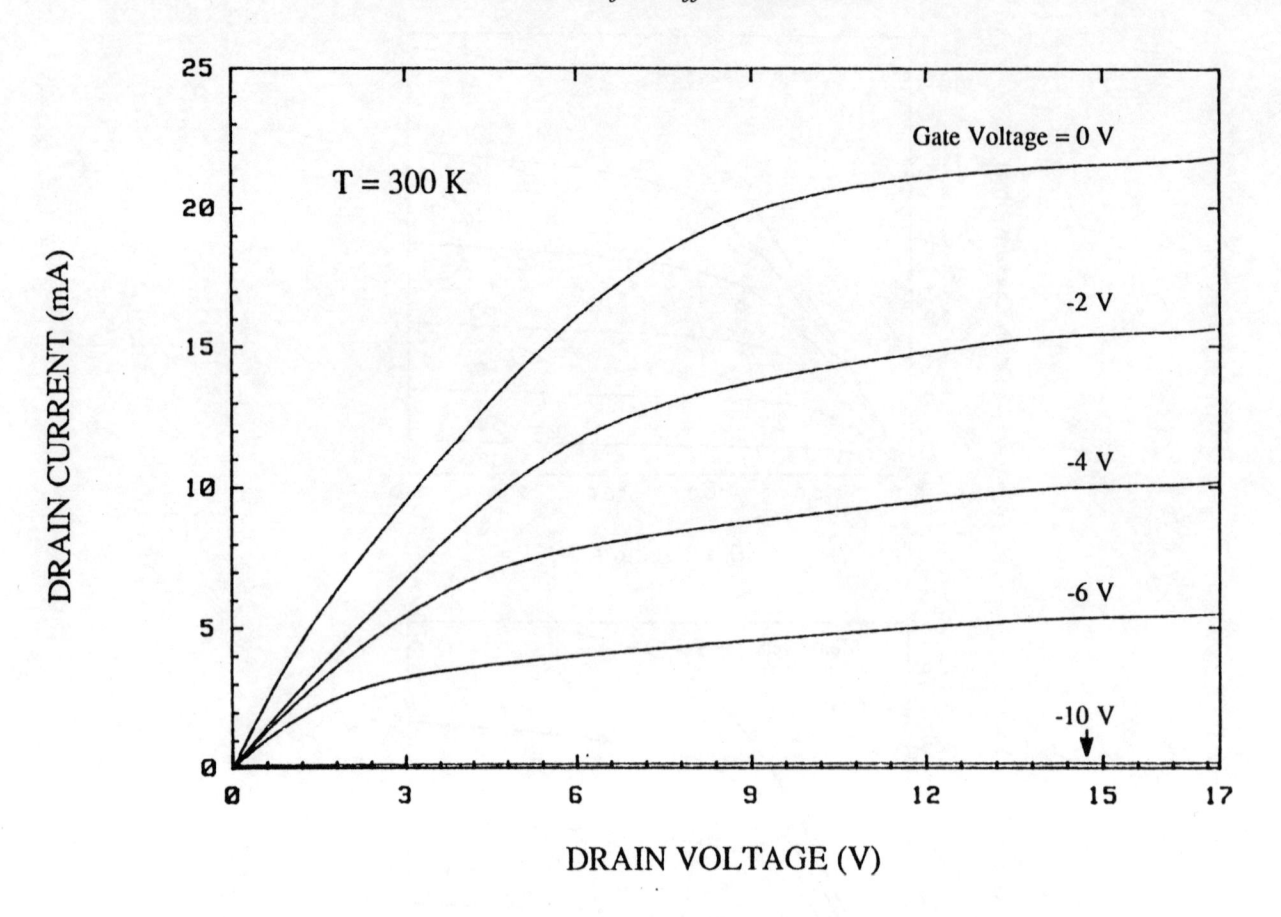

FIGURE 7 Current-voltage characteristics of a depletion mode n-channel α-SiC MOSFET [7].

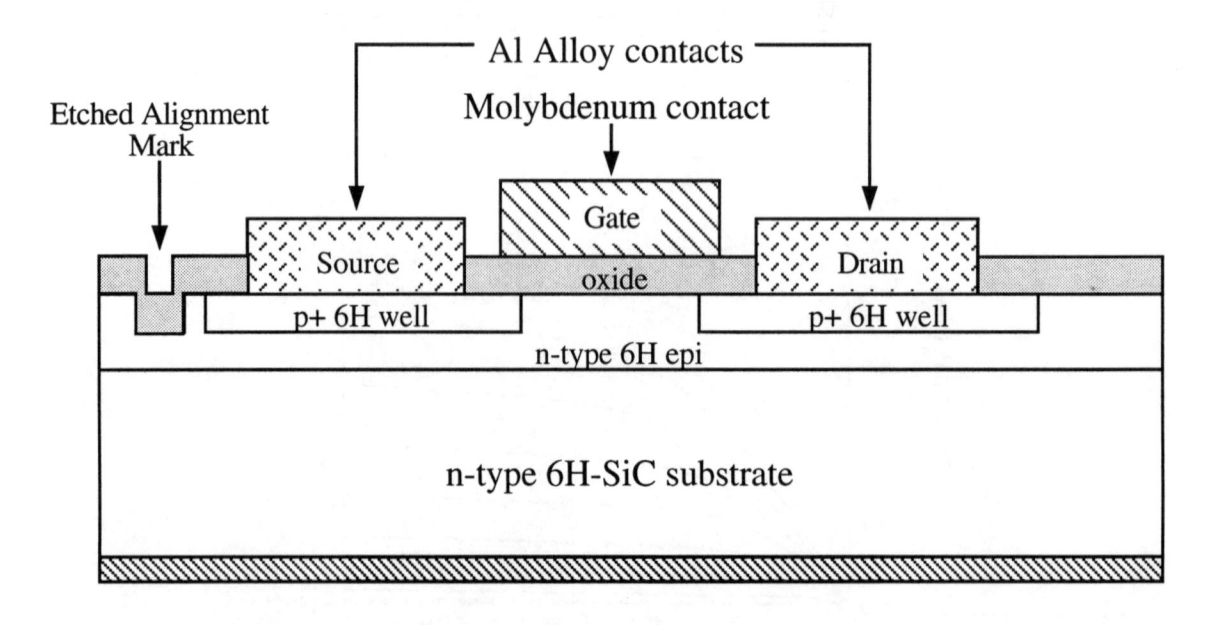

FIGURE 8 Cross-section of enhancement mode 6H-SiC p-channel MOSFET at room temperature [6].

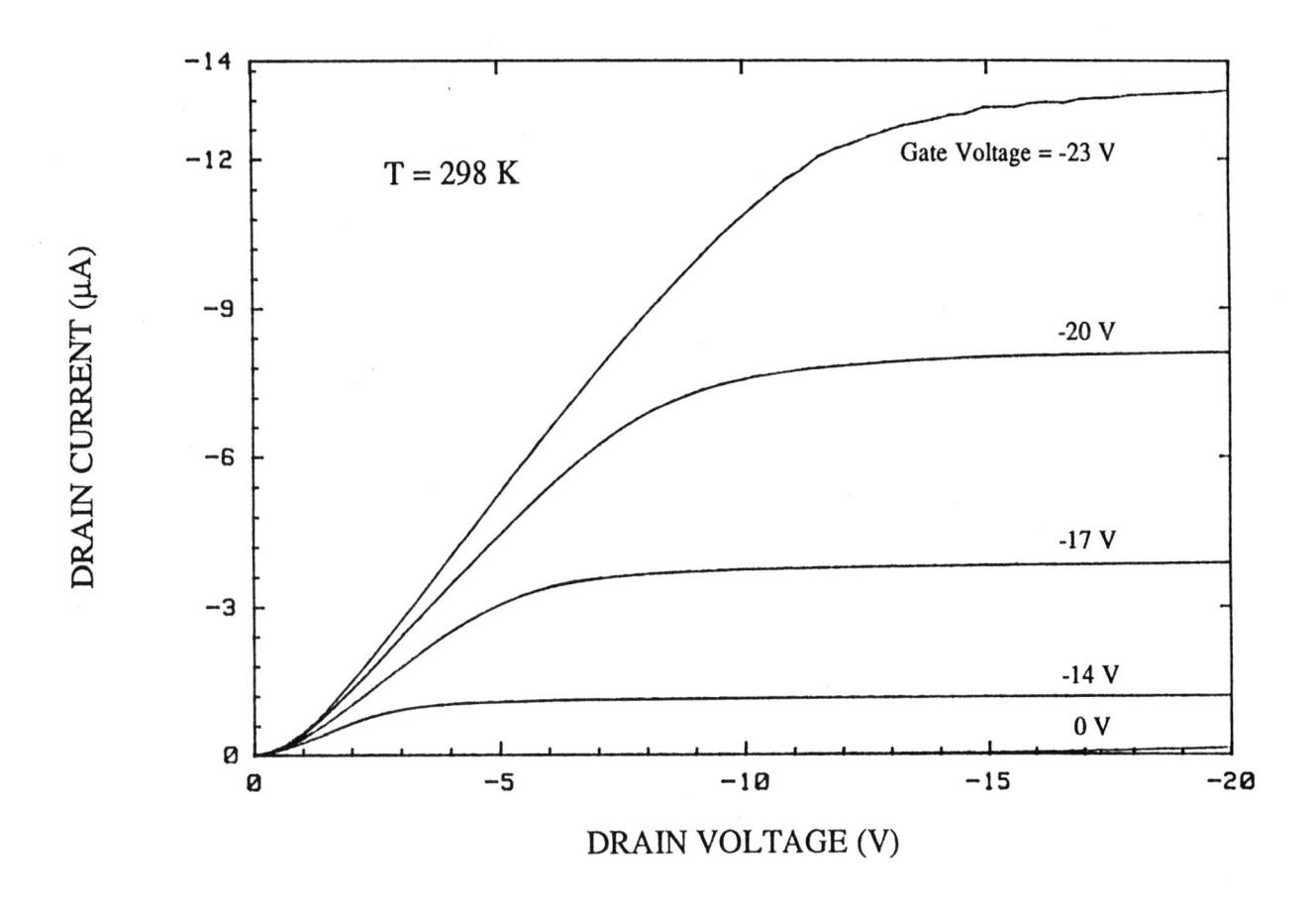

FIGURE 9 Current-voltage characteristics of enhancement mode 6H-SiC p-channel MOSFET at room temperature [6].

α-SiC MESFETs are the only SiC FETs which have achieved microwave operation [13,14,24-26]. A 6H-SiC MESFET with a 0.4 µm gate length demonstrated a cutoff frequency of 5 GHz, 12 dB gain at 2 GHz and breakdown voltage of 200 V (see [14]). This device had a transconductance of 30 mS mm^{-1}. The performance of such devices can be drastically improved if the ohmic contact resistance is reduced.

More recently, the Westinghouse group reported on 6H-SiC MESFETs with a 0.4 µm gate which achieved 2 W mm^{-1} output power at 1 GHz [25]. MESFETs with the output power of 1.8 W mm^{-1} with the gain of 9.7 dB at 0.5 GHz with high drain voltage of 50 V and the total device periphery of 1.9 mm have been reported by Sriram et al [26].

Palmour et al [13] have reported on 4H-SiC microwave MESFETs which exhibited up to 2.8 W mm^{-1} output power at 1.8 GHz for devices with a 0.7 µm gate length. The device transconductance at room temperature was 40 mS mm^{-1}.

9.4 SiC field effect transistors

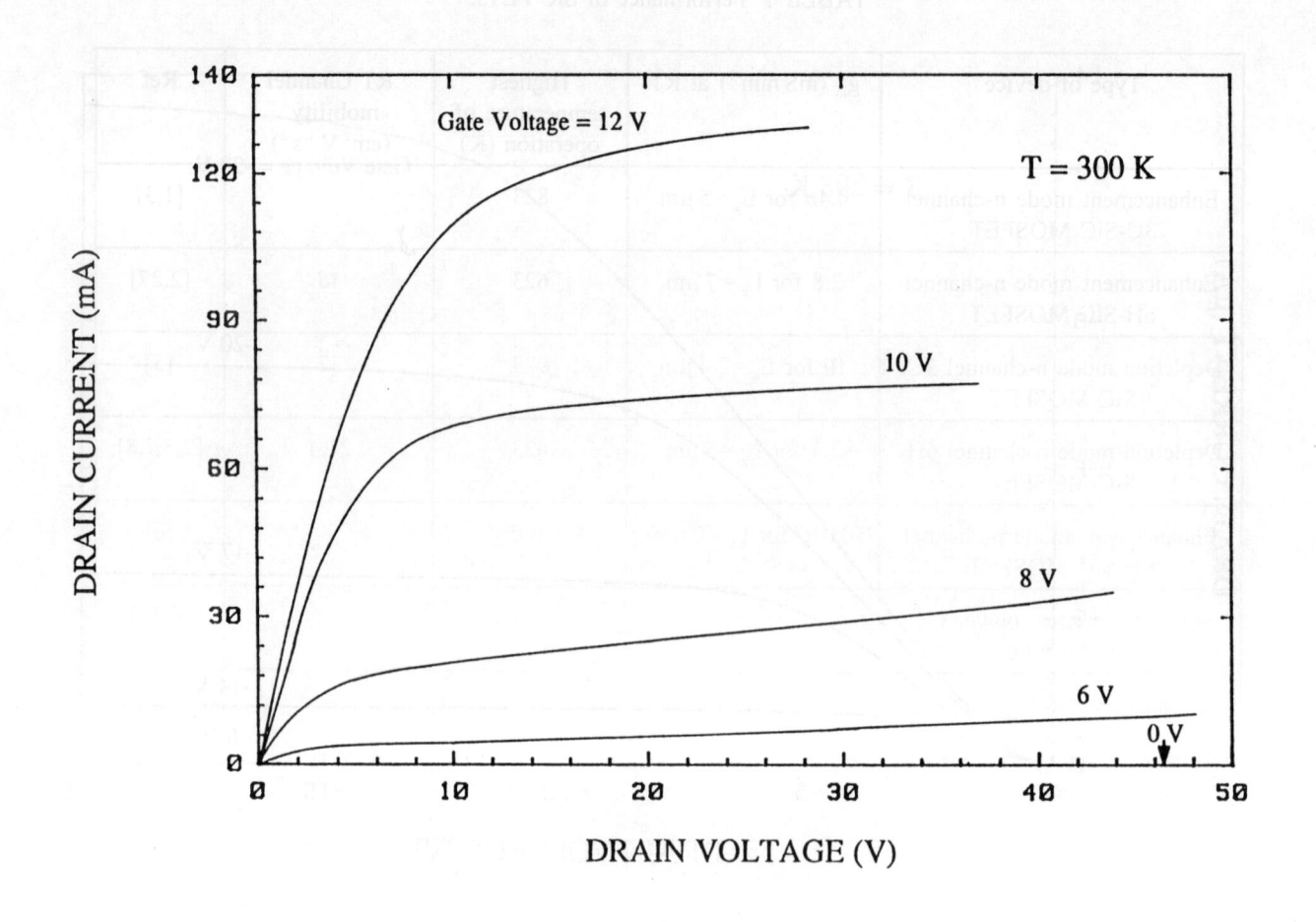

FIGURE 10 Current-voltage characteristics of 6H-SiC n-channel vertical power MOSFET at room temperature [6].

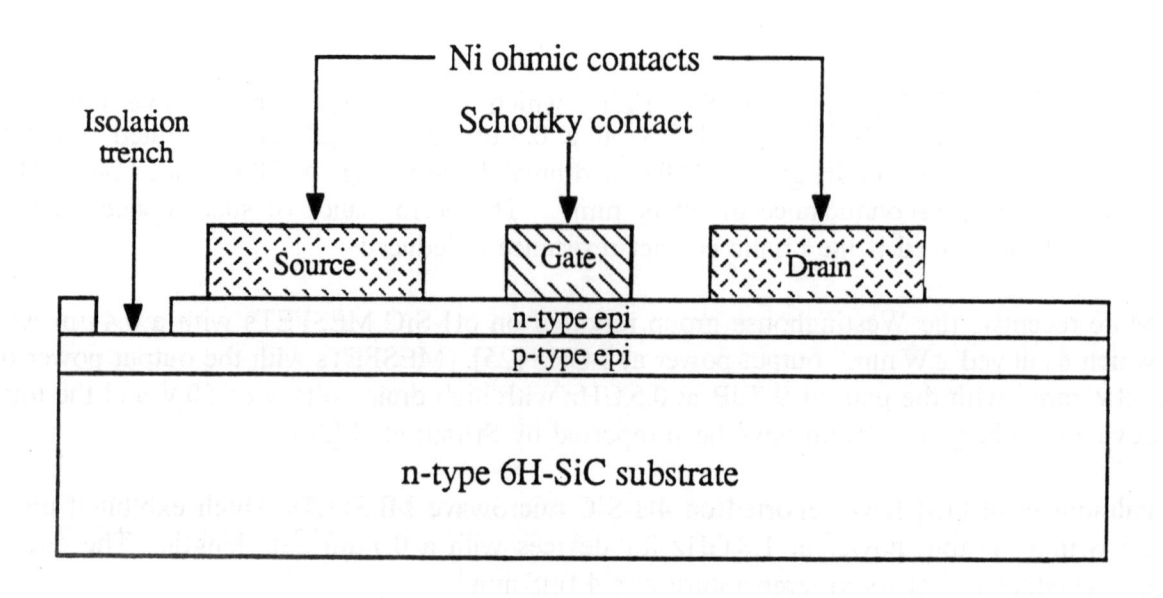

FIGURE 11 Cross-section of α-SiC MESFET [22].

9.4 SiC field effect transistors

TABLE 1 Performance of SiC FETs.

Type of device	g_m (mS mm^{-1}) at RT	Highest temperature of operation (K)	RT Channel mobility (cm^2 V^{-1} s^{-1})	Ref
Enhancement mode n-channel 3C-SiC MOSFET	0.46 for L_g = 5 µm	823		[1,3]
Enhancement mode n-channel 6H-SiC MOSFET	2.8 for L_g = 7 µm	623	46	[2,27]
Depletion mode n-channel 3C-SiC MOSFET	10 for L_g = 2.4 µm	923	37	[3]
Depletion mode n-channel 6H-SiC MOSFET	2.3 for L_g = 5 µm	623	21.4	[2,5,7,8]
Enhancement mode p-channel 6H-SiC MOSFET	0.016 for L_g = 7 µm	300		[6]
n-channel vertical power 6H-SiC MOSFET	6.75	623		[6,12]
n-channel vertical power 4H-SiC MOSFET	10	623		[12]
High temperature 6H-SiC MESFET	4.3 for L_g = 24 µm	623	300	[2,4]
High frequency 6H-SiC MESFET	30 for L_g = 0.4 µm	300		[13,14,23,24]
High frequency 4H-SiC MESFET	40 for L_g = 0.7 µm	300	500	[28,29]
Depletion mode β-SiC JFET	20 for L_g = 4 µm	300	560	[16]
Depletion mode 6H-SiC JFET	17-20 for L_g = 4 µm	800	250	[2,17,18]
Depletion mode 4H-SiC JFET	15 for L_g = 9 µm	700	340	[21]
Enhancement mode 6H-SiC JFET	0.5 for L_g = 10 µm 1.5 for L_g = 5 µm	750		[19,20]

Kelner et al [16] fabricated and evaluated buried gate JFETs in β-SiC grown on an α-SiC substrate. Devices with a 4 µm gate length had g_m = 20 mS mm^{-1}. The data obtained from this study were analyzed using a charge control model. Such an analysis showed that the value of the field-effect mobility ($\approx$ 560 cm^2 V^{-1}s^{-1}) was close to the measured value of the Hall mobility ($\approx$ 470 cm^2 V^{-1}s^{-1}) and that the electron saturation velocity in the channel was consistent with the theoretical value of 2 x 10^7 cm s^{-1}. Kelner et al [17], Anikin et al [18] and Palmour et al [2] reported on the high temperature operation of α-SiC buried gate JFETs (see FIGURE 13). Devices fabricated with a 4 µm gate length had a maximum g_m in the saturation region of 17 mS mm^{-1} (still a record for such a relatively long gate device) and a maximum drain saturation current of 450 mA mm^{-1} at zero gate voltage at room temperature. The transistors were completely pinched-off at a gate voltage of -40 V (FIGURE 14). The device

9.4 SiC field effect transistors

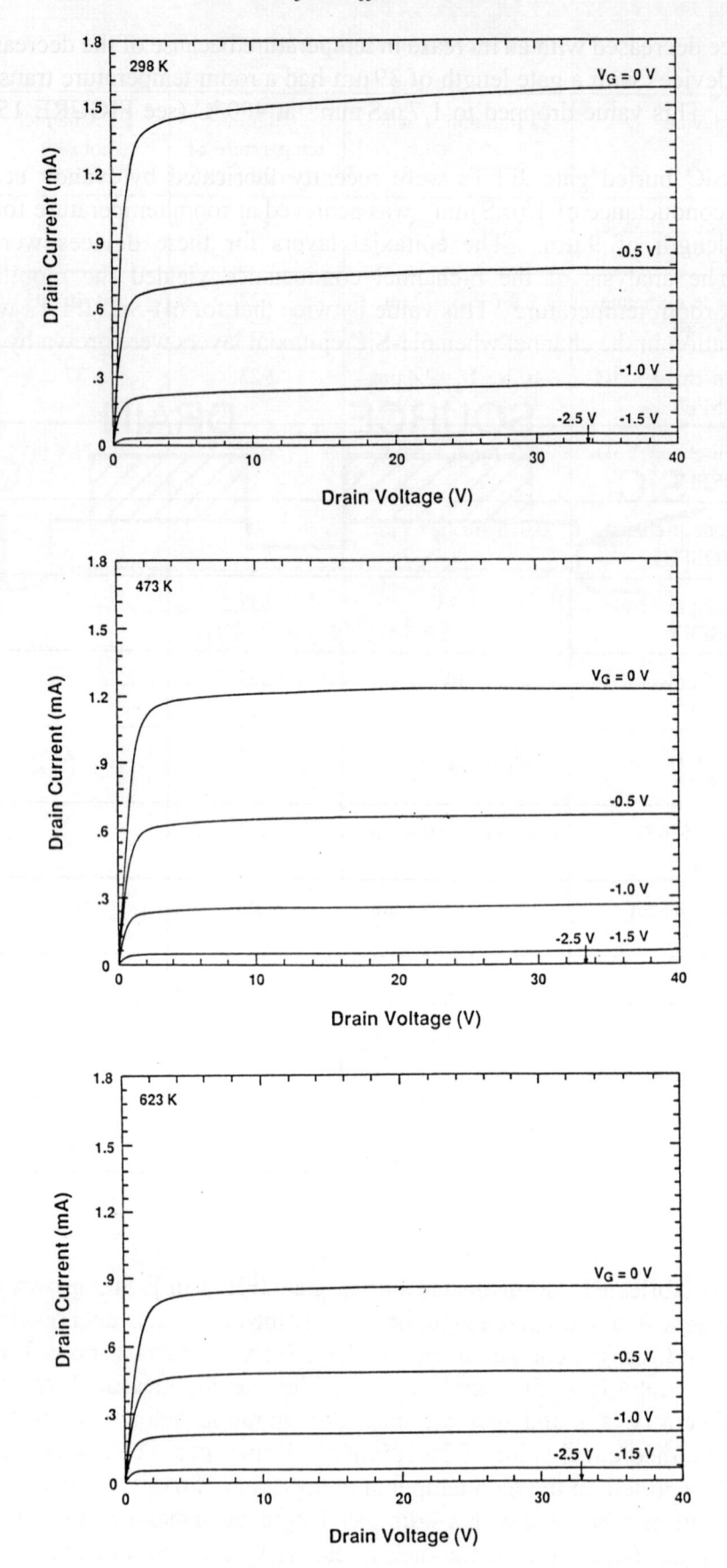

FIGURE 12 Current-voltage characteristics of α-SiC MESFETs at 298 K, 473 K and 623 K [22].

transconductance decreased with an increase in temperature because of the decreasing electron mobility. The devices with a gate length of 39 μm had a room temperature transconductance of 5.4 mS mm^{-1}. This value dropped to 1.7 mS mm^{-1} at 400 °C (see FIGURE 15).

n-Channel 4H-SiC buried gate JFETs were recently fabricated by Ivanov et al [21]. A maximum transconductance of 15 mS mm^{-1} was achieved at room temperature for the devices with the gate length of 9 μm. The epitaxial layers for these devices were grown by sublimation. The analysis of the n-channel conductance yielded the mobility value of 340 cm^2 V^{-1}s^{-1} at room temperature. This value is twice that for 6H-SiC JFETs with the same carrier concentration in the channel when 6H-SiC epitaxial layers were grown by sublimation.

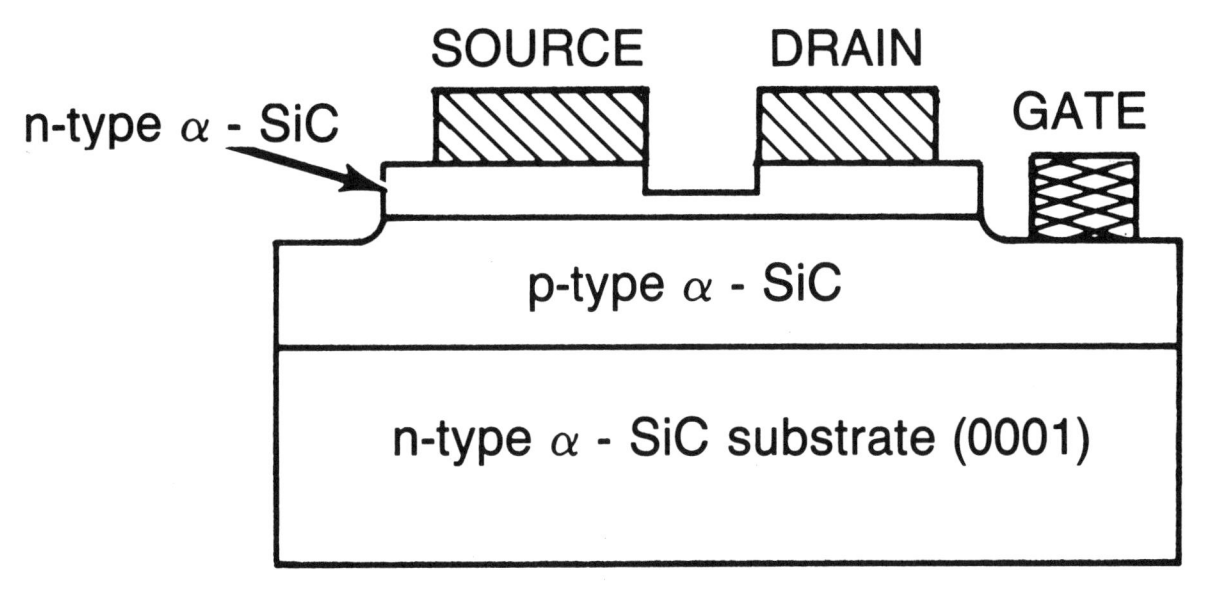

FIGURE 13 Cross-section of α-SiC buried gate JFET [14].

The 4H-SiC buried gate JFETs were operational up to 700 K.

Dmitriev et al [19] and Rupp et al [20] reported on the enhancement mode JFETs. JFETs described in [20] had the saturation current of approximately 1.2 mA mm^{-1} at the gate bias of 2.4 V and transconductance of 1.5 mS mm^{-1}. The devices operated up to 450 °C.

Silicon carbide has a relatively low mobility but is expected to have a relatively high electron saturation drift velocity. The performance of a field effect transistor made from such a material can be substantially improved when the electric field distribution along the channel (in the direction from the source to the drain) is made more uniform. Recently, a new device structure aimed at the generation of a more uniform field along the channel was proposed [30-32]. In this device, the gate voltage swing varies along the channel, with the highest swing near the drain. A possible JFET implementation of this concept is shown in FIGURE 16.

Practical applications of SiC MOSFETs hinge on a substantial improvement in the quality of the SiC-SiO$_2$ interface. α-SiC MESFETs are especially promising for microwave and high-power applications. These devices have reached a cut-off frequency of 10 GHz [12,13,25,33]. SiC buried gate JFETs have the potential for high-power and high-temperature

9.4 SiC field effect transistors

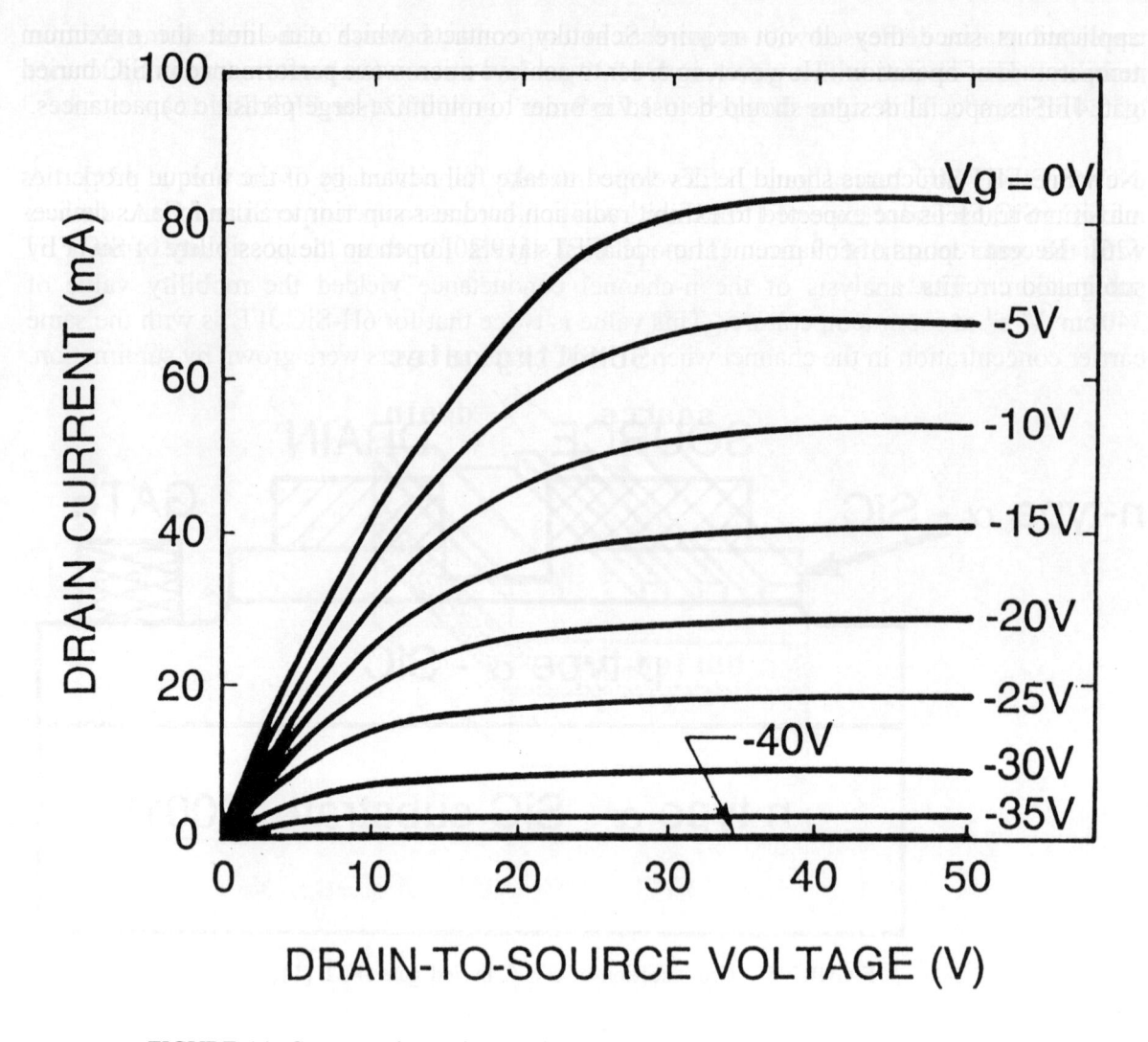

FIGURE 14 Current-voltage characteristics of α-SiC JFETs at room temperature [16].

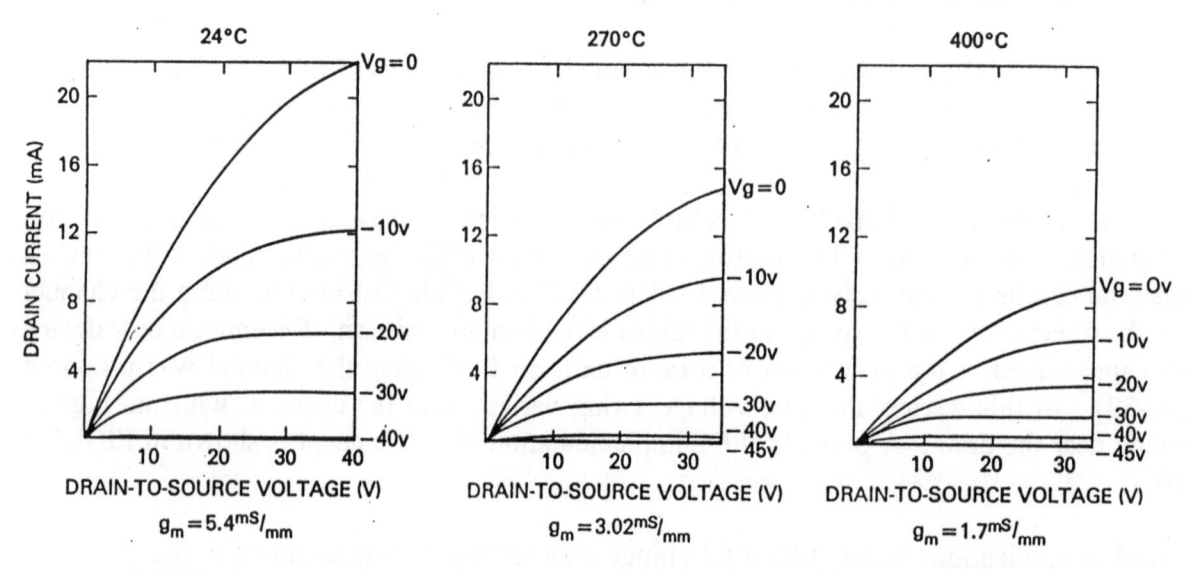

FIGURE 15 Current-voltage characteristics of α-SiC JFETs at different temperatures [16].

9.4 SiC field effect transistors

applications since they do not require Schottky contacts which can limit the maximum temperature of operation. However, in order to achieve microwave performance in SiC buried gate JFETs, special designs should be used in order to minimize large parasitic capacitances.

New SiC FET structures should be developed to take full advantage of the unique properties of SiC. SiC FETs are expected to exhibit radiation hardness superior to Si and GaAs devices [26]. Recent reports of enhancement mode JFETs [19,20] open up the possibility of SiC FET integrated circuits.

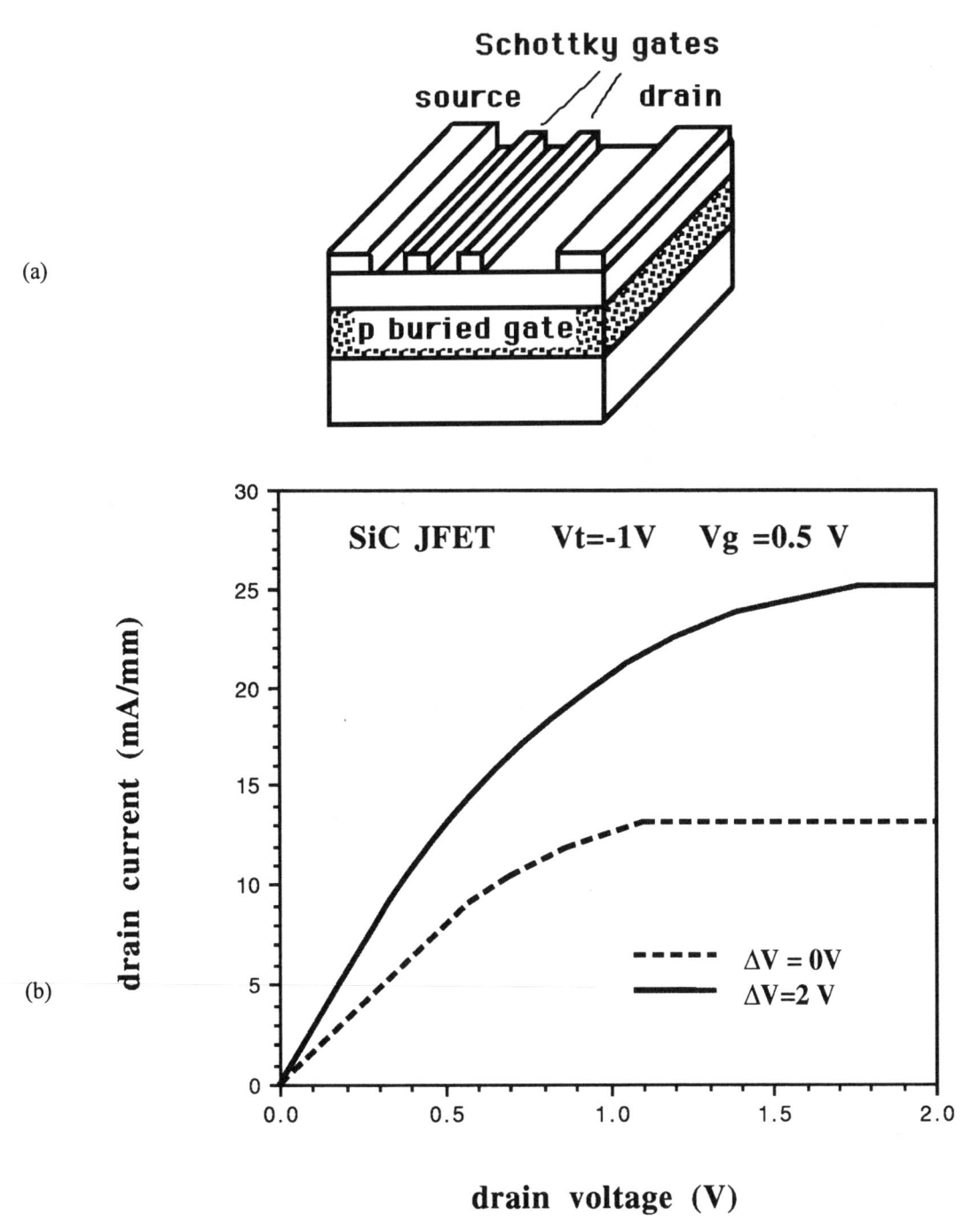

FIGURE 16 Device structure (a) and computed characteristics (b) of α-SiC JFET with several gates for improved field uniformity in the channel [32].

9.4 SiC field effect transistors

The first hybrid 6H-SiC operational amplifier was recently demonstrated by Auburn University in cooperation with Cree [34] (see FIGURE 17). This circuit operated up to temperatures of 365°C.

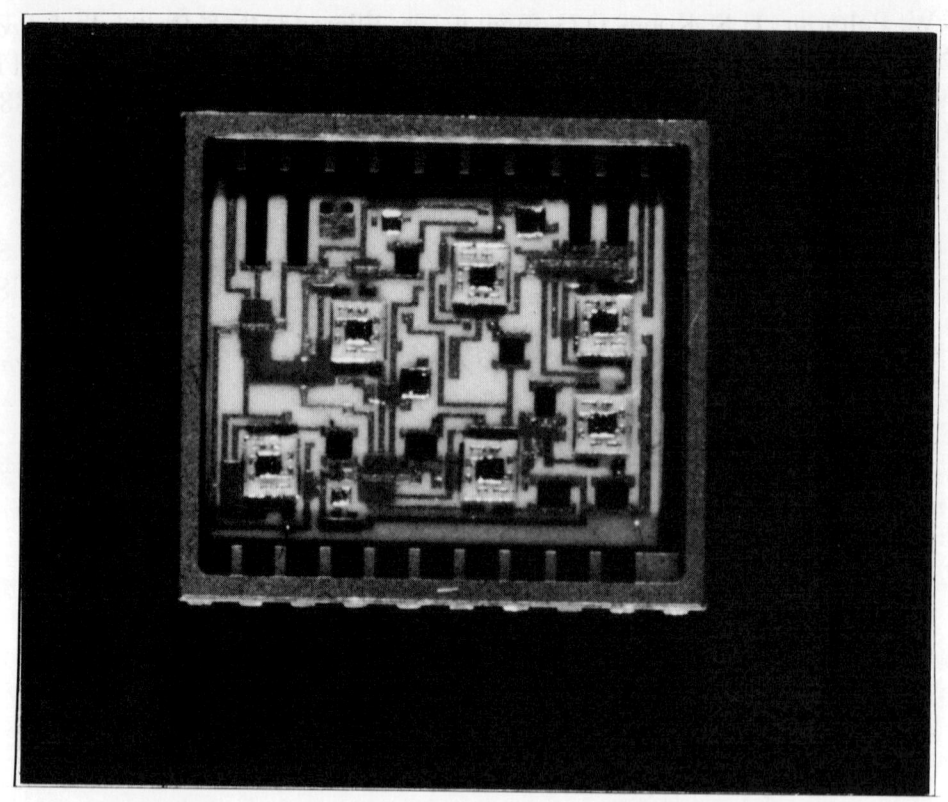

FIGURE 17 Photograph of hybrid SiC operational amplifier [34].

A 3C-SiC MESFET type inverter circuit was recently fabricated at Howard University by Harris et al [35] (see FIGURE 18). This inverter was operational at room temperature with a maximum DC gain of 2.5 and a unity gain at a frequency of 0.143 MHz.

C CONCLUSION

MOSFETs based on SiC require an improvement in the SiC-SiO$_2$ interface to achieve optimum performance. SiC MESFETs and JFETs produce higher field effect mobilities than MOSFETs, with α-SiC giving better device performance than β-SiC. α-SiC MESFETs have also been fabricated which perform in the microwave region. Various additional device concepts/applications are currently being actively researched. Once the problem related to micropipes and 6H-SiC device yield is improved accordingly, power SiC FETs will find many applications.

REFERENCES

[1] A. Suzuki, H. Ashida, N. Furui, K. Mameno, H. Matsunami [*Jpn. J. Appl. Phys. (Japan)* vol.21 (1982) p.579-81]

[2] J.W. Palmour, H.S. Kong, D.G. Waltz, J.A. Edmond, C.H. Carter Jr. [*Proc. 1st Int. High Temperature Electronics Conf.*, Albuquerque, NM, USA, 1991, p.511-18]

[3] J.W. Palmour, H.S. Kong, R.F. Davis [*J. Appl. Phys. (USA)* vol.64 (1988) p.2168-77]

[4] J.W. Palmour, J.A. Edmond [*Proc. 14th IEEE Cornell Conf.*, Ithaca, NY, USA, 1991, p.16-25]

[5] D.L. Barrett, P.G. McMullin, R.G. Seidensticker, R.H. Hopkins, J.A. Spitznagel [*Proc. 1st Int. High Temperature Electronics Conf.*, Albuquerque, NM, USA, 1991, p.180-5]

[6] J.W. Palmour, J.A. Edmond, H.S. Kong, C.H. Carter Jr. [*Abstracts of Workshop on SiC Materials and Devices*, University of Virginia, USA, 1992, p.32]

[7] J. Palmour [private communication, 1992]

[8] V. Krishnamurthy et al [*Proc. 5th Int. Conf. SiC and Related Materials*, Washington, DC, USA, 1 - 3 Nov. 1993, Eds M.G. Spencer et al (IOP Publishing, Bristol, UK, 1994) p.483-6]

[9] M. Shur [*Physics of Semiconductor Devices* (Prentice Hall, New York, 1990)]

[10] J.J. Kopanski [NIST, Maryland, (1990) quoted in [2]]

[11] D.M. Brown et al [*Govern. Microcircuit Appl. Conf.*, Orlando, Fl, USA, 1991]

[12] J.W. Palmour, J.A Edmond, H.S Kong, C.H. Carter [*Proc. 5th Int. Conf. SiC and Related Materials*, Washington, DC, USA, 1 - 3 Nov. 1993, Eds M.G. Spencer et al (IOP Publishing, Bristol, UK, 1994) p.499-502]

[13] J. Palmour [*Proc. ISDRS-93* Charlottesville, VA, USA, 1993, p.695-6]

[14] R.C. Clarke, T.W. O'Keefe, P.G. McMullin, T.J. Smith, S. Sriram, D.L. Barrett [*50th Annual Device Research Conf.*, Cambridge, MA, USA, 22 - 24 June 1992, Abstract VA-5]

[15] S. Yoshida et al [*Mater. Res. Soc. Symp. Proc. (USA)* vol.97 (1987) p.259-64]

[16] G. Kelner, M. Shur, S. Binari, K. Sleger, H.S. Kong [*IEEE Trans. Electron Devices (USA)* vol.36 no.6 (1989) p.1045-9]

[17] G. Kelner, S. Binari, M. Shur, K. Sleger, J. Palmour, H. Kong [*Electron. Lett. (UK)* vol.27, no.12 (1991) p.1038-40]

[18] M.M. Anikin, P.A. Ivanov, A.L. Syrkin, B.V.Tsarenkov, V.E. Chelnokov [*Sov. Tech. Phys. Lett. (USA)* vol.15, no.8 (1989) p.636-7]

[19] V.A. Dmitriev, P.A. Ivanov, V.E. Chelnokov, A.E. Cherenkov [*Proc. 1st Int. Conf. Applications of Diamond Films and Related Materials*, Auburn, Alabama, USA, 1991, p.769-74]

[20] R. Rupp. K. Dohnke, J. Volkl, D. Stephani [*Proc. 5th Int. Conf. SiC and Related Materials*, Washington, DC, USA, 1 - 3 Nov. 1993, Eds M.G. Spencer et al (IOP Publishing, Bristol, UK, 1994) p.625-7]

[21] P.A. Ivanov, N.S. Savkina, V.N. Panteleev, V.E. Chelnokov [*Proc. 5th Int. Conf. SiC and Related Materials*, Washington, DC, USA, 1 - 3 Nov. 1993, Eds M.G. Spencer et al (IOP Publishing, Bristol, UK, 1994) p.593-5]

[22] R.F. Davis, G. Kelner, M. Shur, J.W. Palmour, J.A. Edmond [*Proc. IEEE (USA)* vol.79, no.5 (1991) p.677-701]

[23] G. Kelner, S. Binari, K. Sleger, H. Kong [*IEEE Electron Device Lett. (USA)* vol.8 (1987) p.428-30]
[24] J.W. Palmour, J.A. Edmond, H.S. Kong, C.H. Carter Jr. [*Physica B (Netherlands)* vol.185 (1993) p.461-5]
[25] M.C. Driver et al [*GaAs IC 5th Annual Symp. Tech. Digest* IEEE (1993) p.19-24]
[26] S. Sriram et al [*Proc. 5th Int. Conf. SiC and Related Materials*, Washington, DC, USA, 1 - 3 Nov. 1993, Eds M.G. Spencer et al (IOP Publishing, Bristol, UK, 1994) p.491-4]
[27] J. Palmour, J.A. Edmond, H.S. Kong, C.H. Carter Jr. [*Proc. ICACSC 1991*, Santa Clara, CA, USA, 9 - 11 Oct. 1991, Eds C.Y. Yang, M.M. Rahman, G.L. Harris (Springer-Verlag, Berlin, Germany, 1992) p.289-97]
[28] J. Palmour, C.H. Carter Jr. [*Proc. ISDRS-93* Charlottesville, VA, USA, 1993, p.695-6]
[29] J.W. Palmour, C.E. Weitzel, K. Nordquist, C.H. Carter Jr. [*Proc. 5th Int. Conf. SiC and Related Materials*, Washington, DC, USA, 1 - 3 Nov. 1993, Eds M.G. Spencer et al (IOP Publishing, Bristol, UK, 1994) p.495-8]
[30] M. Shur [*Appl. Phys. Lett. (USA)* vol.54, no.2 (1989) p.162-4]
[31] M. Shur [*Proc. Advanced Material Concepts Conf.*, Denver, Co., USA, Ed. F.W. Smith (Advanced Materials Institute, Colorado, 1989) p.432-40]
[32] G. Kelner, M. Shur [patent no. 5,309,007, 3 May 1994]
[33] R.J. Trew, J.B. Yan, P.M. Mock [*Proc. IEEE (USA)* vol.79, no.5 (1991) p.598-620]
[34] M. Tamana, R.W. Johnson, R.C. Jaeger, J.W. Palmour [*Proc. 42nd Electron. Components and Tech. Conf.*, San Diego, CA, (USA) (1992) p.157-61]
[35] G.L. Harris, K. Wongchotigul, K. Irvine, M.G. Spencer [*Proc. 3rd Int. Conf. on Solid State and Integrated Circuit Technology*, China, (1992)]

9.5 SiC bipolar junction transistors and thyristors

G. Kelner and M. Shur

November 1993

A INTRODUCTION

This Datareview describes recent progress in bipolar junction transistors, thyristors and random access memories made in SiC. The first two device types have been made in both cubic and hexagonal SiC polytypes. The first random access memory made in hexagonal SiC will be described.

B BIPOLAR TRANSISTORS

The highest current gain reported for a 6H-SiC bipolar junction transistor (BJT) is on the order of ten [1]. The cross-section of a BJT fabricated by Cree is shown in FIGURE 1. The devices operated at collector voltages of up to 200 V and exhibited common-emitter current gains as high as 10.4 with the collector leakage current of 350 nA at room temperature and 30 µA at 400°C (see FIGURE 2). The performance of the devices fabricated to date has been limited by recombination losses in the base. Any further attempts to decrease these losses by reducing the base width may lead to unacceptably high base resistances. A promising alternative to a conventional SiC BJT is a heterojunction bipolar transistor with a wider bandgap emitter. This will make it possible to have a high emitter injection efficiency and a low base resistance. Dmitriev et al [3] reported preliminary results on the liquid phase epitaxial growth of a wider bandgap 6H-SiC on 3C-SiC. Further development of the heterojunction epitaxy may lead to more progress in this area. Recently, Yoder [4] proposed the use of an induced base HBT structure. This idea may allow one to alleviate difficulties related to the relatively poor quality of a highly doped SiC heterojunction.

C THYRISTORS

Vainshtein et al [5] and Dmitriev et al [6] described the first SiC dynistor diode. This diode is an n-p-n-p thyristor-like structure but without gate contacts. For this device, an n-p$^+$-n$^+$ structure was epitaxially grown on a p$^+$ 6H-SiC substrate using container-free liquid phase epitaxy in a continuous epitaxial process. The contacts were made from Al. The turn-on time constant varied from 1 ns to 10 ns. The recovery time varied from 150 to 200 ns. The static switching voltage varied between 10 V and 50 V with an inverse voltage of up to 90 V. Recently, Edmond et al [7] have fabricated the first α-SiC thyristor which operated up to 350°C.

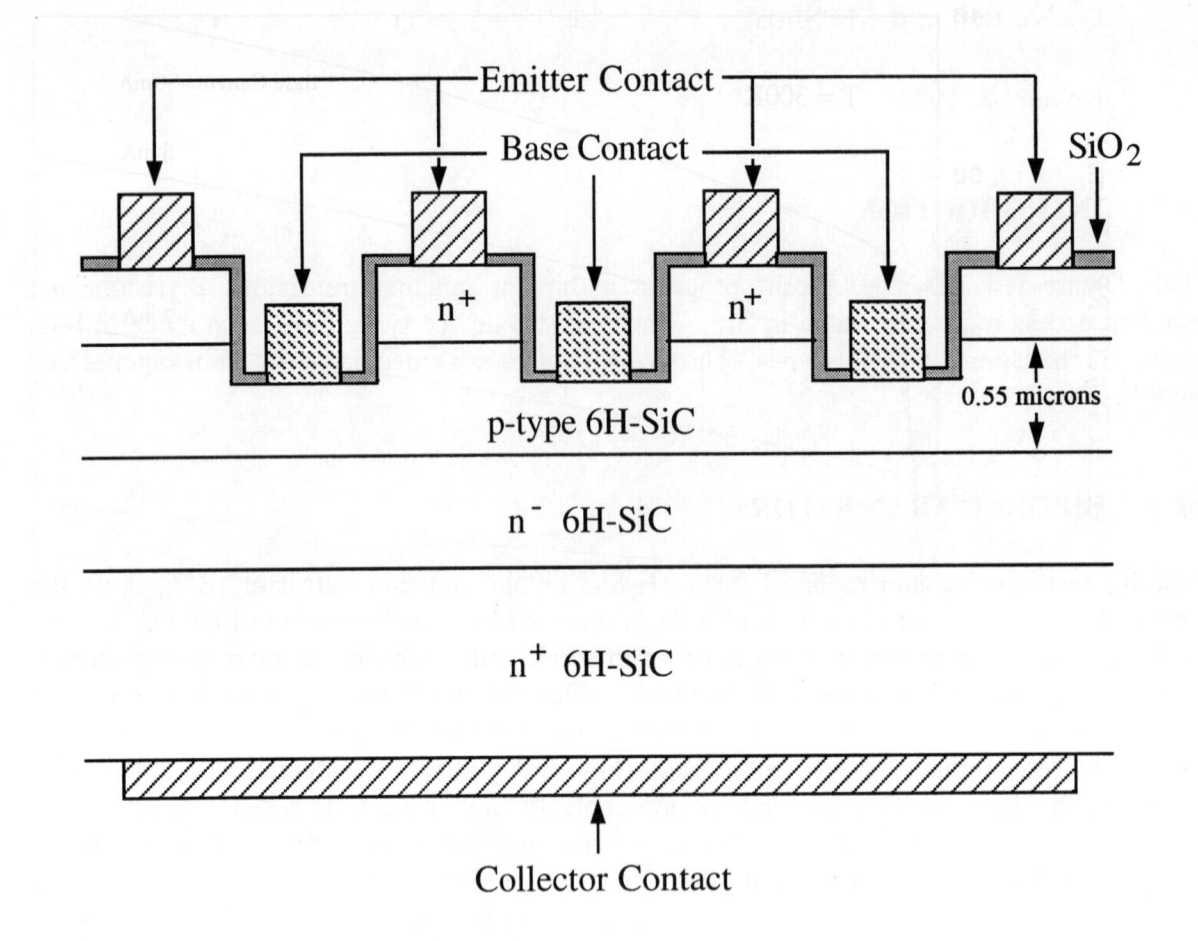

FIGURE 1 Cross-section of SiC BJT [1].

D   RANDOM ACCESS MEMORIES

FIGURE 3 illustrates a random access memory (RAM) structure proposed and demonstrated by researchers from Purdue University and Cree [8,9]. This RAM utilizes either p-n-i-p [9] (FIGURE 3) or n-i-p-i-n [8] α-SiC structures. The charge is stored in the central region of this device (which is floating) by reverse biasing this layer with respect to the surrounding regions. The charge recovery occurs due to the thermal generation in the depletion region. The activation energy for this charge recovery is approximately half of the energy gap (1.45 eV for the studied structures). In the fabricated devices, the room temperature recovery time was too long to measure. The recovery time at 300 °C was longer than 10 min. The project is now under way to demonstrate the feasibility of a high speed non-volatile RAM concept by fabricating a 1 kbit SiC NVRAM within 3 years.

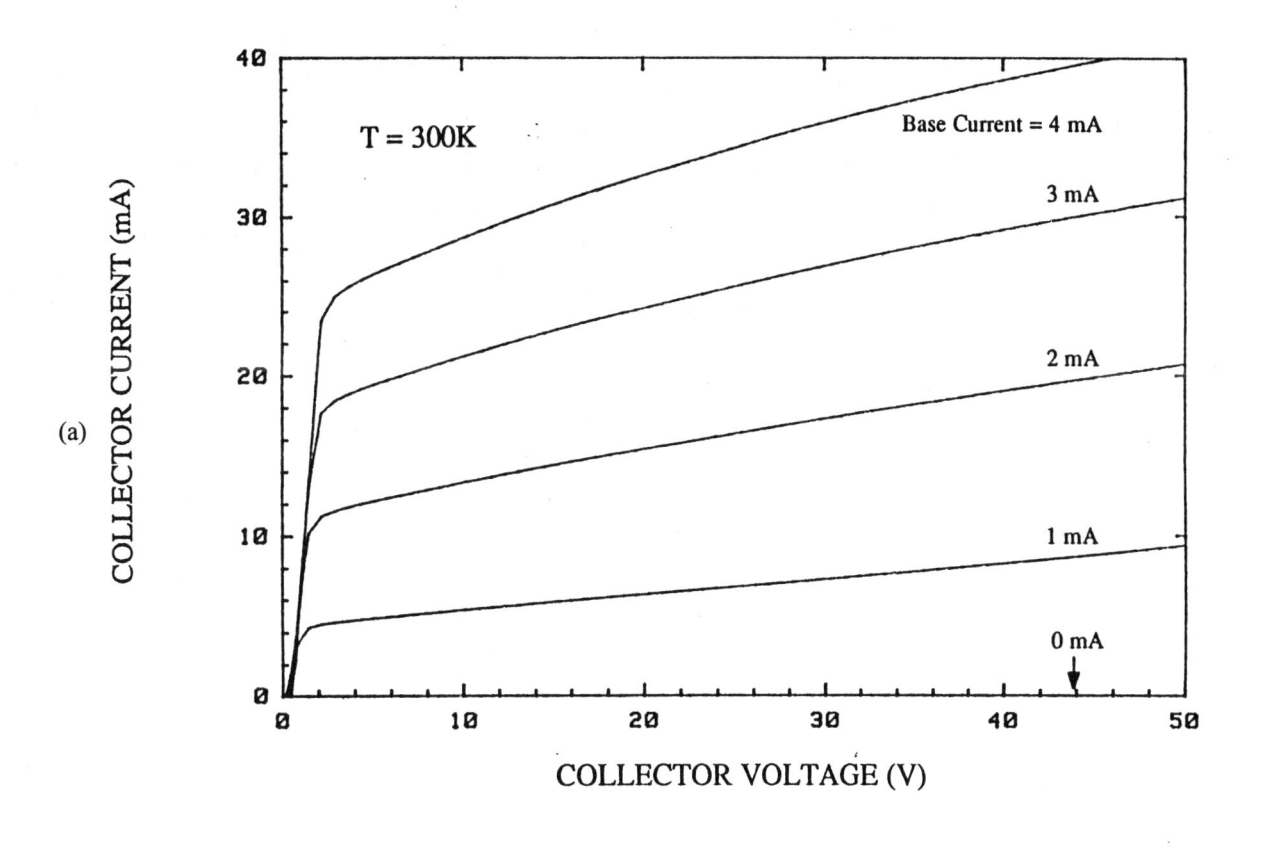

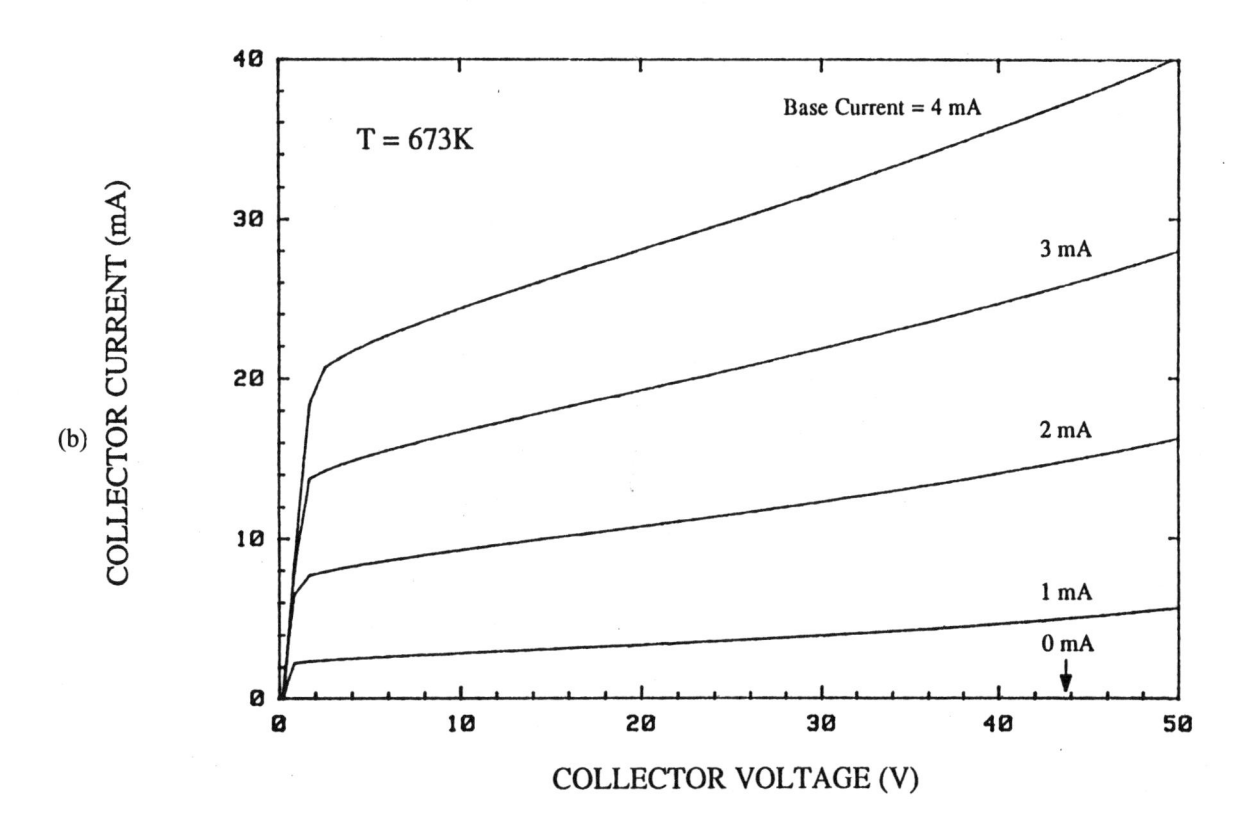

9.5 SiC bipolar junction transistors and thyristors

FIGURE 2 Common-emitter current-voltage characteristics of SiC BJT [2].

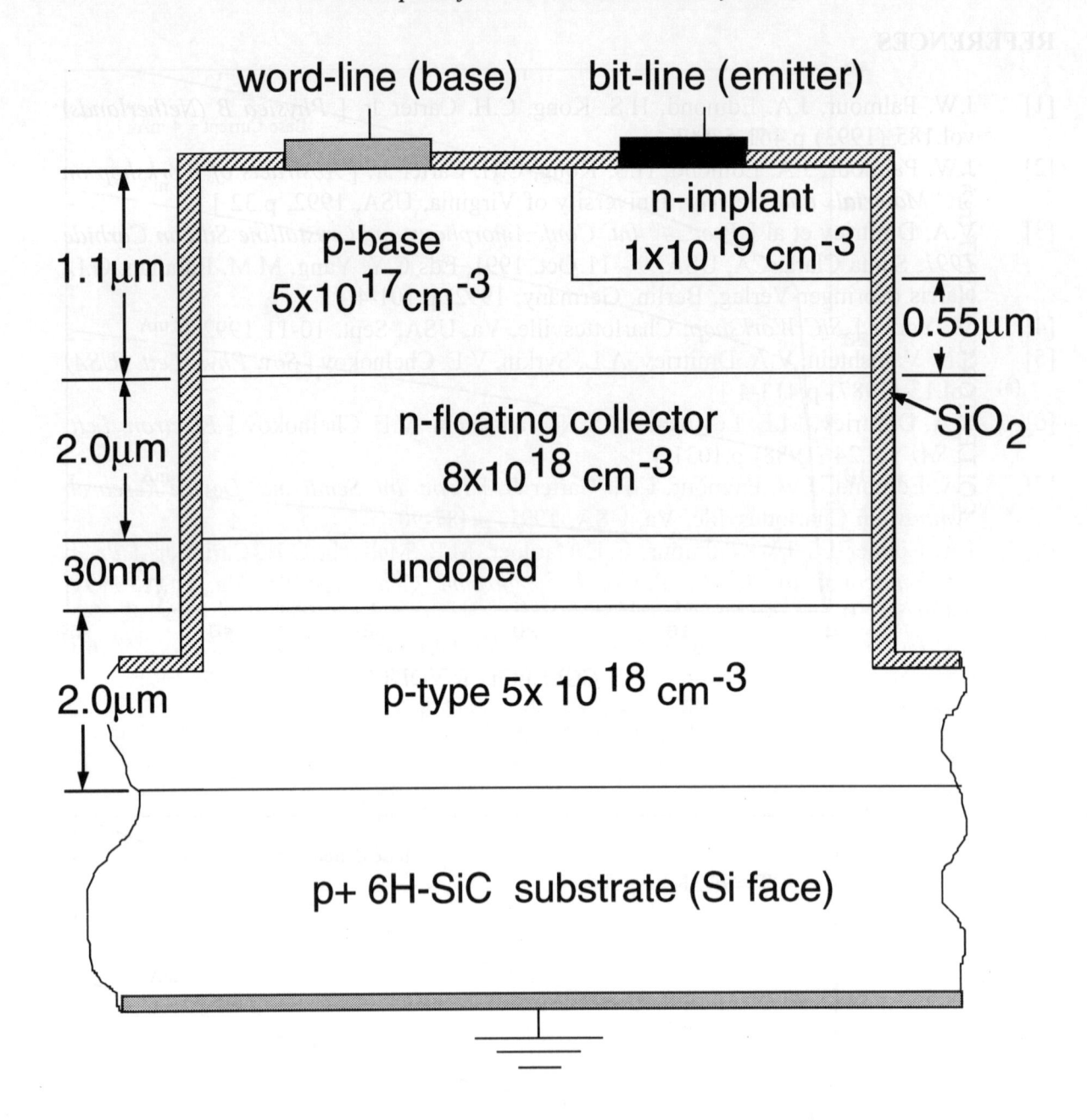

FIGURE 3 Cross-section of the prototype ion implanted bipolar NVRAM cell [9].

E   CONCLUSION

SiC bipolar transistors have been fabricated and measured. Improved performance is to be expected in heterojunction devices with a wider bandgap emitter. A feasibility study is underway to make a non-volatile random access memory in SiC.

REFERENCES

[1] J.W. Palmour, J.A. Edmond, H.S. Kong, C.H. Carter Jr. [*Physica B (Netherlands)* vol.185 (1993) p.461-5]

[2] J.W. Palmour, J.A. Edmond, H.S. Kong, C.H. Carter Jr. [*Abstracts of Workshop on SiC Materials and Devices*, University of Virginia, USA, 1992, p.32]

[3] V.A. Dmitriev et al [*Proc. 4th Int. Conf. Amorphous and Crystalline Silicon Carbide 1991*, Santa Clara, CA, USA, 9 - 11 Oct. 1991, Eds C.Y. Yang, M.M. Rahman, G.L. Harris (Springer-Verlag, Berlin, Germany, 1992) p.101-4]

[4] M. Yoder [*SiC Workshop*, Charlottesville, Va, USA, Sept. 10-11 1992]

[5] S.N. Vainshtein, V.A. Dmitriev, A.L. Syrkin, V.E. Chelnokov [*Sov. Phys. Lett. (USA)* vol.13 (1987) p.413-4]

[6] V.A. Dmitriev, M.E. Levinshtein, S.N. Vainshtein, V.E. Chelnokov [*Electron. Lett. (USA)* vol.24 (1988) p.1031]

[7] J.A. Edmond, J.W. Palmour, C.H. Carter Jr. [*Proc. Int. Semicond. Device Research Symposium* Charlottesville, Va, USA, 1991, p.487-90]

[8] J.A. Cooper Jr., J.W. Palmour, C.T. Gardner, M.R. Melloch, C.H. Carter Jr. [*Proc. Int. Semiconductor Device Research Symposium*, Charlottesville, Va, USA, 1991, p.499-502]

[9] J.A. Cooper Jr., M.R. Melloch, W. Xie, J.W. Palmour, C.H. Carter [*Proc. 5th Int. Conf. SiC and Related Materials*, Washington, DC, USA, 1 - 3 Nov. 1993, Eds M.G. Spencer et al (IOP Publishing, Bristol, UK, 1994) p.711-14]

9.6 SiC optoelectronic devices

G. Kelner and M. Shur

November 1993

A   INTRODUCTION

SiC is an indirect gap material. Nevertheless, it has several luminescence bands involving transitions between impurities (for example, between N and Al levels) and between free and bound exciton states. As a consequence, SiC light emitting diodes (LED) can cover the whole visible spectrum. The wide bandgap of SiC makes it ideal for solar-blind ultra-violet photodetectors (insensitive to visible light). Recent developments in the growth of AlN-SiC solid state solutions, including those with a direct energy gap [1,2], open up new and exciting possibilities for numerous optoelectronic devices.

B   EMISSION DEVICES

The first SiC LED was fabricated by Round in 1907 [3]. He reported on yellow, green and blue emission under forward bias. Blue light emitting diodes have been fabricated by a number of groups and are now available commercially from Siemens, Cree Research, and Sanyo at prices as low as $0.20 [4-6]. Blue LEDs fabricated in 6H-SiC utilize a donor-acceptor pair recombination in an epitaxial layer containing both Al and N dopants. The emission has a peak wavelength of about 470 nm. As mentioned above, the large energy gap of SiC makes it suitable for green [1], blue [2,5] and violet [7] light emitting diodes (see FIGURE 1). Boron implantation into 6H-SiC produces 'yellow' LEDs with the electroluminescence maximum at 580 nm. Green LEDs can also be fabricated by implanting boron into n-type nitrogen-doped 4H-SiC. Thus, as Tairov and Vodakov [8] pointed out, SiC LEDs can cover the entire visible range from red to violet. Dmitriev et al [9] reported a three colour blue-green-red SiC single crystal display. They made an array of 300 - 500 µm diameter diodes which formed a mosaic of the blue, red and green light-emitting diodes with intensity maxima corresponding to 470 nm, 510 nm and 650 nm, respectively. The display dimension was limited by the substrate size ($\approx 1 \text{ cm}^2$). The external power efficiency was fairly low ($\approx 0.03 \%$). We expect that this efficiency can be dramatically improved in AlN/SiC direct bandgap LEDs.

C   PHOTODETECTORS

Both SiC p-n diodes and Schottky barrier diodes are utilized as UV photodetectors operating in the 200 nm to 450 nm range. They can be used up to temperatures of at least 700 K and are expected to be radiation tolerant.

State-of-the-art p-n 6H-SiC photodiodes have exhibited quantum efficiencies higher than 80 % with responsivities of up to 175 mA W^{-1} with dark currents as low as 10^{-11} A cm^{-2} at - 1 V at 473 K. The dark current increased to approximately 10^{-9} to 10^{-8} A cm^{-2} at 623 K. These low values of the dark current are still higher than those in rectifier SiC diodes [10,11] pointing

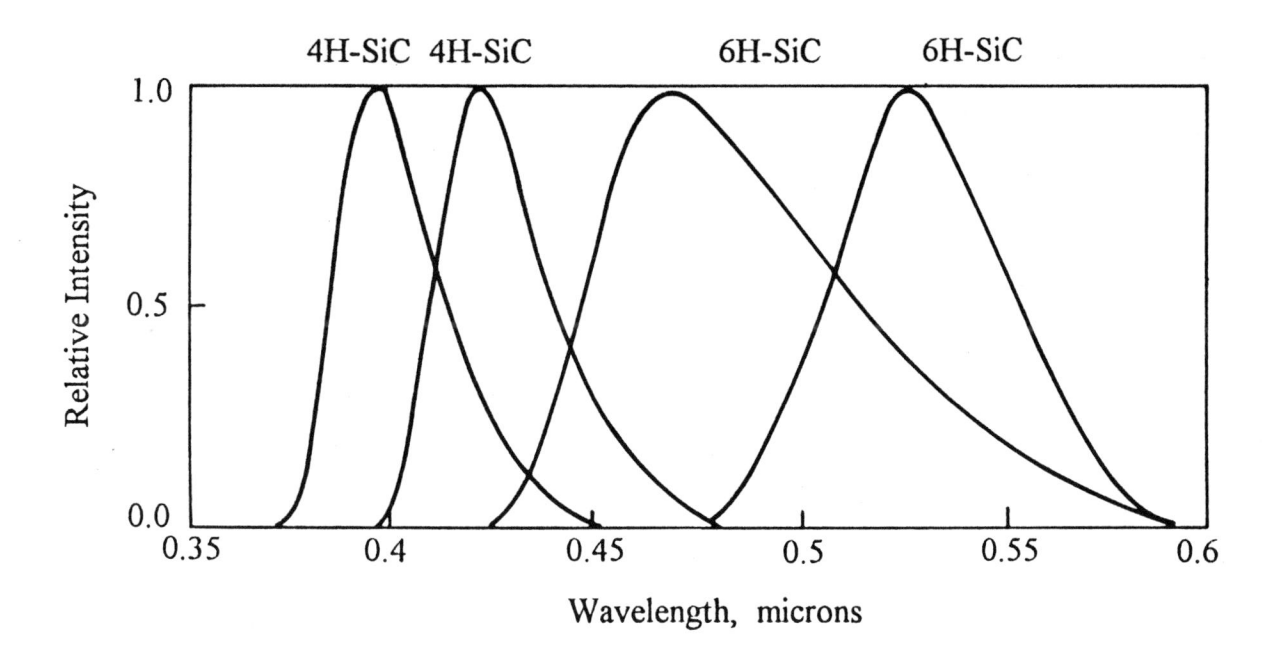

FIGURE 1 SiC light emitting diode emission spectra at room temperature [6].

out the potential improvements in photodetector characteristics. The frequency response depended on the thickness of the p-type layer. When the p-type layer thickness was changed from 3.0 μm to 1.2 μm to 0.4 μm, the peak response shifted from 280 nm to 260 nm to 250 nm. (The junction depth was approximately 0.1 μm [10].)

Silicon carbide (6H and 4H polytypes) Schottky diodes have been used for ultraviolet photodetectors [12]. A thin semi-transparent layer of chromium was used as a Schottky contact. The diffusion length of light-generated holes was estimated to be in the range from 0.1 to 0.3 μm. The earlier work on SiC UV photodiodes [13] used diffusion of Al into n-type substrates. These devices have a high leakage current and a low quantum efficiency. Later they were improved by using nitrogen implantation to form a shallow n^+-p junction in a 5 μm p-type epilayer grown on a p-type substrate. This resulted in devices with the maximum quantum efficiency of 75% at a peak wavelength of 280 nm at room temperature. However, the reverse bias dark current density was excessive, on the order of 10^{-5} A cm^{-2} at -10 V at room temperature [14].

State-of-the art Cree Research, Inc. UV SiC photodiodes are more sensitive than commercial Si photodiodes. The reverse dark current at room temperature in these devices is less than 1 pA cm^{-2} and a few nA cm^{-2} at 400 °C.

D   CONCLUSION

Blue LEDs made in the 6H-SiC polytype, utilising a donor-acceptor pair recombination, are now commercially available. Boron-doped 6H-SiC produces a yellow colour while green

LEDs are made from boron-doped 4H-SiC. Three colour displays have been demonstrated. SiC ultraviolet photodetectors made from p-n junction and Schottky barrier diodes can be used up to temperatures of 700 K and are expected to be radiation tolerant. These photodiodes are more sensitive than their silicon counterparts.

REFERENCES

[1] V.M. Gusev, K.D. Demakov [*Sov. Phys.-Semicond. (USA)* vol.15 (1981) p.2430-1]
[2] B.I. Vishnevskaya et al [*Sov. Phys.-Semicond. (USA)* vol.22 (1988) p.664-5]
[3] H.J. Round [*Electr. World (USA)* vol.19 (1907) p.308-9]
[4] L.A. Tang, J.A. Edmond, J.W. Palmour, C.H. Carter [*Proc. Electrochem. Soc. Meeting*, Hollywood, Fl., USA, 1989, p.705]
[5] T. Nakata, K. Koga, Y. Matsushita, Y. Ueda, T. Niina [*Amorphous and Crystal. SiC* vol.43, Eds M.M. Rahman, C.Y.W. Yang, G.L. Harris (Springer Verlag, Berlin, 1988) p.26-34]
[6] V.A. Dmitriev, P.A. Ivanov, Ya.V. Morozenko, I.V. Popov, V.E. Chelnokov [*Sov. Tech. Phys. Lett. (USA)* vol.11, no.12 (1985) p.101-2]
[7] V.A. Dmitriev, P.A. Ivanov, V.E. Chelnokov, A.E. Cherenkov [*Proc. 1st Int. Conf. Applications of Diamond Films and Related Materials*, Auburn, Alabama, USA, 1991, p.769-74]
[8] Yu.M. Tairov, Yu.A. Vodakov [*Top. Appl. Phys. (Germany)* vol.17 (1977)]
[9] V.A. Dmitriev, Ya.V. Morozenko, I.V. Popov, A.V. Suvorov, A.L. Syrkin, V.E. Chelnokov [*Sov. Tech. Phys. Lett. (USA)* vol.12, no.5 (1986) p.221-2]
[10] J.A. Edmond, J.W. Palmour, C.H. Carter Jr. [*Proc. Int. Semicond. Device Research Symposium* Charlottesville, Va, 1991, p.487-90]
[11] J.A. Edmond, H.S. Kong, C.H. Carter Jr. [*Proc. 4th Int. Conf. Amorphous and Crystalline Silicon Carbide 1991*, Santa Clara, CA, USA, 9 - 11 Oct. 1991, Eds C.Y. Yang, M.M. Rahman, G.L. Harris (Springer-Verlag, Berlin, Germany, 1992) p.344-51]
[12] R.G. Verenchikova, V.I. Sankin [*Sov. Tech. Phys. Lett. (USA)* vol.14, no.10 (1987) p.756-8]
[13] P. Glasow, G. Ziegler, W. Suttrop, G. Pensl, R. Helbig [*Proc. SPIE (USA)* vol.898 (1987) p.40-5]
[14] 6H-SiC Study and Developments at the Corporate Research Laboratory of Siemens AG and the Institute for Applied Physics of the University in Erlangen (Germany) [*Proc. Phys. (Germany)* vol.34 (1989) p.13-33]

9.7 Potential performance/applications of SiC devices and integrated circuits

G. Kelner and M. Shur

November 1993

Calculations by Trew et al [1] showed that SiC MESFETs are capable of generating up to 65 W of power at 10 GHz (4 W mm^{-1}). They also predicted that SiC IMPATT devices should be capable of achieving about 4 W of power at 20 to 30 GHz with conversion efficiencies of 15 to 20 %. SiC devices have successfully demonstrated excellent performance at high temperature and in a harsh environment.

SiC also has potential for applications in integrated circuits. As the device size is scaled down, the disadvantages related to a relatively low mobility diminish, and advantages related to a high breakdown voltage and a high saturation velocity become more important. In particular, these features are important for SiC non-volatile memory integrated circuits. Finally, emergence of a new material system based on SiC/AlN/GaN/InN points to new exciting opportunities for device applications using heterostructures which utilize the superior optoelectronic properties of the direct gap materials.

High ohmic contact resistances in SiC devices present a serious limitation for high frequency performance. Furthermore, the problem of ohmic and Schottky contact thermal stability has not been solved. Contacts (and, sometimes, packages) usually limit the maximum SiC device operating temperatures. The existence of micropipes in 6H- and 4H-SiC material leads to a low yield for power devices. Improvements in material quality, the development of bulk 3C-SiC for 3C-SiC homoepitaxy, and the development of better contacts are of primary importance for the advancement of the SiC device technology.

REFERENCE

[1]   R.J. Trew, J.B. Yan, P.M. Mock [*Proc. IEEE (USA)* vol.79, no.5 (1991) p.598-620]

SUBJECT INDEX

α-SiC: see hexagonal SiC
acceptor binding energy 33, 34
acceptor concentration 187, 206, 214
acceptor ionisation energy 34, 89, 94
acceptors in SiC
    Al 33, 34, 46, 55, 56, 88-90, 94
    B 33, 34, 45-47, 55, 56, 88-90, 94, 95
    Be 34, 90, 94
    Ga 34, 46, 47, 55, 56, 88-90, 94
    O 90
    Sc 34, 55, 56, 94, 96
Acherson process 170
acoustic velocity 7
    temperature dependence 7
AlN-SiC: see SiC-AlN
AlN/SiC/InN 235
amorphous SiC, reactive ion etching 138
antiphase boundaries 134, 205
antiphase disorder 205
antisite defects 36
applicance 10
Avogadro's constant 3

β-SiC: see cubic SiC
band structure 74-79, 81-83
    conduction band minimum 74-77, 79
    pressure effects 81-83
    valence band maximum 74-76, 79
bandgap 29, 31, 33, 34, 74-79, 81-83, 141
    direct 74, 76, 77, 82
    indirect 29, 74-77, 79, 81, 82
    pressure coefficients 81-83
    temperature dependence 74
barrier height, thermal oxide/SiC 125, 126
bipolar transistors 136, 141, 223, 265-267
    current gain 265
    I-V characteristics 265, 267
BJT: see bipolar transistors
bound excitons 29, 31, 32, 57
breakdown field 9, 141
bulk growth 163-166, 174, 175, 183, 191
    growth rate 183
    polytype control 165, 166
    single crystal 163-165
    source temperature 165
    substrate temperature 165
    substrates 165
bulk modulus 11

carrier concentration 63-67, 206
    temperature dependence 63
carrier concentration, intrinsic 222, 238
carrier mobility 63-67
cathodoluminescence spectra of porous SiC 148

chemical etching 134, 135
    gases 134, 135
    molten salts 134
chemical vapour deposition: see CVD growth
CMOS devices 141, 250
colour, polytypes 15
common phonon spectrum 24
compensation ratio 235
container-free LPE 218-221
crystal structure 21-24
cubic SiC
    acoustic velocity 7
    atoms per unit cell 22
    band structure 75, 76, 79, 81-83
    bandgap 29, 31, 33, 74-76, 81-83
    breakdown field 9
    bulk growth 165, 166
    carrier concentration 63, 64
    carrier mobility 63, 64
    colour 15
    CVD growth 32, 204, 205
    Debye temperature 11
    deep levels 93, 94, 96, 97
    density 3
    dielectric constant 9
    diffusion coefficients of impurities 153-156
    electron effective mass 69-71
    electron mobility 235
    electro-optic coefficient 10
    electropolishing 145, 146
    ESR spectra 43, 44, 46-48
    etching 134, 138, 141-145, 147
    gauge factor 10
    hole effective mass 70, 71
    impurity energy levels 88-90, 93, 94, 96, 97
    interface structure 110-116
    lattice parameters 4
    LPE growth 224, 225
    ODMR 51, 53, 54, 58
    ohmic contacts 233
    optical absorption coefficient 16, 17
    oxidation 121-127
    phonon mode frequencies 25, 30
    photoluminescence spectra 29-36
    Raman spectra 23
    refractive index 18, 19
    Schottky barrier heights 243
    space group 22, 23
    specific heat 11
    stacking sequences 21, 22, 87
    stiffness 10
    sublimation growth 190, 193, 194

SUBJECT INDEX

    surface cleaning 102-104
    surface structure 104-109
    thermal conductivity 5, 6
    work function 9
    Young's modulus 8
CVD growth 32, 204-206
    buffer layers 204, 205
    substrates 204-206
    temperature profile 205

D centre 94, 95
Debye temperature 11
deep levels 34, 35, 93-97
    irradiation effects 96
defect characterisation 29-36, 42-49, 51-58
defect energy levels 93-97
defect formation in sublimation growth 189-193
defects
    antisite 36
    dislocations 134
    electron traps 94-96
    hole traps 94, 96
    ion implantation induced 35, 36
    micropipes 165, 236, 237, 262, 273
    radiation induced 35, 36, 48, 58
    self defects 95
    stacking faults 134, 174, 191, 205
    twinning 190, 205
density 3
    temperature dependence 3
density-of-states effective mass 69, 70
device technology 273
dichroism 15
dielectric constant 9
diffraction gratings 142
diffusion coefficients of impurities 153-156
    Al 155
    B 155
    Be 155
    Ga 155
    N 155
    O 155
    Sc 155
diffusion of impurities 153-156, 197-199
    Al 197
    B 197
diffusivity
    Al 195
    Be 188
dimer pairs 104-107, 116
diodes 215, 223, 238-245
dislocation density 220
dislocations 134
donor-acceptor pair recombination 56
donor-acceptor pairs 33, 34, 51, 53, 54, 56

donor binding energy 33, 34, 45
donor concentration 187, 206, 220
donor ionisation energy 29-31, 89, 94
    donor concentration dependence 89
donors in SiC
    N 29-32, 42-46, 51-54, 88-90, 94
    P 90
dopant capture coefficients: see impurity capture coefficients
doping 157-159, 186-189, 206
doping efficiency 186, 187
dry etching 136-138
    plasma etching 136, 138
    reactive ion etching 137, 138
dynistors 214, 223, 265

elastic coefficients 10
electrochemical etching 141-148
electrochemical polishing 145, 146
    surface roughness 146
electroluminescence 218, 223-225, 270
electromechanical coupling coefficient 10
electron diffusion length 222
electron effective mass 69-72
    density-of-states 69, 70
    longitudinal 69-71
    transverse 69-71
electron-hole liquid 31, 32
electron-hole liquid binding energy 31
electron-hole plasma 31, 32
electron irradiation effects 48, 58, 94, 95
electron lifetime 222
electron mobility 63-67, 221, 235
electron traps 94-96
electro-optic coefficient 10
electropolishing 145, 146
    surface roughness 146
energy gap: see bandgap
epitaxial growth 166, 180, 191, 193, 204-206, 214-225
EPR spectra: see ESR spectra
ESR spectra 36, 42-49
    doping effects 42-47
    irradiation effects 48
    temperature dependence 43, 44, 46, 47
    transition metal impurities 47, 48
etch pits 134
etch stops 141, 146-148
etchants 134
etching 133-138, 141-148
    chemical 134, 135
    dry 136-138
    electrochemical 141-148
    plasma 136, 138
    reactive ion 137, 138
etching rate 136-138, 141-145, 148

SUBJECT INDEX

exciton binding energy 31, 89
exciton energy gap 31

Faust-Henry coefficient 26
FET 136, 223, 247-263: see also JFET, MESFET, MOSFET
figures of merit 235, 236
    GaAs 236
    Si 236
free excitons 31
free-to-bound transitions 29, 34
furnaces for sublimation growth 171-173
    induction 171-173
    resistive 171-173

GaAs
    Johnson's figure of merit 236
    Keyes' figure of merit 236
gauge factor 10
Ge, order-disorder transitions 105
grain boundaries 134
gram formula weight 3
growth rate
    bulk growth 183
    impurity effects 180
    LPE growth 214, 216, 220, 222, 224, 225
    pressure dependence 163, 180, 181
    Si vapour pressure dependence 181-183
    sublimation growth 163, 174, 180-183
    temperature dependence 163, 180, 181, 183
g-tensors 43, 46-48, 51, 53, 55-58

Hall mobility 63-67, 220, 221
    doping effects 63-67
    temperature dependence
HBT: see bipolar transistors
heteroepitaxial growth 204, 205
heterojunction devices 205, 265
heteropolytype epitaxy 224, 225
hexagonal SiC
    acoustic velocity 7
    applicance 10
    atoms per unit cell 22
    band structure 76-79
    bandgap 31, 74, 77-79
    breakdown field 9
    bulk growth 163, 165, 166
    bulk modulus 11
    carrier concentration 63, 65, 66
    carrier mobility 63, 65, 66
    colour 15
    CVD growth 32, 206
    Debye temperature 11

deep levels 93-97
density 3
dielectric constant 9
diffusion coefficients of impurities 153-156
electromechanical coupling coefficient 10
electron effective mass 69-71
electron mobility 235
electropolishing 145, 146
ESR spectra 45-48
etching 134, 137, 138, 142, 148
gauge factor 10
hole effective mass 69, 70
impurity energy levels 88-90, 93-97
interface structure 110-116
lattice parameters 4
LPE growth 217, 218, 220, 222
magnetic susceptibility 11
minority carrier diffusion length 9
ODMR 53, 55-58
ohmic contacts 231, 232
optical absorption coefficient 16
oxidation 121-127
phonon mode frequencies 25, 30
photoluminescence spectra 29-36
piezoelectric coefficient 10
Raman spectra 23, 24
refractive index 18, 19
saturation velocity 9
Schottky barrier heights 243
solubility limits 155
solubility of impurities 153-155
space group 22, 23
specific heat 11
stacking sequences 21, 22, 87
stiffness 10
sublimation growth 189, 193, 194
surface cleaning 102-104
surface structure 108, 109
thermal conductivity 5, 6, 237
work function 9
hexagonality percentage of polytypes 22, 74
hole concentration 222
hole diffusion length 220
hole effective mass 69-72
    heavy 70
    light 70
hole mobility 63-67, 221, 222
hole traps 94, 96
homoepitaxial growth 204, 206

IMPATT diodes 136, 273
impurity binding energy
    Al 34
    B 34
    Be 34

SUBJECT INDEX

    Ga 34
    N 31
    Sc 34
impurity capture coefficients 186, 187
    Al 187
    B 187
    Ga 187
    N 187
impurity energy levels 87-90, 93-97
    Al 88, 89, 94, 96
    B 88, 89, 94, 95
    Be 90, 94
    deep levels 93-97
    Ga 88, 89, 94
    N 88, 89, 94
    O 90
    P 90
    Sc 94, 96
impurity ionisation energy 88-90, 94, 95
inequivalent sites in polytypes 87, 89
insulating layers 124
    aluminium oxide 124
    silicon dioxide 124
    silicon nitride 124
integrated circuits 136, 261, 273
interface growth 194-199
interface structure 101, 109-116
    Ag/SiC 110, 115
    Al/SiC 110, 111
    annealing effects 111-116
    Au/SiC 110, 112
    Cr/SiC 110, 111
    Cu/SiC 110, 112
    Fe/SiC 110, 112, 113
    Mn/SiC 110, 113
    Mo/SiC 110, 113
    Ni/SiC 110, 113, 114
    Pd/SiC 110, 114
    Pt/SiC 110, 114, 115
    Sn/SiC 110, 115
    Ta/SiC 110, 115
    Ti/SiC 110, 115, 116
    W/SiC 110, 116
intrinsic carrier concentration 222, 238
inverter circuit 262, 263
    gain 262
ion implantation effects 35, 36
ion implantation of SiC 157-159
    Al 157, 159
    annealing effects 157-159
    B 157, 159
    Ga 158, 159
    Kr 159
    N 158, 159
    P 158, 159
    Sb 159

irradiation effects 36, 48, 58, 94-96
    electron irradiation effects 48, 58, 94-96
    neutron irradiation effects 58, 94-96
    proton irradiation effects 48, 95

Jahn-Teller effect 46, 57
JFET 214, 247, 252, 257, 259-262
    depletion mode 257
    enhancement mode 257, 259, 261
    field effect mobility 257, 259
    I-V characteristics 260
    operating temperature 257, 259
    transconductance 257, 259
Johnson's figure of merit 235, 236
    GaAs 236
    Si 236
junction diodes 238-242
    I-V characteristics 239
    leakage current 239
    reverse saturation current density 238, 239

Keyes' figure of merit 235, 236
    GaAs 236
    Si 236

lapping rate 133
large zone 23, 24
laser etching 143, 144
lasers 235
lattice dynamics 21-27
lattice mismatch
    SiC/Si 24, 204
    Ti/SiC 116
lattice parameters 4, 83
    Si 83
    temperature dependence 4
lattice phonons 23-25
LED 136, 214, 216-218, 223, 225, 236, 270-272
    quantum efficiency 216
Lely method 170, 174, 175, 189, 193, 194
liquid phase epitaxy: see LPE growth
localised vibrational modes 26
low pressure CVD growth 205
low temperature growth 205
low temperature LPE 223-225
LPE growth 214-225
    container-free LPE 218-221
    crucibles 215-217
    doping profile control 220, 225
    growth rate 214, 216, 220, 222, 224, 225
    growth temperature 215-225
    heteropolytype 224, 225
    layer thickness 215, 217, 220, 223-225
    solvents 223, 224

SUBJECT INDEX

vertical dipping technique 217, 218

magnetic susceptibility 11
mechanical polishing 133, 146, 166
    surface roughness 146
melting point 141
    Si 220, 221
MESFET 247, 252, 255-259, 262, 273
    breakdown voltage 255
    cut-off frequency 259
    field effect mobility 252, 257
    gain 255
    I-V characteristics 258
    microwave operation 255, 259
    operating temperature 252, 257
    output power 255
    transconductance 252, 255, 257
metallization 101, 109-116
metastability 189
microcracks 205
microelectromechanical sensors 141, 147
micropipe density 236, 237
micropipes 165, 236, 237, 262, 273
    annealing 165
microplasmas 196, 197, 199, 223
microwave devices 255, 259, 261
minority carrier diffusion length 9
MIS capacitors 124
MOS capacitors 124-127, 247
    breakdown voltage 126
    C-V characteristics 126
    doping effects 126, 127
    fixed charge density 125, 126
    interface trapped charge at midgap 125, 126
MOSFET 247-257, 259, 262
    depletion mode 247, 250, 252-254, 257
    enhancement mode 247-251, 254, 255, 257
    field effect mobility 247, 257
    I-V characteristics 249, 251, 253-256
    leakage current 250
    operating temperature 247, 250, 257
    power MOSFET 250, 252, 256, 257
    transconductance 247, 250, 257

neutron irradiation effects 58, 94-96
neutron scattering spectroscopy 21
NVRAM 266, 268, 273

ODMR 46, 51-58
    doping effects 51-56
    impurity effects 57, 58
    irradiation effects 58
ohmic contacts to SiC 231-233
    contact resistance 231-233

operational amplifier 262
optical absorption 15-17, 143
    doping effects 15
optical absorption coefficient 15-17
    doping effects 17
    wavelength dependence 16, 17
optical modes 23-25
optoelectronic devices 270-273
order-disorder transitions 105
    Ge 105
    Si 105
overcompensation method 216
oxidation 121-127
    dopant redistribution 123
    dry 121-123
    temperature effects 127
    thermal 121-123
    wet 121-123, 127
oxidation rate 121-123
    temperature dependence 122, 123
oxide layers on SiC 121, 122, 124-127
    barrier height 125, 126
    breakdown field 126
    dielectric strength 125
    fixed charge density 125
    interfacial charge density 125
    refractive index 125
    resistivity 125, 126

phase diagrams, Si-C 214, 215
phase equilibria 175-179
    evaporation product activation energy 176
    impurity effects 179
    SiC-C 176, 177
    SiC-H 179
    SiC-N 179
    SiC-Si 176, 177
    temperature dependence 176, 177
    vapour pressure 176-179
phase transitions 189, 190, 193, 194
    solid phase mechanism 190
    vapour phase transport 190
phonon dispersion 21, 24
phonon mode frequencies 25, 30
    acoustic 25, 30
    optical 25, 30
phonons 21-27
photodetectors 270-272
photodiodes 270-272
    dark current 270, 271
    frequency response 270
    quantum efficiency 270, 271
    responsivity 270
photoelectrochemical etching 141-148
    dopant selectivity 146-148

SUBJECT INDEX

electrolytes 141-143
    surface roughness 144
photoelectrochemical polishing 145, 146
    surface roughness 146
photoluminescence spectra 29-36
    doping effects 29-34, 95
    implantation-induced defects 35, 36
    irradiation effects 36
    phonon replicas 29, 30
    SiC/Si 32
    transition metal impurities 34, 35
    zero phonon line 29, 30
piezoelectric coefficient 10
piezoresistive effect 10
plasma CVD growth 205
plasma etching 136, 138
    mask materials 138
    selectivity ratio 138
plasmariton 26
plasmon 26
p-n junction diodes 238-242, 245
    breakdown voltage 240-242, 245
    ideality factor 240, 241
    I-V characteristics 240
    leakage current 240-242, 245
    operating temperature 245
    reverse bias current density 240
    reverse recovery time 240
    series resistance 240
p-n junction growth 192, 194-199, 214-219, 222-225
    container-free LPE 218, 219
    LPE 214-219, 222-224
    overcompensation method 216
    SiC-AlN 225, 235
    temperature effects 196, 197
    two-step epitaxy 216
    vertical dipping technique 217
p-n junctions 171
    breakdown voltage 195, 196, 199, 223, 224
    doping profile 196, 197
    electroluminescence 218, 223-225
    ideality factor 222, 239
    I-V characteristics 222-224
    leakage current 195, 196, 223
    microplasmas 196, 197, 199, 223
polaritons 25, 26
polishing techniques 133, 166
polytype notations 22
polytype transformations 189, 190, 193, 194
    impurity effects 194
    substrate orientation dependence 193
porous SiC 148
power SiC MOSFET 250, 252, 256, 257, 262

proton irradiation effects 48, 95

R centre 94, 95
radiation hardness 261
RAM 266, 268
Raman spectra 21, 23-27
    ion implantation effects 26
    irradiation effects 26
reactive ion etching 137, 138, 148
    mask materials 138, 148
    selectivity ratio 138
recombination processes 56
rectifier diodes 240, 243, 270
refractive index 17-19
    extraordinary 17-19
    ordinary 17-19
    wavelength dependence 18, 19
rhombohedral SiC
    acoustic velocity 7
    atoms per unit cell 22
    band structure 78
    bandgap 31, 74
    carrier density 67
    carrier mobility 67
    colour 15
    electron effective mass 69-71
    electron mobility 235
    ESR spectra 45, 46
    impurity energy levels 88, 89
    lattice parameters 4
    ODMR 57
    ohmic contacts 231
    optical absorption 15
    phonon mode frequencies 25, 30
    photoluminescence spectra 30, 31, 34-36
    Raman spectra 24
    refractive index 19
    space group 22
    stacking sequences 22, 87
    stiffness 10
    sublimation growth 189, 193, 194

S centre 94, 95
sandwich growth method 166, 170, 173-175, 180, 186, 195
saturation velocity 9
scattering mechanisms 63, 67
Schottky barrier diodes 243-245
    breakdown voltage 243, 245
    ideality factor 243
    operating temperature 243-245
    reverse leakage current 243, 245
Schottky barrier heights
    Au/SiC 243, 245
    Co/SiC 243, 245
    Hf/SiC 243, 245

SUBJECT INDEX

Pd/SiC 245
Pt/SiC 243, 245
thermal oxide/SiC 125, 126
Ti/SiC 243, 245
SCR devices 141
self defects 95
Si
    Johnson's figure of merit 236
    Keyes' figure of merit 236
    lattice parameter 83
    melting point 220, 221
    order-disorder transitions 105
SiC-AlN 225, 235, 270
    direct-indirect bandgap transition 235
SiC-AlN growth 225, 235
    AlN concentration 225
    growth temperature 225
    layer thickness 225
SiC/HF interface 141-143, 147
SiC/Si
    biaxial stress 83
    lattice mismatch 24, 204
    photoluminescence spectra 32
    thermal expansion coefficient mismatch 24, 204
SiC/SiO$_2$ interface 247, 250, 262
site competition epitaxy 206
solubility limits for impurities in SiC 153, 155, 187-189
    Al 153, 155, 187, 189
    As 155, 189
    Au 155, 189
    B 153, 155, 188, 189
    Be 155, 188, 189
    Cr 155, 189
    Cu 155, 189
    Ga 153, 155, 187, 189
    Ge 155, 189
    Ho 155, 189
    In 155, 189
    Li 155, 189
    Mn 155, 189
    N 153, 155, 189
    P 153, 155, 189
    Sb 155, 189
    Sc 155, 189
    Sn 155, 189
    Ta 155, 189
    temperature dependence 188
    Ti 155, 189
    W 155, 189
    Y 155, 189
solubility of C
    in Al 223, 224
    in Pb 223, 224

solubility of impurities 153-155, 184, 185, 187
    crystal face dependence 154
    growth rate dependence 184, 185
    partial pressure dependence 187
    temperature dependence 154, 185, 187
solubility of SiC
    in Ga 223, 224
    in Ge 223, 224
    in Si 223, 224
    in Sn 223, 224
space group 22
speaker diaphragms 7
specific heat 11
spectral emissivity 11
stacking faults 134, 174, 191, 205
stacking sequences of polytypes 21, 22, 87
step-controlled epitaxy 204, 206
stiffness 10
sublimation growth 163-166, 170-200
    background impurities 184-186
    crucibles 163, 164, 171, 173, 177, 179, 184, 185
    defect formation 189-193
    doping 186-189
    doping efficiency 186, 187
    furnaces 171-173
    growth cavities 174, 175
    growth rate 163, 174, 180-183
    mass transfer 179-183
    phase equilibria 175-179
    p-n junctions 194-199
    polytype transformations 193, 194
    source materials 174, 175, 177, 178
    temperature range 171, 176
surface characterization techniques 101-103, 134
surface cleaning 102-104
surface composition 108, 109
surface contamination 191
surface domains 105
surface modification 190-192
surface phases 104-109
surface reconstructions 101, 104-109, 191, 192
surface recrystallisation 191, 192, 195
surface roughness 144-146, 205
surface structure 101, 104-109

thermal conductivity 5, 6, 237
    doping effects 5, 6
    temperature dependence 5, 6
thermal EMF 11
thermal expansion coefficient mismatch
    SiC/Si 204

SUBJECT INDEX

thermal oxidation 121-123
thermal oxide layers on SiC 121, 122, 124-127
    barrier height 125, 126
    breakdown field 126
    dielectric strength 125
    dry 122, 124
    fixed charge density 125
    interfacial charge density 125
    refractive index 125
    resistivity 125, 126
    wet 122, 124, 125
thermal stress 205
thyristors 265
transition metal impurities in SiC
    Ti 34, 35, 47, 48, 57, 58
    V 34, 35, 47, 48
tunnel diodes 136
twinning 190, 205
two-step epitaxy 216

UV photodetectors 270-272

vacuum plasma etching 136-138
    plasma etching 136, 138
    reactive ion etching 137, 138
valence band offset 77, 79
voids 165

wafer polishing 166
work function 9
wurtzite structure: see hexagonal SiC

X-ray rocking curve half-width 220, 225

Young's modulus 8
    temperature coefficient 8

zinc blende structure: see cubic SiC
zone-folding 23, 24